Contemporary Issues in Technology Education

Contemporary Issues in Technology Education - About this series

Technology education is a developing field, new issues keep arising and timely, relevant research is continually being conducted. The aim of this series is to draw on the latest research to focus on contemporary issues, create debate and push the boundaries in order to expand the field of technology education and explore new paradigms. Maybe more than any other subject, technology education has strong links with other learning areas, including the humanities and the sciences, and exploring these boundaries and the gaps between them will be a focus of this series. Much of the literature from other disciplines has applicability to technology education, and harnessing this diversity of research and ideas with a focus on technology will strengthen the field.

More information about this series at http://www.springer.com/series/13336

P. John Williams • David Barlex
Editors

Pedagogy for Technology Education in Secondary Schools

Research Informed Perspectives for Classroom Teachers

Springer

Editors
P. John Williams
Curtin University
Perth, WA, Australia

David Barlex
University of Exeter Associate
Brighton, UK

ISSN 2510-0327 ISSN 2510-0335 (electronic)
Contemporary Issues in Technology Education
ISBN 978-3-030-41547-1 ISBN 978-3-030-41548-8 (eBook)
https://doi.org/10.1007/978-3-030-41548-8

This Springer imprint is published by the registered company Springer Nature Switzerland AG
The registered company address is: Gewerbestrasse 11, 6330 Cham, Switzerland

Contents

About the Authors

Moshe Barak is a Full Professor in the Department of Science and Technology Education, Ben-Gurion University of the Negev, Israel, and formerly the Department Chair. His background is in electrical and electronics engineering, and he received his PhD in 1987 from the Department of Science and Technology Education at the Technion – Israel Institute of Technology. He has over 40 years of experience in teaching technology and engineering in high schools and college and teacher training colleges. His research interests focus on fostering higher-order cognitive skills, such as problem-solving and creativity in technology and engineering education, and teaching advanced interdisciplinary technological subjects such as control systems and robotics to children. Barak has published over 120 papers in international refereed scientific journals, conferences and chapters in collective volumes.

David Barlex is an acknowledged leader in design and technology education, curriculum design and curriculum materials development. He taught in comprehensive schools for 15 years before taking university positions in teacher education. He directed the Nuffield Design and Technology Project. He is well-known for developing curriculum materials that support pupil learning from a constructivist perspective. He uses this approach to develop young peoples' ability to understand and critique the design decisions made by professional designers and those they make themselves. This informed the Nuffield Design and Technology publications, which have been widely used in the UK and emulated abroad – in Russia, Sweden, Canada, South Africa, Australia and New Zealand.

Marilyn Fleer (PhD, MEd, MA, BEd) holds the Foundation Chair of Early Childhood Education and is a Laureate Professor at Monash University, Australia, and is the President of the International Society for Cultural and Activity Research (ISCAR). She is also an Honorary Research Fellow, Faculty of Education, University of Oxford, has the 2019 Ashley Goldsworthy Award for Outstanding leadership in university-business collaboration and is also a Professor at Western Norway University of Applied Sciences. Her research interests focus on early years learning and development, with special attention on pedagogy, culture, science and design and technology.

Mishack T. Gumbo is a Full Professor in the Department of Science and Technology Education at the University of South Africa's College of Education. He is also a Coordinator of the Master's and Doctoral Programme in the College. His research specialisation fields include technology teachers' continuing professional development, indigenous knowledge and the curriculum, decolonisation of the curriculum, distance education and e-learning. He has published numerous scholarly articles, conference papers, book chapters and three coedited scholarly books. He has given keynote addresses at conferences and other academic occasions and mentored developing researchers and postdoctoral fellows. He leads a community engagement on Strategic Intervention in Mathematics, Science and Technology Education.

Jonas Hallström, PhD, is Professor of Technology Education at Linköping University, Sweden, where he heads the unit Technology and Science Education Research, TESER. His research mainly concerns the history and epistemology of technology (education), attitudes to technology, gender and technology education, modelling in technology education and teaching and learning about technological systems. Two of his latest publications are Hallström, J., & Klasander, C. (2017), Visible parts, invisible whole: Swedish technology student teachers' conceptions about technological systems, *International Journal of Technology and Design Education*, 27(3), 387–405, and Hallström, J., & Schönborn, K. J. (2019), Models and modelling for authentic STEM education: reinforcing the argument, *International Journal of STEM Education*, 6:22.

Janet Hanson is an Education Researcher with the Centre for Real-World Learning (CRL) at the University of Winchester. After working in higher education for over 25 years, she joined CRL in 2013 to undertake research projects with a focus on engineering education, mostly on behalf of the Royal Academy of Engineering. Arising from CRL's research, the concept of engineering habits of mind (EHoM) has been adopted as one of the 'Big Ideas' for reinvigorating engineering education and has inspired teachers to introduce engineering into their STEM teaching. She has also investigated the role of school leadership in delivering engineering within the curriculum and collaborated with colleagues at the University of Winchester to introduce EHoM to students training to be primary teachers in technology, science and computing.

David Hills-Taylor has over 12 years' experience of teaching design and technology at secondary level. He has been both a key stage 3 and key stage 4 coordinator, as well as an SSAT Lead Practitioner for the subject. In 2012, he won the D&T Association IET National Award for Digital Design and Technology in Schools. He has expertise in digital and online learning and has held whole school responsibility for this area of the curriculum.

He has acted as a Consultant and Author and has written or cowritten several books and online resources, including the new AQA-accredited Hodder GCSE Engineering textbook, and has significant examining experience. He is a Co-founder and Director of Attainment in Education, an organisation that provides high-quality resources and support for teaching the STEM subjects in schools.

Steve Keirl is a Curriculum Theorist whose research advocates technological and design, literacy that is critically, ethically and democratically focused. As such, he strongly resists STEM as a curriculum construct. His research helped inform the 2001 (birth age 18) South Australian Design and Technology Curriculum for which he was Chair of the Technology Expert Working Group and subsequently Principal Author. Critiquing, as one of, the three strands running throughout that curriculum, was then a 'world rst' and has, subsequently been recognised internationally for its value to design and technology education. He is currently Reader in Design Education at Goldsmiths, University of London.

Richard Kimbell special interest is in design learning and its interaction with assessment. He founded the Technology Education Research Unit (TERU) at Goldsmiths, University of London, in 1990, and over 25 years ran research projects for research councils (e.g., ESRC, NSF [USA]), for industry (e.g., LEGO, BP), for government departments (e.g., DfES, DfID) as well as for professional and charitable organisations (e.g., Engineering Council, Royal Society of Arts, Design Museum). He has published widely in the field of technology education, including five single-authored books, many chapters for edited compilations as well as reports commissioned by UK Government Departments, the Congress of the USA, UNESCO and NATO. He has written and presented television programmes and regularly lectures internationally. He has been a Consultant to the National Academy of Engineering and the National Science Foundation in the USA and a Visiting Professor at the University of British Columbia, the University of Stockholm, Edith Cowan University in Perth Australia and Texas A&M University.

Claes Klasander has been the director of CETIS (the Swedish National Centre for School Technology Education at Linköping University) since 2014. He was a member of the group who in 2011 wrote the Swedish national curriculum for the school subject Technology. As the director of CETIS he has a broad knowledge on the situation for technology education, a large network, and he cooperates with many other bodies, for example the Swedish National Agency for Education. In 2010 he wrote his PhD thesis on how different actors in the Swedish school system have been talking about teaching subject content relating to technological systems. He has been teacher educator at Linköping University for more than 20 years. Prior to that he has worked as a secondary school teacher in technology, science and mathematics for 15 years.

Bill Lucas is Director of the Centre for Real-World Learning (CRL) and Professor of Learning at the University of Winchester. In 2017, he was appointed to be the Co-chair of the new PISA 2021 test of creative thinking and with Ellen Spencer has published an authoritative handbook for teachers, *Teaching Creative Thinking: Developing Learners Who Have Fresh Ideas and Think Critically*. His research with Janet Hanson for the Royal Academy of Engineering into engineering habits of mind has been described as a big idea in engineering in schools. His critique of the education system in England, *Educating Ruby: What Our Children Really Need to Learn*, written with Guy Claxton, will be followed up with *Zest for Learning: Developing Curious Learners Who Relish Real-World Challenges in 2020.*

Stephen Petrina is a Professor in the Department of Curriculum and Pedagogy at the University of British Columbia. He specialises in science, technology, engineering and mathematics education (STEM), design and technology education, media and technology, science and technology studies (STS) and curriculum studies. For over a decade, Stephen and a team of researchers have been focusing on how we learn (HWL) media and technology across the lifespan. HWL focuses especially on how children, youth and adults innovate in classrooms, labs, workshops, makerspaces, virtual spaces, home spaces and workplaces. As an Environmental Activist, he also researches learning by ecodesign and natural spaces.

Elwin Savelsbergh is Professor of Science and Technology Education at Utrecht University of Applied Sciences and Associate Professor at Utrecht University. His research programme is about curriculum development and teacher education in STEM, with a particular focus on STEM education for sustainability.

Niall Seery, BTech(Ed), PhD, is currently the Vice President of Academic Affairs and Registrar at Athlone Institute of Technology. He is a qualified Secondary School Teacher of Engineering, Technology and Design and Communication Graphics. Niall has a background in Technology Teacher Education, where he spent 15 years as an academic with a special interest in pedagogical practice. He has served as Director of Studies at undergraduate and master's level while also developing an emerging research agenda. In 2010, he founded and continues to direct the Technology Education Research Group, where he is still active in research supervision and mentorship. Niall also served as a Visiting Associate Professor of Technology Education at the Royal Institute of Technology, KTH, in Stockholm. He has experience in hosting international education conferences and participation in international research projects. He remains committed to advocating for technology and engineering education research and supporting the development of associated policy and practice.

Kay Stables is Emeritus Professor of Design Education at Goldsmiths, University of London. A Founder Member of the Technology Education Research Unit (TERU), she has directed and contributed to projects in primary and secondary education in the UK and overseas. With Richard Kimbell, she authored the TERU retrospective: *Researching Design Learning* (2007). Together with Steve Keirl, she published *Environment, Ethics and Cultures: Design and Technology Education's Contribution to Sustainable Global Futures (2015)*, an edited collection of theoretical and practice-based approaches. More recently, with John Williams, she published a further edited collection titled *Critique in Design and Technology Education (2017)*. Her recent research has focused on design, creativity and sustainable development, digital tools in assessment (the e-scape project), the concept of designerly wellbeing and creating dialogic frameworks for supporting the development of D&T capability, including in digital environments.

Torben Steeg is an Education Consultant with a background in secondary teaching of D&T science and computing, in teacher education, in CPD and in curriculum development. He is particularly interested in new technologies and the maker movement. He is a Fellow of the Royal Society of Arts (FRSA), a Consultant Member of the D&T Association, a Member of the editorial board for the D&T Association's *Design and Technology Education: An International Journal* and an Owner and Editor of the online journal *ECT Education*. He is a part-time Visiting Lecturer on the D&T, Computing and Physics ITE programmes at Manchester Metropolitan University.

Gerald van Dijk started his career as a Science and Technology Teacher in Botswana and the Netherlands. He now works as a Researcher and Senior Teacher Educator in the field of STEM education. In his research, he focuses on design education, maker education and language-sensitive STEM teaching. He is also an active Member of several advisory boards for the improvement of national technology curricula in the Netherlands. Gerald works at the University of Applied Sciences Utrecht.

Didier Van de Velde works in Brussels as a pedagogical advisor for the school boards associated with the Catholic school's network in Flanders, which represents approximately 70% of secondary schools in the Flemish region of Belgium. He is responsible for coordinating the STEM curriculum and is involved in the development of curricula for technology, sciences and mathematics. He supports also pedagogical initiatives that enhance more coherence in STEM education. Before that, he worked as an educational advisor for the curriculum agency of the Flemish Government and worked for 16 years as a Lecturer and Researcher in technology education at the Teacher Education Department of Arteveldehogeschool University College (Ghent University Association). Earlier, he started his career as an engineer in the plastics processing industry.

Arjan van der Meij has a Master's degree in Physics. He has taught at a high school in the Hague since 1995. Around 2009, he joined the emerging maker movement as a 'maker', and right away, he saw the great potential that this movement could provide for education. Together with a group of enthusiastic colleagues, he introduced making in his school and soon after that became one of the trail blazers of maker education in the Netherlands. His goal is to introduce as many children as possible to the maker mindset.

Belinda von Mengersen is Discipline Leader for Design and Technologies at the National School of Arts, Australian Catholic University. As Senior Lecturer in textiles, she specialises in traditional, current and emerging textile technologies. A background in textiles, visual art, fashion and theatre design informs her teaching and research. Her research interests include reflective, creative and speculative writing in design practice, practice-led research in design, comparisons between signature pedagogies in visual arts, design and technology and the inherently interdisciplinary and dynamic nature of design related fields. She is curious to see how the rapid emergence of new textiles technologies will continue to influence all fields of design and influence material thinking and inform ethical sustainable design strategies.

John G. Wells is a Professor in the School of Education at Virginia Tech in Blacksburg, Virginia (USA), and the Program Leader and Co-creator of the Integrative STEM Education (I-STEM ED) Graduate Programme (https://tinyurl.com/VT-I-STEM-ED). He earned a BS degree in Biological Sciences and a BS degree in Industrial Arts from Florida State University, as well as both an MS degree and PhD in Technology Education from Virginia Tech. Dr. Wells is an internationally recognised scholar in the field of Technology and Engineering Education on Integrative STEM Education and Technological/Engineering Design-Based Biotechnical Learning (T/E DBBL). Over the past three decades, his research has been focused along two distinct and closely related areas of study: (1) establishing viable I-STEM ED pedagogical models (PIRPOSAL) that represent the habits-of-mind instructional strategies of T/E DBL and (2) evidencing the efficacy of T/E DBBL practices for intentionally having students 'design to understand' (D2U); promoting designerly ways of knowing and validating such pedagogical practices fosters critical thinking.

Terry Wilkinson (BFA, BEd, MEd) is currently a Course Director in the Bachelor of Education, Technological Education programme, at York University (Toronto, Canada). Her doctoral research focuses on teacher advocacy work in curriculum policy reform. Throughout her working life, Terry has taught and learned from children and adolescents with a wide range of cognitive and behavioural exceptionalities. As a former D&T Teacher with a visual arts degree, she has a keen interest in designing and technological problem-solving processes. She is also a strong proponent of hands-on, project-based learning – especially in school subjects like design and technology, science and technology and mathematics.

P. John Williams is a Professor of Education and the Director of Graduate Research in the School of Education at Curtin University in Perth, Western Australia, where he teaches and supervises research students in STEM and technology education. Apart from Australia, he has worked and studied in a number of African and Indian Ocean countries and in New Zealand and the USA. He is a longstanding Member of eight professional associations, is the Series Editor of the Springer Contemporary Issues in Technology Education and is on the editorial board of six professional journals. He has authored or contributed to over 250 publications and is elected to the International Technology and Engineering Education Association's Academy of Fellows for prominence in the profession.

Deborah Winn is a Practising Teacher and has taught Design Technology in secondary schools in England for 20 years, 15 years of this has been teaching in Neale-Wade Academy, which is part of the Active Learning Trust. Having developed a particular interest in digital technologies, she completed a PhD thesis through the Open University to investigate ways in which younger students could be aided to be creative in the use of complex technologies in the classroom. She continues to explore the issues surrounding these technologies through practical application in the classroom.

An Introduction to Effective Pedagogies of Design and Technology Education

P. John Williams

Abstract The chapters in this book reveal a rich tapestry of pedagogical options from which educators can choose, based on a rationale that is appropriate to their educational purpose. The rationale may derive from theory (constructivism, socio-cultural theory, critical theory, critical pedagogy, activity theory), it may derive from an element of design and technology education (design, making, systems, digital technology, critiquing), or it may derive from a related issue (STEM, problem-solving, the material world, poorly resources environments, vocational-general technology education, pupil teacher interactions). Regardless of its derivation, the discussions throughout the book begin with a particular rationale for design and technology education, and from that base, explore aligned pedagogies.

A core theme running through all the chapters is the notion of authentic pedagogy, activities which are perceived to be real and meaningful by students. Again, we are fortunate in design and technology because we do not have to contrive to be authentic, the area of education lends itself to such an approach. The strands of authentic pedagogy which weave themselves through the chapters in this book do not represent a call for a radical shift in pedagogical philosophy but an encouragement to capitalise on the natural opportunities that are afforded by working toward technological literacy through design and technology education.

The contemporary prevailing wisdom about the constituents of effective education are embedded in design and technology education – active knowledge making, student centredness, learning by design, constructive and construction – ism. In an uncontrived way, such notions appear repeatedly throughout this book, attesting to the potential of design and technology education as an effective form of education. Many of the chapters utilise vignettes to illustrate the discussion and prompts for reflection, which can be used by the reader to guide their contemplation.

P. J. Williams (✉)
Curtin University, Perth, WA, Australia
e-mail: pjohn.williams@curtin.edu.au

P. J. Williams, D. Barlex (eds.), *Pedagogy for Technology Education in Secondary Schools*, Contemporary Issues in Technology Education,
https://doi.org/10.1007/978-3-030-41548-8_1

How fortunate we are to be involved with design and technology education, with its rich history of technology, its current ubiquitous application, its broad epistemology, its diverse pedagogies, and the resultant endless opportunities to use it to engage students in authentic learning opportunities.

A foundation of this book is appropriately set in chapter "Technology Education: The Promise of Cultural-Historical Theory for Advancing the Field" by Marilyn Fleer through her examination of the theoretical frames adopted by educators, which determine what counts as knowledge in technology education, how this is thought to develop over time, and what might constitute valued teaching practices for realising this development. The theoretical frame speaks directly to how secondary teachers engage learners and organise learning experiences in technology education.

The chapter asks, "How has technology education been theorised in a quest for better understanding how teachers teach and young people learn in technology education in secondary classrooms?" In technology education, the focus is not so much on discovering laws or developing propositions about *"reality"*, but rather it is on what should be. Therefore, the teaching of technological knowledge brings a futures lens where values, ethics, and aesthetics are brought to the fore. This demands a very different way of teaching to other disciplines and a different way of conceptualising learning.

Technology education appears in the literature to be theorised through the lenses of either constructivism, or sociocultural theory, or critical theory. A range of teaching models cluster around each of these approaches:

- Constructivism (models of student-centred practice; concept mapping, misconceptions, technological, pedagogical, and content knowledge framework/TPCK, content representation).
- Critical and post-structuralist perspectives (disruptive technologies, political ecology).
- Cultural-historical theory (communities of learners, with Verillon's trio, funds of knowledge), activity theory (content representation).
- Cultural historical activity theory (education for sustainability).
- Physical, intellectual, emotional, and social (Vygotskian every day and scientific concept formation, zone of proximal development).

Marilyn suggests that the profile of teaching models found could benefit from an additional exploration of the way young people learn through Vygotsky's cultural-historical theory of development because his central concepts discuss the unity of social and material environment as a unit of technological practice and cognition. This theoretical frame speaks directly to the unique nature of learning in technology education, and advances on other theories, such as constructivism, because they appear to only capture half of the secondary students' learning experience.

A proposed cultural-historical model of technology education encompasses knowing the learner, supporting creativity, teaching for learner agency, addressing change and new motive orientation, developing professional concepts, and imagining through design.

In chapter "The Case for Technology Habits of Mind", Janet Hanson and Bill Lucas discuss how habits of mind for technology might be conceived and developed to offer technology teachers an alternative lens through which to explore their pedagogy. Habits of mind are ways of thinking or behaving intelligently when meeting new learning challenges. They include dispositions linked to academic success such as perseverance and curiosity and capabilities important for employability and long-term well-being such as creativity and sociability. Disciplinary habits of mind emphasise distinctive ways of thinking and behaving which support learning. In STEM subjects, they can help to bring learning to life for children by demonstrating links between the subject and its application in the world outside school. Technology habits of mind will complement those already developed for science, engineering, and mathematics.

The term habits of mind (HoM) was purposely selected to reinforce the idea that as habits, they can be cultivated concurrently with subject content through appropriate teaching and learning conditions and that as powerful thinking tools they can be used habitually by individuals in many different contexts, both within and outside education. Literature may refer to them as "capabilities", "personal competencies", or "essential skills".

Janet and Bill's proposal, based on an exploration of conceptions of technology, is that there are six habits of mind which are important to technology and, although aligned with other disciplines, they are sufficiently distinct to constitute a unique differentiator for technology.

Pragmatic thinking	Moving from observations to possible explanations, generating the best possible explanation, challenging the norm
Critiquing	Observing, asking questions, testing ideas out on others, risk-taking, accepting feedback, evaluating, making judgements
Imagining	Flexible thinking, creative experimenting, developing multiple alternatives, accepting ambiguity, reflecting, transforming
Making	Sketching, modelling, proto-typing, considering the properties of materials, generating solutions, accepting failure
Human-centred designing	Seeking to understand people, their culture and their interaction with technology, empathising, multidisciplinary collaboration
Maximising contexts	Appreciating the affordances and constraints of different situations, recognising the need to achieve balance, function, and sustainability

They identified four important factors that teachers should focus on when cultivating habits of mind:

- When learners are introduced to a new way of thinking, they need to understand what it is, what it looks like when it is being used successfully, and why it is important.
- The learning climate within the classroom should encourage and reward the new way of thinking.
- Teaching approaches should expect learners to use the new way of thinking.
- Learners should be encouraged to take ownership of the new way of thinking.

Jonas Hallström and Claes Klasander in chapter "Making the Invisible Visible: Pedagogies Related to Teaching and Learning About Technological Systems" examine the characteristics of technological systems in relation to teaching and learning, the rationale being the ubiquitous nature of systems within our everyday lives, and the consequent need to explore these in the development of technological literacy.

Systems thinking, in the context of technology, is the ability to think about, analyse, and even design technology in terms of systems. This is based on systems theory, which is concerned with an interdisciplinary study of systems, and has developed in different scientific fields such as mathematics, biology, economy, and engineering. The most prominent technological aspect of a system is its physical-technical, human-made core, a difference made obvious in comparing, for example, the human digestion system with the national power grid. There are 11 groups of concepts related to systems: the technical core of a system; hierarchies, subsystems, components; connections and wholeness; system boundary and surrounding; isolated, closed, or open systems; control, feedback, flow of information; systems' functions and behaviour, processes, models; scale and complexity; dynamics, development, change; socio-technological perspectives; systems for innovation, conditions for production.

Based on the somewhat limited research available on technological systems in technology education, some conclusions can be made about the teaching and learning of systems:

1. Students understand systems better when they are scaffolded, either by an interviewer or by teaching interventions.
2. Students gain a deeper understanding of systems as they grow older, especially regarding the included components.
3. Both students and teachers are better at understanding structure, input and output of a system than its behaviour, and control mechanisms and flows of information are particularly difficult to grasp.
4. The role of humans in and around a technological system is difficult to understand, probably because humans fulfil so many different roles as designers, users and operators and thereby function as crucial but multifaceted components of the system (Vermaas et al., 2011).

When teaching technological systems to secondary students, research shows that teachers as well as textbooks and curricula can apply four different pedagogies or strategies for making systems conceivable to the students. These pedagogies are underpinned by both practical and theoretical aspects. The pedagogies are interface pedagogy, holistic pedagogy, historical pedagogy, and design pedagogy.

When planning their pedagogy, teachers should ask questions about how students should deal with technological systems. There are three questions that are particularly important:

1. How can students *describe* a certain technological system?
2. How can students *design*, *control* or *regulate* a certain technological system?

3. How can students *understand* changes in a certain technological system over time?

Over the last decade, maker education and, more broadly, the maker movement have attained worldwide social popularity. Makers' enthusiasm for creative design and construction, using old and new tools has proven contagious, and in chapter "Maker Education: Opportunities and Threats for Engineering and Technology Education", Gerald van Dijk, Elwin Savelsbergh, and Arjan van der Meij explore how this movement is related to technology education.

The maker movement is not a single movement but a diverse coalition of entrepreneurs, hackers, sustainability activists, crafts hobbyists, and so on. But there are characteristics which bring them together, for example, the innovative use of tools and playful creativity. The availability of relatively cheap equipment such as laser cutters and 3D printers enabled individuals to prototype artefacts that could only be made in specialised facilities before. The maker movement has been linked to the rising popularity of entrepreneurship, characterised by Make, Share, Give, Learn, Tool Up, Play, Participate, Support and Change. Democratic and entrepreneurial values are attributed to the maker movement, because it gives individuals the capacity to make sophisticated products that can change the world, in a way that was previously unthinkable.

Making is obviously is an important part of technology education with its focus on technological literacy, making is an inherent part of the iterative process from conceptualizing an idea based on some human need to realizing and testing the product. Experiencing and reflecting on this process teaches children how the designed world comes into existence and is therefore an inherent part of a curriculum that targets technological literacy.

The theory of constructionism has been used to shape a maker education pedagogy and describe its foundations. Constructionism asserts that embodied experiences and the production of artefacts are central to the ways people learn and that pedagogies should be interest-driven rather than content-driven. A key constructionists' notion is that education should not stifle children's natural propensity to learn and has resulted in a critique of the "design models" commonly used in schools for containing too much instruction, which would lead to too many interruptions of the learning process. The authors propose "Think, Make, Improve" as an appropriate pedagogical approach which enables design and making.

Maker education is often considered as contributing to STEM learning objectives, and its openness and emphasis on student-centred learning resonates with inquiry-based STEM approaches. Knowledge and skills from a variety of disciplines such as physics, chemistry, mathematics, and art are used to make products, but generally it is not explicated what the relation is with these disciplines. Maker education is often located outside schools, for instance, in libraries and museums. Where it is school-based, it is often part of extracurricular activities, whereby students voluntarily participate and without high stakes formal assessment. The learning environment is often called a "maker space". The space and the tools reflect the characteristics of maker education as described above. However, regular technology workshops in schools vary widely and some can doubtlessly be used as maker spaces.

It is asserted that replacement of a regular technology and engineering curriculum with a radical form of maker education is likely to result in a misfit with generally accepted learning objectives. This has been acknowledged from within the maker education community, but nevertheless, maker education can potentially strengthen more traditional forms of engineering and technology education.

Kay Stables focuses on pedagogies appropriate for teaching design in chapter "Signature Pedagogies for Designing: A Speculative Framework for Supporting Learning and Teaching in Design and Technology Education". One of the common, ubiquitous elements of curriculum in many countries is the centrality of processes of designing. Despite this centrality, there is general recognition that teaching and learning designing is a challenge, not least because of confusion over definitions and models of designing. The aim of this chapter is to provide support for developing approaches to overcome this challenge by proposing a pedagogic framework that crosses borders and enables teachers to focus on the why, what, and how of teaching and learning designing in a flexible and creative way.

Although everyone has *potential* design capability, for the potential to be realised, it needs to be developed through learning experiences. These experiences are not simple and are not easy to learn or teach.

Kay develops the case for design being a particular kind of intelligence, a cognitive function that humans have whether they are professional designers or not. "Designerly ways of knowing" is not an imaginary academic construct, but an everyday reality; it is an activity that constantly provokes a need for new learning, whether it is a specific skill or particular understanding.

The theoretical foundation for the pedagogical suggestions in this chapter are similar to the Habits of Mind discussion in chapter "The Case for Technology Habits of Mind" , based on concepts of signature pedagogies, pedagogies that prepare people for their profession, in ways of thinking, performing, and acting with integrity that have their own "signature" methods of teaching and learning.

The ubiquitous linear notion of "the" design process has been seen by some as creating a systematic process that learners can learn and then apply to design problems. Despite no evidence to support this either in reality or in the classroom, its existence has, by default, created prescriptive approaches that could be viewed as current "signature pedagogies" of design and technology education: pedagogies of identifying a problem, conducting research, generating an idea, making and evaluating.

An alternative approach is suggested which provides a way of considering options for developing designerly thinking, which enable a teacher, and potentially a learner, to make a decision about the facet of designing that needs to be focused at any given time. These pedagogies are pedagogies of: speculation, imaging and modelling, materiality, need-to-know, critiquing, and collaboration.

Design capability is at the heart of technological activity but teaching and learning designing in the context of mainstream schooling is challenging. By moving from a prescribed approach, following a linear set of steps, to a responsive approach, where there is recognition that, as a project and its needs become clearer, the ground will shift, a teacher's pedagogic practices will inevitably be unsettled. A consideration of signature pedagogies, within a framework of pedagogic purpose has the potential to scaffold both a teacher and their learners.

While in some jurisdictions, digital technologies represent a separate curriculum to design and technology, there is an integral relationship between these two areas, and many digital technologies are utilised in the practice of design and technology, for example, in image manipulation, animation, robotics, coding, product designing, manufacture, simulation, and planning. In chapter "Pedagogies for Enabling the Use of Digital Technology", Deborah Winn teaching digital technologies is seen by some teachers as an onerous task, and others simply follow a scheme of work or a set sequence that focuses on a specific outcome. A consideration of what is most useful to students will conclude that the important aspects are to encourage an interest and confidence in the subject, learn how to learn, and develop strategic knowledge. Students then become more confident to try new technologies as they emerge and develop, they also become more informed consumers able to make reasoned choices and enjoy digital technologies for multiple purposes – hobbies, entertainment, work, etc. The least important skill would be for the student to possess the knowledge and have the ability to complete a specific outcome, such as a 3D CAD model, from memory. It is the process, and sometimes failures, which have the most value.

The evidence laid out in this chapter suggests that the most important factors to consider when developing a pedagogical approach to digital technologies is:

- To inspire the students to have an interest in using digital tools.
- To teach students to know what digital tools can be used for and therefore have a context for learning.
- To teach students to have an understanding of the language used in the programs so as to instil confidence when they use the program.
- To teach students to recognise and solve the common errors that result in failure.
- To have tasks that focus on individual progress rather than a successful end product.

Steve Keirl's focus in chapter "Developing a Pedagogy of Critiquing as a Key Dimension of Design and Technology Education" is on the pedagogy of critiquing, and well complements Kay's chapter "Signature Pedagogies for Designing: A Speculative Framework for Supporting Learning and Teaching in Design and Technology Education" on design. All technologies have been designed and with designs come consequences. No technology is neutral, values-free, or universally good. In fact, all technologies are problematic and, like education, a matter of politics. Our local and global challenges such as climate alteration, pollution, resource depletion, and inequitable economic systems are all technological in nature by design. In order to ensure comprehensive engagement with ethics, sustainability, and democratic politics, being critical is essential.

Critiquing means far more than offering tokenistic platitudes which can amount to being demeaning or patronising. The judgments that are sought will be of the articulate, informed, and reasoned kind. Critiquing can be viewed as being both an inward-directed reflective practice and an outward-directed intentional act addressing phenomena and the actions of others. It embraces a lexicon of related terms such a critique (as noun and verb), criticism, critical reflection, critical thinking, imagination, and interpretation. To these can be added critical stance, critical distance, critical disposition, scepticism, empathy, and ambiguity.

Critical theory, critical literacy, and critical pedagogy are contexts for critiquing. Critical theory is fundamentally tied to a struggle for a qualitatively better life for all through the construction of a society based on non-exploitative relations and social justice. The critical educator does not believe that there are two sides to every question but that there are *many* sides to a problem. Critical literacy leads students to challenge and interrogate the status quo as "critical thinkers" when engaged in inquiry and creative transformation. And critical pedagogy places dialectical thought, critique, and conflict as central to significant student experience.

Steve provides the following guidance in developing pedagogies for critiquing:

- Just as there is no one way (or process) to design, so critiquing must eschew formulaic or prescriptive approaches.
- Critiquing helps clarify needs-wants issues, values issues, highlights the contestable nature of technologies and their multiple effects, and heightens student designerly and technological consciousness.
- Critiquing always *responds to* something that exists or has happened – critiquing raises and clarifies questions through reformulation and reassessment.
- Critiquing is deconstructive but not destructive, it has excellent problem-finding or fallacy-exposing capacities.
- Critiquing calls for an understanding of the audience for the critique.
- Critiquing aids *selection* of thinking styles, it is a form of metacognition.
- In being reflective and deconstructive, critiquing may involve discomfort but that is an aspect of critical purpose.
- Critiquing contributes to ethical education.

While both critiquing and designing can enhance student consciousness, thinking styles and confidence, such traits remain undervalued in many situations. While both invite students to be (positively) critical, both also reject rote-learning, fact-learning, or linear/staged procedures. However, there are important differences between the two. While critiquing happens *after* an idea, event, argument, or product, designing *brings into being* these circumstances. While designing is pro-active, critiquing is re-active, and while critiquing is focused, designing is holistic and dynamic. So far as their working arrangement is concerned, critiquing is a tool that serves the design enterprise. Good designing demands good critiquing.

Because material selection in design is complex, Belinda von Mengersen and Terry Wilkinson in chapter "Question Think Learn: A Pedagogy for Understanding the Material World" present a framework of key elements and strategies that can support teaching and learning in design and technology education. In particular, they focus on the pedagogical role of dialogue, peer collaboration, and media resources to foster critical awareness of current production-consumption-throwaway practices and their effects. They propose that critical life-cycle thinking can enable students not only to choose appropriate materials for school projects but also encourage them to think and act carefully about the ways they use, consume, and design technologies.

The two guiding principles of environmental stewardship and social justice should inform teaching in order to develop a "planetary ethic" of care and respect that takes into account the effects of our actions on the lives and well-being of all. From an ecological sustainability standpoint, a functionalist model of teaching and learning fails to account for *all* the costs of design decisions – not simply economic or fit for purpose, but environmental, cultural, political, and social consequences. The question must be asked what do students really *need to know* about materials? Accordingly, the pedagogy advanced requires different questions to seek out a much wider range of material-related factors and their consequences.

Collaborative talk developed through a question-think-learn approach can foster critical thinking and facilitate a deep discussion of all aspects of materials. Students must perceive problems as real or they will not "own" them. Debate, stories as dialogues, argumentation, and reflection can all be utilised to this end. An important aim of D&T in its general educational role is to develop a critical awareness and understanding that materials are intricately linked with how we make and use things. When young people understand that our technological world is not a given but has been *designed* this way, they can begin to imagine how things could be done differently.

In chapter "Pedagogy for Technical Understanding", Torben Steeg and David Hills-Taylor focus on pedagogies for technical understanding, "technical" referring to the elements of a technology that together make technological systems work. These relate to relate to pupils' understanding of how things work and how to make things work, including understanding about the materials they use, scientific and mathematical concepts related to materials and methods, and understanding of mechanical, structural, and control systems.

The purpose of technical understanding is very different to scientific understanding, being to inform designer-maker capability and to underpin technological perspectives. This means that very often scientific concepts have to be reworked or transformed to be directly useful as technical ideas – that is ideas that can be used in the service of practical capability.

It is common to categorise the knowledge that leads to understanding into factual, conceptual, or procedural. Factual knowledge ('knowing that') "includes knowledge of terminology, names or symbols of components, technical vocabulary or names of processes" (ibid). Conceptual knowledge encompasses the abstractions that tell us how facts are related (classifications, theories, definitions, equations, models, etc.). Procedural knowledge is "knowing how" to do things (i.e. how to carry out procedures) and includes, at a strategic level, knowing which procedures are applicable to a situation and when they should be employed.

Some of the perspectives on learning and teaching that underpin the development of technical understanding in design and technology are constructivism, constructionism, situated cognition, and some of the recent findings of cognitive psychology and neuroscience.

A mixture of just-in-time and just-in-case approaches is supported by the idea that four broad kinds of activity are required to make up an appropriate pedagogy for design and technology: making without designing, designing without making, designing and making, and considering consequences. These activities can be supported by three approaches. Firstly, the use of methods to encapsulate (re-conceptualise) explicit scientific knowledge to make it useful technical knowledge. Secondly, product analysis where examining closely the designing and making of others reveals the application of technical knowledge in the development of products and systems. Thirdly, the use of systems thinking to focus attention on the relevant level of detail when designing and making. These are not the only ways to approach the teaching of technical understanding, but they have particular utility.

The conundrum of assessment lies at the heart of any effective pedagogy for design and technology. This is the position from which Richard Kimbell, in chapter "Pedagogy for Technical Understanding", discusses assessment as a core element of teachers' pedagogy. The current practice of summative high stakes assessment is deeply flawed because when teachers are required to be coaches maximising the achievements in externally constructed and atomised examinations, it is very difficult to see how they can simultaneously be developing learners' autonomous capability. The solution presented is the process of comparative judgement, which for learners can be seen as assessment **as** learning, and in the realm of summative assessment (for awarding) it can operate to build communities of teachers that share and disseminate constructs of quality. This enables high reliability in the assessments but additionally it develops cohesion within the community of teachers.

If the curriculum seeks to develop imagination, emotional subtlety, and toughness, and if task-based learning is the chosen vehicle, and if these tasks contain degrees of uncertainty so that we quite deliberately require learners to struggle to find appropriate direction and gain a hand-hold on what they should do next, then the very least that is required by them of us is that they ***trust*** us to deal fairly with them. A capability curriculum is all about the *affective* – and specifically about trust. This is the domain where measurement struggles to get a bearing. It is not just that these qualities are difficult to assess, it is that, since capability acts are essentially uncertain, learners will not go out on a limb and take chances if they believe that – should they fail – they will suffer serious penalties.

The reconciliation of assessment with pedagogy, for examiners and teachers as well as for learners, lies in the constructs of quality that they hold. *Developing* this construct is the key for learners beginning to build their autonomous technological capability. And perversely, seeing the construct as variable between individuals and becoming familiar with the "changing hats" routine has the potential actually to *strengthen* the constructs that learners hold.

So, using a comparative judgement tool, the act of engaging in the judgement process empowers learners progressively to articulate a personal view about what they mean by "good", "ok", "better", "weaker", and "outstanding" work. In this way, they develop a rich sense of quality that enables them to ground their autonomous technological capability in their own judgements. Seen within the requirements of a capability curriculum, the assessment process is, in itself, a learning process. The two are inseparable.

Science, technology, engineering, and mathematics (STEM) education is a relatively recent phenomena in which technology education can play a key role. In chapter "Teaching Problem Solving in the Digital Era", John Wells and Didier Van de Velde discuss the pedagogical contribution technology education can make to integrative approaches to education such as STEM.

The pedagogy of technology education embraces and capitalises on an integration of multiple disciplinary content and practices demanded of the learner as they work toward a plausible design solution with reference to authentic practices and contexts. As opposed to the traditional silo approach, the integration of content and practices in technology education is not contrived. To the contrary, it is predicated on authentic problem scenarios that impose on a learner the "need-to-know", requiring them to use their knowledge of STEM disciplines to formulate questions and/or seek answers for that which they do not yet know. In this way, technology education has the potential to reflect the "nature of technology" through integration of STEM content and practice. Conceptually, based on grade levels and curricular goals, integration is a move away from the traditional mono-disciplinary approach toward a progression of integrative strategies from multidisciplinary, interdisciplinary, and finally to a transdisciplinary approach. When implemented through an integrative technology education pedagogy, interdisciplinary and transdisciplinary are found to be those which best enhance STEM learning.

Fundamental to the teaching of technology education is a pedagogy that capitalises on the integrative nature of technological design-based learning as an instructional strategy for teaching about technology. Technological design-based learning aligns with the pedagogical framework of Integrative STEM Education where learners construct knowledge through engagement in the phases of technological design.

The process of design is guided by the questions posed by the learner at any given point throughout the phases of design. Whether it be a technological question (function, behaviour, structure) or one of content (bioprocessing, liquid/gas flow rates, chemical conversion), in all phases of design students are confronted by designerly questions that reflect a genuine need-to-know. In responding to designerly questions, learners oscillate seamlessly between convergent thinking and divergent thinking in building the body of knowledge necessary for producing a viable design solution. These rapid, continuous transitions between what I know (knowledge domain) and what I need to know (concept domain) lead to informed design decisions that foster habits of mind and higher order thinking skills characteristic of problem solvers.

The value of integration through technological design is realised by the opportunities it provides students to utilise their knowledge and skills, recognise content connections, develop blended disciplinary perspectives, achieve depth and breadth of understanding, develop positive attitudes toward learning, and more thoroughly explore the curriculum.

The imposed cognitive demands of designing serve as the glue for engaging learners in multiple STEM subjects from an authentic experiential learning approach. In particular, the experiential nature of technological design-based learning engages learners in discipline-specific content/practices at varying levels of complexity throughout the phases of design.

In chapter "Technology Education Pedagogy: Enhancing STEM Learning", Moshe Barak demonstrates that the impacts of new technologies means that, in order to ensure relevance, teaching technological problem-solving cannot just be about dealing with peoples' specific needs or technical issues, but also must be about preparing students to integrate into the sophisticated technological world in which new innovative products and services appear rapidly.

In elaborating on this challenge, Moshe proposes four aspects of teaching problem-solving in this digital era: developing new products and services that create new needs and push the market forward, systematic inventive problem-solving, computational thinking, and project-based learning.

In the modern area, inventing new products and services is often designed to create new uses and needs in ways people had not thought of before. This is an innovative approach, beyond the traditional view of problem-solving as identifying and answering people's needs and desires.

Using methods for "systematic inventing thinking" means carrying out systematic manipulations with a system's attributes or components to solve a problem or create a new product. This is an opposite, and complementary, approach to the conventional method of randomly searching for new ideas by methods such as brainstorming.

Fostering students' computational thinking (CT), which is an important target of technology education today, relates to solving problems, designing systems and understanding human behaviour by drawing on concepts fundamental to computer science. This may include, for example, data collection, analysis and representation, problem decomposition or abstraction, and using algorithms, procedures, automation, and simulation for problem-solving.

A rationalised approach to applying the project-based learning (PBL) methodology in the technological class takes into account that many students could benefit from PBL only after they have gained some basic knowledge and working skills in learning a new subject. The P3 scale – practice, small-scale problem-solving, and broad open-ended projects – could help educators in designing effective curricula and project-based learning in the technological class.

The challenge for technology teachers and students today is to design products and services for the future, and to use new and emerging technologies in their design proposals. The new technologies are often virtual, global, and dynamic, and they rapidly produce new products and services extensively affecting individuals' lives, society, and the economy.

Many schools offer both vocational technology education and general technology education, and while they are both forms of technology education, the goals, pedagogies, and assessments are different in each. In chapter "Pedagogical Approaches to Vocational Education", John Williams focuses on the misalignment between the goals of vocational education and the pedagogies employed to achieve those goals.

The vocational instructional approach most commonly involves the demonstration of the application of knowledge to a skill, followed by the students practice until mastery is achieved. The curriculum therefore consists of a sequence of tasks, sometimes related, and the pedagogy applied is skill demonstration followed by individual practice and support.

On the other hand, many employers report the need for new employees who have developed a range of transversal competencies and adaptable occupational knowledge, such as thinking creatively, problem-solving, and collaboration, which can best be achieved through a focus on learning processes. They argue that specific competencies related to machines and manipulative skills can be taught on the job, and these change within short periods of time anyway. Whereas the transversal competencies are required to enable employees to be versatile and flexible within an organization.

The pedagogical approach required to address these transversal competencies would tend to be more student centred and project or problem based, involving a focus on a design process (Billett, 2016), rather than teacher centred with a focus on competencies. This would enable the students to have a degree of creative input into their studies, which would in turn facilitate the development of the transversal competencies.

The thesis of this chapter is that a disconnect exists between the goals of vocational education and the pedagogies most effective in the achievement of those goals. The chapter describes the current situation, proposes frameworks within which a critique of the current situation can be made and then discusses the elements of appropriate pedagogical practice. It concludes that appropriate pedagogies are those which are student centred, implemented in authentic contexts, social, and cross disciplinary; really just sound pedagogical principles of general education. So it would seem that in order for vocational education to fulfil its potential, better satisfy the needs of employers and develop students to their potential, there is no need to develop unique pedagogies but to subscribe to those which are proposed as suitable for sound general education.

The level at which school technology education departments are resourced varies greatly. In chapter "Teaching Technology in "Poorly Resourced" Contexts", Mishack Gumbo focuses on poorly resourced contexts and suggests a paradigm to support transformative and creative approaches to teaching technology in such environments.

This chapter critiques the notion of poorly resourced and emphasises the richness that resides in contexts perceived as poorly resourced – tools, equipment, consumable materials, and curriculum materials. It introduces the notion of "resource sensitive teaching" as it outlines a culturally responsive teaching approach. While it is a government responsibility to resource schools, the fact remains that many are not adequately resourced, and teachers need to react to this situation by taking advantage of the resources commonly used by communities, and so contextualizing the teaching of technology.

Students come to class with technological world views from their communities that can facilitate learning new concepts. The incorporation of these funds of knowledge into school technology practices may require teachers to critically examine their own views about knowledge, human nature, values, and society, so they can adopt a culturally relevant form of teaching.

In rural and poorly resourced communities, blacksmiths, architects, creative industrialists, food design and processing specialists, agriculturalists, expert poets, musicians, dancers, historians, cultural interpreters, all maybe skilled in traditional forms of knowledge and are local human resources that can be consulted to enrich the teaching and learning of technology.

Effective teachers in indigenous contexts possess several qualities that are aligned with ubuntu. Ubuntu is an African value system that means humanness or being human, a world view characterised by such values as caring, sharing, compassion, communalism, communocracy, and related predispositions. This philosophy can be applied universally. For instance, the Aboriginal world view is framed on the unity and coherence of people, nature, and time. Caring, sharing, compassion, communalism, and communocracy could be used to promote cooperative learning in knowledge co-construction, co-investigation of existing solutions or designs, co-design new solutions. Unity and coherence of people, nature, and time can be used in students' design projects, and in this way, a teacher becomes a caring project manager and advisor.

Mishack uses the food technology context as an example of utilizing local resources to achieve culturally relevant teaching. In addition to indigenous food systems being important for human sustenance, they also form a treasure trove of knowledge, which plays a role in the well-being and health, environmental sustainability, and cosmic balance of the ecosystem.

Creative teachers can demystify the notion of "poorly resourced" by creating an awareness about local contexts being rich in resources – energy forms, modes of transport, cooking technics, etc. Teachers who teach in such schools need to make their teaching relevant by recognising the wealth of resources in such environments. Being "resource-poor" can also mean being "technology-rich".

To support pedagogical decision-making in design and technology, in this chapter Niall Seery considers the relationship between cognition and social interactions, framed by the notions of capability and the nature of knowledge which underpin design and technology education.

At the core of effective teaching is the interaction and exchange between the teacher and pupil and by extension, pupil-to-pupil transactions. Understanding the confounding dimensions to these transactions is fundamental to advancing practice. The pedagogical challenge is to examine the relations between cognition and social interactions focusing on:

- The nature of practice – e.g. the visual, the make, the enquiry, and the designerly.
- The relationship between knowledge and authentic D&T activity.
- Unpacking the complexity of language in articulating the disposition (relationship between critical and speculative) of enquiry.
- The educational transaction and the interactions between pupil, material, and teacher as core to cognitive development.

The nature of D&T education requires engagement with a dialectic process, where conversations with the self, the medium, and with others identify the need for discourse, as a means of pushing thinking and performance forward. The generative process at the centre of the educational transaction in the classroom becomes even more critical as we consider the complexity of design education. Design education supports interactions that see social construction as the cornerstone of negotiating new meaning and as such positions designing as exploratory and conditional. However, building an iterative paradigm of speculation and critique requires clarity of understanding as to what is success.

The relationship and balance between knowledge and the designerly can be somewhat managed in younger design classes, where the need for foundational knowledge is apparent and serves the purpose of a starting position, (e.g. the material properties of wood as a consideration for designing a handle) and an explanatory position for further exploration (what else are these characteristics useful for? What characteristics have other materials that also may be useful? etc.). Therefore, the emphasis on the delivery of knowledge associated with a design task is directly related to how permeable the task design is. Considering the role of knowledge as a means of speculating and synthesising new conceptions as a creative endeavour is further enforced when knowledge is used to validate and confirm the viability and utility of the proposed. The more permeable, the less likely students are to rely on a specific knowledge base, therefore relying more on a collaborative process, where there is a negotiated meaning and an evolving knowledge base. In addition, permeable tasks that push the boundaries of the collective knowledge result in a generative process to create the necessary insights, which relies heavily on biological primary knowledge and innate human capacity. The objective is governed by the search, appraisal and application of knowledge in a "lean" or "just-in-time" model. From the perspective of practice, this can appear to be chaotic, as pupils work on different interpretations of the design task and try to seek out knowledge through an experiential and experimental practice, loosely linked by "hunch" and "half-knowing". Once understood, the organised chaos can be respected for its sophistication and benefit.

Niall hopes that through this chapter, D&T teachers will be encouraged to embrace the chaos that is the exploration of new realities and focus on an environment that celebrates the breadth and ingenuity of human capacity.

Stephen Petrina's discussion of the philosophy of technology for children and youth in chapter "Philosophy of Technology for Children and Youth" is an appropriate penultimate chapter for this book as it brings together many of the concepts discussed in other chapters through the book. Stephen wonders what exactly education can offer children if they are natural designers, engineers, inventors, makers, and technologists, is there anything that pedagogy and philosophy can offer them that they do not already have or know? It has become common sense that schools – especially secondary schools – disrupt and stifle children's natural development and quash their innate gifts of creativity and criticism. Why then is it a paradox that in a transformation of youngster and youth to adult, is the loss of the gifts and wisdom necessary to obligations towards future generations?

Philosophy for students "means the habit of always seeing an alternative, of not taking the usual for granted, of making conventionalities fluid again, of imagining foreign states of mind" (James, 1876, p. 178). Dewey defended arguments against teaching philosophy in schools if this amounted to "conscious moralizing". Qualifying the argument, he reasoned that if ethics was defined as human relationships in action, then it is "not only teachable, but is necessary to include in all curriculum". Dewey concluded that "philosophy is *love of wisdom*" if wisdom is understood as "knowledge-plus" (1949, p. 713). In turn, for this chapter, pedagogy is defined as translating or rendering knowledge-plus teachable and learnable.

The Philosophy for Children (P4C) movement largely omits technology from consideration in its classroom materials. For technology educators, this is perceived as a real problem with the pedagogy and discourages technology educators from engaging with philosophy, even though children's literature is awash with themes related to the philosophy of technology. If pedagogy is rendering knowledge-plus teachable and learnable, then of course it is inseparable from philosophy *and* technology; however, school philosophy curriculum have generally not included considerations of technology.

Nearly each day we hear about the speed of technology and get reminded that kids operate faster than any generation that has come before. Kids and technology are fast and impulsive while pedagogy and philosophy are slow and contemplative, conventional wisdom holds. Pedagogy's and philosophy's slow adoption of kids' and new technologies' spontaneous adaptation to one another is proof positive. Philosophers and teachers grew up pulling wagons around, just like medieval children, while kids now grew up on the fast technology. We often marvel at the achievements of kids and technology in spite of the laborious nature of pedagogy and philosophy. Kids and technology roll with Zuckerberg's (2010) wisdom, "move fast and break things", whereas pedagogy and philosophy are preoccupied tinkering with what cannot be fixed.

In the final chapter of the book, David Barlex selects key elements of each chapter and extends the pedagogical discussion to consider other aspects, and to make links between the chapters.

This is a rich book. The argued approaches to design and technology education and the logic of implementing pedagogies matched with those approaches presents powerful ways for teachers to think about their practice. The diversity in the book responds to the often cited need for teachers to vary their practice, ideas about which can be sourced from this rich compendium of practical grounded advice.

References

Billett, S. (2016) Beyond competence: an essay on a process approach to organising and enacting vocational education, *International Journal of Training Research, 14*(3), 197–214. https://doi.org/10.1080/14480220.2016.1254365

James, W. (1876, September 21). The teaching of philosophy in our colleges. The Nation, 178–179.

Vermaas, P., Kroes, P., van de Poel, I., Franssen, M., & Houkes, W. (2011). A Philosophy of Technology: From Technical Artefacts to Sociotechnical Systems. San Rafael, CA: Morgan & Claypool Publishers.

Zuckerberg, M. (2010, October 14). Mark Zuckerberg, moving fast and breaking things [interview with Henry Blodget]. Business Insider. Retrieved from: http://www.businessinsider.com/mark-zuckerberg-2010-10.

Technology Education: The Promise of Cultural-Historical Theory for Advancing the Field

Marilyn Fleer

Abstract Although technology education has a long history in practice (i.e. apprenticeship model, guilds), the theorisation of learning technology and design is relatively recent when compared with other disciplines, such as science education. Consequently, this chapter examines how scholars of technology education in contemporary times have theorised their work in their quest for better understanding how teachers teach and young people learn in technology education. To achieve this goal, this chapter conceptualises the outcomes of this focused theoretical review by examining the essence of what constitutes constructivism, social-constructivism, sociocultural theory and cultural-historical theory. In drawing upon primary sources, the questions posed for critique are: How does each inform design and technology education? What is unique to each? What is the same? What might be the gaps? A critique of the grey zone between social constructivism and cultural-historical theory will be made in the context of theories placed on a continuum, which moves from 'in the head' and 'in the hand', to a dialectical relation between cultural and societal contexts and the cultural development of the person through technology education. It is argued that through examining the alignment, contradictions and movement of thought practice, a theoretically informed discussion of learning the practice and knowledge of technology and design is possible. The chapter concludes by returning to the theoretical review to present a discussion on the place of theory for informing a dialectical understanding of how young people learn and teachers teach in technology and design education.

1 Introduction

Technology education brings to the student new ways of thinking about how their technological world works, how the designed environment is and what it could be. But importantly, technology education gives students different lenses for

M. Fleer (✉)
Monash University, Melbourne, VIC, Australia
e-mail: marilyn.fleer@monash.edu

P. J. Williams, D. Barlex (eds.), *Pedagogy for Technology Education in Secondary Schools*, Contemporary Issues in Technology Education,
https://doi.org/10.1007/978-3-030-41548-8_2

interrogating and critiquing what they see, experience and imagine. This mirrors government and industry discourses of twenty-first-century skills with a renewed interest in STEM education (Wells and van de Velde, 2019; Chap. 13, this volume). Technology education is important for engaging with twenty-first century skills and projecting forward. But in the current political context of reduced numbers of students choosing STEM careers (Education Council, 2018), the field of technology education is increasingly asked what can it bring to general education (Kimbell, 2019; Chap. 12, this volume). As secondary teachers, we know of the relevance and unique contribution of technology education (Hanson, Lucas, 2019; Chap. 4, this volume). We also know that technology education can support general education through critique and futures thinking. Helping students to develop insights, to constructively critique, to push against technological determinism and to build new systems and products for supporting society are clear goals in technology education. Now more than ever, secondary technology teachers need a robust theoretical foundation from which to argue, and an armoury of concepts to speak back. Without a theoretical and conceptual foundation, the field could reinforce, rather than de-bunk, the stereotype that technology education is simply skills-based practice (Chap. 6, Stables, this Volume 7, calls this stereotype busting).

Research reflection 3.1 What theoretical frames do you believe inform the teaching of technology education in secondary schools?

Williams (2016) has shown in his extensive international review of the literature and of the activities of professional associations that the field of technology education has seen major "Shifts in views of learning from a cognitive constructivist perspective to a more sociological view which considers the cultural context and interactions between people" (p. 156). This represents a significant theoretical change in thinking about practices in technology education.

The theoretical frame adopted by the educator draws attention to what counts as knowledge in technology education, how this is thought to develop over time and what might constitute valued teaching practices for realising this development. Therefore, the theoretical frame speaks directly to how secondary teachers engage learners and organise learning experiences in technology education. Williams (2016) has argued that a:

> … more **social constructivist perspective aligns** well with the essentially **social manner** in which technology is developed through **design teams**, for example, and so further supports a **collaborative classroom environment in technology education**. (Williams, 2016, pp. 156–157; my emphasis)

A change in theoretical direction, as noted by Williams, can be best described as moving from a focus on the individual learner constructing knowledge for him or herself, to groups of learners who collectively engage in activity for valued and shared outcomes. The latter theoretical framing is thought to better align with how technology teachers and their students work.

But how do teachers of technology conceptualise their practices and how have researchers used theory to develop models of teaching that support student learning? This chapter seeks to examine what is known about the models and pedagogical practices of technology education from the lens of theory. The chapter asks, "How has technology education been theorised in a quest for better understanding how teachers teach and young people learn in technology education in secondary classrooms?"

To answer this big question, this chapter begins with a discussion of the theoretical frames that currently exist within the field of technology education. An overview of the different ways theory has shaped our worldview of the nature of technological knowledge, the positioning of the learner and what this means for the role of the secondary teacher is given. This is followed by a discussion of different teaching models that show how different theoretical lenses give different readings of technological practices, and therefore shape what might constitute technology education now and into the future. Finally, the central concepts of cultural-historical theory are advanced for technology education where a discussion of *why this approach is particularly suited to the learning requirements of technology education* is given. The chapter closes with a clarification of what Jones, Buntting and deVries (2013) identified as the domination of constructivist approaches and sociocultural views, and Liu and Matthews (2005) suggested are theories incorrectly presented in the technology education literature.

2 Considering the Theoretical Foundations of Technology Education

In many fields of study, researchers seek to determine the laws or develop propositions to explain the world. However, in technology education the focus is not so much on discovering laws or developing propositions about *'reality'*, but rather it is on "what should be, not about what is" (de Vries, 2014, p. 37). Therefore, the teaching of technological knowledge brings a futures lens where values, ethics and aesthetics are brought to the fore. This demands a very different way of teaching to other disciplines, and a different way of conceptualising learning. But, by looking closely at the technology education literature, how the theoretical frame is named and conceptualised appears to be rather muddled and, at times, incorrectly presented (Liu & Matthews, 2005). Therefore, some teasing out of theory in the context of technological practices and pedagogy is important.

Research reflection 3.2: In your teaching, do you privilege technological artefacts, systems or materials that already exit, or do you focus on what has not yet been imagined and created?

Technology education is a relatively young research and curriculum field. However, it does sit on a longer more established philosophical tradition of technology (Scharff

& Dusek, 2003), where debates surrounding the nature of knowledge and reality abound. The philosophical foundations and their related theoretical frames give particular ways of thinking about the nature of knowledge and reality (Kotezee, 2010), and this in turn shapes how technology education practices are theorised, resulting in specific ways of teaching and learning (de Vries, 2014).

Broadly speaking, technology education appears in the literature to be theorised through the lenses of either constructivism, or sociocultural theory or critical theory. Each of these broad terms includes other theoretical frames or names, as we see in cultural-historical theory which is also called CHAT (cultural historical activity theory), or sociocultural theory or activity theory. From within sociocultural theory, references have been made to funds of knowledge or communities of practice, when theorising technology education. Similarly, according to Kotezee (2010), constructivism is also a broad church. Critical constructivism, social constructivism and constructivist teaching models, such as TPCK (Technological, Pedagogical Content Knowledge), have also emerged as terms to capture and theorise practices in technology education.

In linking terms with constructivism, we give different ways of thinking about what might constitute valued forms of technological knowledge and practice. For instance, Fleury and Garrison (2014) draw attention to 'critical' by linking this to constructivism. That is, they discuss 'critical-constructivism' as a way of talking about the individual who is building a relationship to knowledge, where local knowledge, including indigenous knowledge, is brought to the fore. Here they argue that critical constructivism positions the technological learner as someone in the process of "resisting the imposition of global forms of knowledge associated with oppressive and exploitative forms of global political economy, including domestic oppression as well as colonialism" (pp. 21–22). Critical constructivism is also evident in the work of Barlex (2017), who writes about disruptive technologies, and Petrina (2000) in drawing attention to the political ecology of technologies.

The introduction of 'social' to the more traditional theoretical lens of constructivism has opened up other ways of talking about technology education. For instance, Fleury and Garrison (2014) suggest that,

> Accepting a genuine social constructivism would mean rejecting such common assumptions as atomistic individualism, complete autonomy, innate rationality, innate free will as the ultimate essence of all human beings. A thoroughgoing social constructivism involves a philosophical anthropology wherein mind and self are contingent emergent, and evolving, even if relatively stable, social constructions. (p. 20)

Different worldviews bring different solutions to established problems, as noted by Djordjevic, Spirtovic and Acimovic (2016) when they suggest "The social constructivist perspective is meant, therefore, as a strong antidote to technological determinism" (p. 178).

Poststructuralist theories appear in a completely different part of the literature (e.g. Wajcman's (2004, 2010) theoretical work on cyborgs and TechnoFeminism) and, to the best of my knowledge, do not appear to sit centrally within the technology education community. However, important work by Pavlova (2015) on pedagogies that develop, activate and utilise capabilities for global citizenship draw upon Freire – a poststructuralist. Similarly, von Mengersen and Wilkinson (Chap. 10, this Volume) open up new thinking in materiality through the development of a pedagogical model that supports the consciousness of students' materials selection and design for ecological sustainability and social justice. Longstanding researchers such as Pavlova, and more recent contributions, such as von Mengersen and Wilkinson, appear to be setting new directions and generating a new kind of technological consciousness through their use of poststructuralist theories (see also Kierl, 2019, Chap. 9, this Volume).

There appear to be many different theoretical ways of writing about technology education, and these theories inform how we talk about what matters in technology education. In Table 1, a summary of worldviews or theoretical frames are presented that loosely capture and sort out what is generally in the technology education literature.

Research reflection 3.3:

1. Use Table 1 to identify the theoretical frame that informs how you conceptualise technology education.
2. Column 1: Which theory do you believe informs your work?
3. Column 2: When you think about learning, what theory informs your thinking? Column 2 draws out how reality is framed for the learner, and therefore how learning is conceptualised.
4. Column 3: When you consider the learner, does the theory you have chosen explain your view of the learner and your role as the teacher?

Theory underpins the teaching models for technology education that are used in secondary schools. Use Table 2 (further below) to back-track/check which theoretical foundations inform the complexity of your work in technology education. It is possible you may draw upon different models at different times.

Research reflection 3.4:

1. Did you select one or many? Why/why not?
2. What might this allow for your thinking and research?
3. What might be missing?
4. How do these theories speak to the models of teaching developed, and how have they informed teaching practices in the field?

3 Theories Informing Models of Teaching in Technology Education

De Vries (2014) in discussing the teaching of technological knowledge and technical artefacts notes that "most learners are much more aware of the artefacts dimension than of the knowledge dimension" but technology also "comprises knowledge and that technology is something one can study, is much less obvious" to students

Table 1 Theories that inform and shape our view of technology education

Theories	Worldview of young person's learning and development	Worldview of technology education	Most like me
Constructivism	*Reality* exists independent of the observer and is knowable. *Learning* is a process in which the individual constructs meaning for themselves	*When applied* to technology education, knowledge is viewed as a human product that is individually constructed	
		This means the learner has an active role in discovering principles, concepts and practices in technology education	
		The teacher has a facilitation role, creating engaging conditions and providing situations that support the learner to construct understandings for themselves of technological knowledge and practices	
Social constructivism	*Reality* is constructed through human activity	*When applied* to technology education, knowledge is viewed as a human product that is socially and culturally constructed with others	
	Learning is viewed as a social process, where the mind is not an individual entity, but rather a part of an expanded social system	*This means* that others help shape how the learner thinks and acts technologically in their world	
		The teacher has an active role in organising learning contexts that support the young person's learning through and with others. The teacher's attention is on the learners and their engagement with the content	

Cultural-historical theory or socio-cultural theory (also includes communities of practice; funds of knowledge)	*Reality* is always a cultural construction, shaped and developed through interactions between people who also shape and give new meaning to social practices	*When applied* to technology education, this means that technology is defined and developed historically by specific groups in relation to societal needs and what is meaningful for them. Practices leave historical traces within evolving and socially dynamic communities which in turn shape and are shaped by contemporary societal, institutional and personal needs	
	Learning is dynamically captured as the unity of the person, environment and the social collective	*This means* that there is a dramatic relation between the person and their social and material environment that simultaneously includes the past, present and future	
	But also, development of the learner is key and is centred on dramatic moments in everyday life, where new motive orientations to valued community/society practices are engineered. The learner is shaped by but also shapes this dynamic unity	*The teacher*, as part of young peoples' social world, has an active role in developing their motive orientations to concepts and practices in technology that support learners to read but also shape their own activity and participation in the world/cultural practices	
Critical theories	*Reality* is produced through societal structures which have been developed historically and through political processes	*When applied* to technology education, this means examining structures and processes in the designing and making of technologies that reinforce discrimination that make it difficult to support equality and diversity	
	Learning is centred on noticing, analysing, and critiquing societal structures and discourses that support particular readings of reality	*This means* that an individual learner has an active role in analysing and determining existing social, political and historical structures that unfairly discriminate, and which have social and environmental consequences on the human condition	
		The teacher has an active role in supporting the young learner by providing tools that support the analysis of existing and evolved structures, discourses and discriminatory systems	

(continued)

Table 1 (continued)

Theories	Worldview of young person's learning and development	Worldview of technology education	Most like me
Post-postmodern theories	*Reality* is a moving target and is relative. This theory offers insights into issues of power, equity and social justice	*When applied* to a view of technology education, this means there will never be a single and regular ways of defining or enacting technology education. Progress is seen as a 'master narrative' that is to be subverted	
	Learning and the content of learning is always localised, contingent and plural. There is no truth	*This means* that it is not possible to profile what is technology or valued forms of technological knowledge to be learned	
		The teacher has an active role in supporting the young learner to look for multiple ways that reality can be constructed and push against a master narrative acting and performing technologically in the world. Multiple perspectives are valued and supported, as there is no one truth to be taught	

Table 2 Models of technological practice

Models to guide practice	Researchers	Explanation	This model of practice informs my work
Constructivism			
Student-centred education and constructivism	Krahenbuhl (2016)	*Teaching model*: Learners construct their own meaning; social interaction plays a key role; authentic learning tasks are crucial for meaningful learning; learning is dependent on existing understanding	
Concept mapping	Koycu and de Vries (2016)	*Research model*: Attitudes and understandings of engineering are mapped and used as the basis for teaching	
Misconceptions model	Firat (2017)	*Teaching and research model*: Finding out students' misconceptions of reasoning about artefacts as foundational to technological literacy	
Technological pedagogical content knowledge framework TPACK	Harvey and Caro (2017)	*Teaching model*: Technological content knowledge captures the idea that understandings develop from interactions among content, pedagogy and technology knowledge	
TPCK	Mapotse (2017)	*Research and teaching model*: Using TPCK in community engagement projects	
CoRe	Williams, Eames, Hume and Lockley (2012)	*Research and teaching model*: CoRe (content representation) – Subject knowledge, knowledge of curriculum, assessment standards, teaching and learning plan (pedagogical content knowledge) are the focus	
Constructionism:	Papert (1991) and Steeg and Hills-Taylor 2019, Chap. 11, this volume)	*Practice*: Child-centred activities using Lego blocks to build and program a vehicle	
Maker movement/ maker education	van Dijk, Savelsbergh and van der Meij (2019, Chap. 6, this volume)	*Model of practice* (e.g. *out of school clubs)*: Originally digital fabrication in after-school contexts, developed into the use of scratch, MaKey, little bits, etc. the making of physical objects in maker Spces, FabLab, Maker's blogs, etc. think, make, improve model often adopted	
Systematic inventive thinking (SIT)	Barak (2019, Chap. 14, this volume)	Problem-solving approach in which students find solutions to problems by making systematic alterations or they manipulate components (can include computational thinking)	

(continued)

Table 2 (continued)

Models to guide practice	Researchers	Explanation	This model of practice informs my work
Critical and poststructuralist perspectives			
Disruptive technologies	Barlex (2017)	*Teaching model:* Critique tool – Disruptive technology, society, market, people; pedagogy – Life cycle of technology, building scenarios and critiquing technologies that exist in society	
Pedagogies of critiquing	Kierl (2019, Chap. 9, this volume)	In a context of the spectrum of total design avoidance, tokenistic closed design, prescribed but open design briefs and student-determined research-based designing, it has become increasingly important to recognise that all technologies are designed and with this comes consequences. A pedagogy of critiquing foregrounds that technologies are not value-neutral, include ethical, environmental and political dimensions, and can and should be interrogated in all their phases	
Studio teaching as an iterative model of practice	Stables (2019; Chap. 7, this volume)	The many faces of designerly thinking are developed through a pedagogical choreography of: Speculation, imagining and modelling, materiality, need-to-know, critiquing and collaboration	
Political ecology	Petrina (2000)	A teaching model that works as a critiquing tool for considering the design life cycle in technologies. An eco-pedagogy foregrounds the footprint, streams, and wakes of the life cycle of the technology being imagined, created, deployed and disposed. A philosophy for children approach is realised through this teaching model	
Pedagogies of materiality	von Mengersen and Wilkinson (2019, Chap. 10, this volume)	A pedagogical model that supports the development of consciousness of students' materials selection and design for ecological sustainability and social justice	
Cultural-historical approaches			
Lave's community of learners model	Slatter and France (2011a, b)	*Teaching model:* Focus on interactions within communities of learners engaged in authentic design problems for enhanced technological practice	

(continued)

Table 2 (continued)

Models to guide practice	Researchers	Explanation	This model of practice informs my work
Verillon's trio and communities of learners model	Head and Dakers (2005)	*Teaching model:* Pragmatic (tool transforms object), epistemic (instrument affords knowledge) or semiotic (transforms information held by another) nature of technologies is mediated by the teacher to be transformative (challenging use and purpose of technology as presented). Communities of practice capture the engagement beyond objects and through participation in practices with others, where newcomers learn the craft of the old timers (legitimate peripheral participation)	
Funds of knowledge and mediated practices	Fox-Turnbull (2015)	*Teaching model:* Drawing upon the strength of localised knowledges and practices to inform technology education, interactions and conversations support learning in technology education classrooms through engagement with real-world activities and authentic practices	
Activity theory and CoRe	Eames (2016)	*Teaching model:* PCK through CoRe focusing on multivoicedness, the contradictions and the expansive learning through different activity systems (workshop, planning a unit, delivery of unit)	
Education for sustainability through CHAT	Lockley (2016)	*Curriculum model:* Local curriculum development for sustainability (education for sustainability: EfS) – OBJECT teacher's perceptions of EfS – OUTCOMES enacted local curriculum in EfS: TOOLS teacher's perception of national curriculum; SUBJECT teacher's perception of sustainability; RULES school culture of education success; COMMUITY, shareholders in local curriculum development; DIVISION OF LABOUR local curriculum practices	

(continued)

Table 2 (continued)

Models to guide practice	Researchers	Explanation	This model of practice informs my work
PIES	Barlex (2012)	*Curriculum model:* Physical, intellectual, emotional and social (PIES) as a framework for classifying human needs and conceptualising practices in technology education. Draws upon an interpretation of a Vygotskian approach that is centred on social relations for conceptualising peer collaboration and teacher guidance, including taking account of students' everyday concepts, the teaching of abstract (scientific) concepts and working within the student's zone of proximal development (ZPD)	

of technology education (p. 36). A practice-based profession is complex and offers insights into the development of theory and models to support teaching and learning – that go beyond technology education – as Table 1 above shows.

Table 2 illustrates examples of the particular models of teaching and learning in technology education that are found in literature and which are used to inform practice.

Research reflection 3.5 concerning Table 2:

1. Look at column 1. Which model of teaching do you use?
2. Column 3 draws attention to pedagogical practices captured in the model. Which of these relate to your national curriculum? Which do not? What might that say about how the teacher and the learner are positioned in your country?
3. Identify which theory informed your model (or models, if you use more). Can you link these well-known models of teaching in technology education found in the literature to the content of Table 1?
4. Tick which model you currently use. What you captured will always be in the process of development, and a range of models for guiding pedagogical practice in secondary schools will always be present.

You might have noticed that the teaching models shown in Table 2 cluster primarily around constructivism (models of student-centred practice; concept mapping, misconceptions, Technological Pedagogical Content Knowledge framework – TPCK, Content Representation), critical and poststructuralist perspectives (disruptive technologies, political ecology), but also cultural-historical theory (communities of learners, with Verillon's trio, funds of knowledge), Activity theory (Content Representation), Cultural-Historical Activity Theory (education for sustainability) and Physical, Intellectual, Emotional and Social (Vygotskian everyday and scientific concept formation, Zone of Proximal Development).

Research reflection 3.6: Consider what models of technology education dominate in your school. Analyse if the models cluster around constructivism or cultural-historical or critical or poststructuralism.

Whilst there is evidence of movement towards a more cultural-historical approach for teaching in technology education (Williams, 2016), the reality is that no one model of pedagogical practice has dominated. Head and Dakers made mention of this omission back in 2005, stating, "it is our contention that, in mainstream schools (at least in Scotland), and in technology classes in particular, the theories (Activity theory – Engestrom and socio-cultural theory – Bruner) are either largely unknown or have not been interrogated fully in order to explore their implications for teaching and learning" (p. 33).

It would appear that when considering the available cultural-historically informed models, they mostly use Lave's (1988) original *communities of learners model,* Gonzalez, Moll and Amanti's (2005) original *funds of knowledge* framework, Engestrom's (1999) conception of *Activity Theory* and *everyday and scientific concept formation* and the *Zone of Proximal Development* as originally theorised by Vygotsky. In addition, some papers discuss significant cultural issues, such as Seeman's (2015) field work on design for cultural groups and humanisation, and Gumbo's (2015) synthesis of models of teaching to realise a new framework for integrating indigenous technologies into the daily teaching practices of technology teachers.

Taken together, it is suggested that the profile of teaching models found could benefit from an additional exploration of the way young people learn through Vygotsky's cultural-historical theory of development because his central concepts discuss the unity of social and material environment as a unit of technological practice and cognition. This theoretical frame speaks directly to the unique nature of learning in technology education, and advances on other theories, such as constructivism, because they appear to only capture half of the secondary students' learning experience.

4 A Cultural-Historical Conceptualisation of the Unique Nature of Learning in Technology Education

We begin this section with a teaching example that was designed from the cultural-historical model shown further below in this chapter (Fig. 4). *The everyday practice concepts and abstract technological concepts, imagination and creativity, and the social situation of development* frame the lesson sequence that follows.

In the lesson example Helene the teacher looks at the following site for starting point ideas for setting up the design challenge: https://www.data.org.uk/for-education/curriculum/starting-points-for-designing/. In particular she notes that each starting point could engage her students with a wide range of different design responses:

- Playtime
- Communication
- Keeping in Touch
- Staying Safe
- New Worlds
- Thinking Machines

She selects a controversial article from a newsfeed on semi-autonomous car as part of her research from the site on **Going places** – transport at different levels of detail and transport systems and futures thinking about intelligent infrastructure.

Lesson 1: *Helene reads out a newsfeed on a recent fatality involving a semi-autonomous car:*

Are we ready for self-driving cars? On the 19th of March 2018 an autonomous car struck and killed a pedestrian who was pushing their bike at night across a road. The car had not been programmed to recognise this particular scenario of a pedestrian (see AP images, 2018*; http://www.apimages.com/metadata/Index/Self-Driving-Vehicle-Fatality/d7a9837e1275452d84aed6ea70a7f9ec/2/0).*

Helene poses three questions, and then invites small group discussion, followed by a whole group sharing in order to facilitate potential debate amongst students:

1. ***What comes to mind?*** *– After hearing about the newsfeed scenario, discuss your thoughts, consider the design challenge that comes to mind and plan your solution.*
2. ***How many more bodies must be sacrificed in the design of autonomous vehicles****? What are the instructions that a programmer would need to code into an autonomous vehicle in order to keep the driver, pedestrians and oncoming cars and their passenger's safe, and to minimise the carnage? What everyday scenarios would need to be worked through (like the one in the newsfeed where the car did not recognise the bike being pushed as being associated with a human being)? What could be the body count if we don't predict all the possibilities?*
3. ***What are the safety issues the programmer would need to consider in coding autonomous vehicles?*** *Consider the following scenario and discuss what you think the programmer should do? Who decides when confronted with the need to choose between swerving in order to miss hitting two cyclists, or the mother walking with her toddler and infant or a tree which would kill the driver? Who lives and who dies must be decided in this situation – not by the driver, but by the designer, programmer and manufacturer of the car. Prepare a set of instructions to guide each.*

Lessons 2–5: The follow-up lessons are organised to deepen the students thinking about design by introducing a system perspective, ethical coding and an ecological worldview (Himmelreich, 2018, *http://theconversation.com/the-everyday-ethical-challenges-of-self-driving-cars-92710). Helene sets up the following discussion questions and design challenges to research, role-play and for games designing. But she also gives the students the option to do design work to follow their interests – as an open-ended design challenge.*

Design the road rules and environmental features for self-driving cars in your community. *The decisions that are made by an engineer and a programmer now won't be about how to instruct one car but rather it will be about how all cars drive. Cars will be communicating with each other. In the future, will we need crosswalks or traffic lights? Autonomous cars will change what our environments and road systems look like. As self-driving cars emerge in our community, we may have to invent new systems, new road rules and different community features to support them. Design a road system that would support autonomous vehicles.*

Design a code of practice for ecological and ethical algorithms: *As an engineer working out how to move people and goods across cities and country towns, decisions have to be made between the footprint of driving, costs and the needs of humans, particularly those living in remote regions (access and inclusion). All decisions have some kind of environmental impact. How might you design a sustainable transport system for both travel efficiency whilst taking into account the environmental impact of what is designed. Can you design a set of ecological principles into code? Prepare a code of practice for designers. In addition to a 5 star rating for safety, what might be a 5 star rating for ecological sustainability? What might be evidence of a 5 star rating for a company that uses ethical algorithms?*

Open-ended design challenge: *After considering the newsfeed about semi-autonomous cars, what design challenge comes to mind? In small group discuss your ideas, and then, either collectively or as an individual, identify the design challenge and design your own solution.*

Assessment: *Because Helene organised the students into small groups, this gave her the opportunity to move about the classroom observing the students' learning during the process of discussing, designing, playing with systems and researching about transport and semi-autonomous vehicles. She made anecdotal notes on how students engaged with the topic, but also how they designed, made models/processes and how they critiqued and evaluated their solutions.*

Research reflection 3.7: After reading this teaching example, why do you think this might be an example of cultural-historically informed approach to technology education?

1. What concepts did Helene draw upon to frame her pedagogy?
2. How might the choice of starting points:
   (a) Appeal to girls?
   (b) Support learners from different cultural backgrounds?
   (c) Build on students' interests generally?
   (d) Motivate collective designing?
   (e) Make visible the individual's confidence and competence in technology education within a collective context, but where individual assessment is the norm?

What we know about Helene's plan and theoretical approach is that, first, she used the dialectical logic of *cultural-historical* theory. As originally developed by Vygotsky (1987), cultural-historical theory is a system of concepts that work together to explain cultural development. Learning about ethics through designing a code of practice for ethical coding is a cultural practice. *Culture* is not ethnicity or race, but rather is the cultured practices of particular societies and communities, such as those technologies that are invented to solve the new human need of keeping people safe when autonomous vehicles become the norm. These cultural practices are realised through everyday life as well as in specifically created institutions, such as secondary schools. Within these institutions, young people participate in the practices that the institution values and makes time and resources for, such as, technology education. Within these classrooms are activity settings, such as group teaching, bench or lab work, field trips and group discussions about ethical *algorithms*. How a young person enters into these different activity settings, and how they participate, is determined by their social situation of development (their motives and the demands made upon them) (Vygotsky, 1994). But also, the demands the learner makes upon the activity setting, where their agency and competencies, shape the activity setting (Hedegaard, 2012). In line with this, the term *historical* in 'cultural-historical' is not a description of the past practices or technological inventions, but rather the focus is on how current practices carry with them the genesis of their development, and this shapes the secondary student's current technological experience in everyday life. Knowing that the existing road system is based on human driving abilities, and could become a relic from the past, is an example of how history shapes existing and future practices. Therefore, dialectical logic that underpins cultural-historical theory is captured in its naming, as a *developmental dynamic*, always in motion, and always as a relation between practice and people. This *developmental dynamic* can be seen in the teaching example, because it examines the past and projects into the future through what it will mean for design and production in the car industry.

From a cultural-historical perspective, technology education that explores this ethical dilemma places the secondary student into a new relationship with reality as this problem is debated with other learners. In this scenario, students engaged in technology education will think differently about an everyday technological practice such as stopping at a traffic light and thinking about this as a designed system, because they have come to understand it in a new way, with new conceptual lenses – such as an ethical algorithm. The student through being introduced to technological concepts is positioned to consciously relate to this technological practice differently. Technology "education creates a completely new relationship" between the student's "motives and actions" and this "forms the conditions for the emergence of new structures and functional features of psychological processes" (Leontiev & Luria, 2005, p.44) – to think ethically, to realise the power of design and to consciously design in new ways. In the teaching example of an ethical dilemma, brings to the students' attention deeper considerations, than simply physical features of aerodynamics or aesthetics or even a business lens of increased market share through being able to design a popular car shape to increase sales. This dilemma is emotionally charged, has no simple solution, and invites collective reasoning that goes beyond the car and engages with everyday life and the human condition.

4.1 Everyday Practice Concepts and Abstract Technological Concepts

The pedagogical practices associated with this scenario can be explained through the dialectical relations between everyday practice concepts and abstract technological concepts. Vygotsky (1987) used the term scientific concepts to reflect what I am calling *abstract technological concepts*. Vygotsky also used the term everyday or spontaneous concepts to capture what I am calling *technological practice concepts*. In cultural-historical theory, technological practice concepts and abstract technological concepts work together, as noted by Vygotsky (1987): "...there is a *mutual dependence* between these two paths of development" (p. 221; original emphasis). Through exploring the design, development and even the mass production of autonomous vehicles (whether as simulations or actual designed and developed models), students are located in a practice context that requires a team approach. Students work practically to realise their design solutions, and in so doing, pave the way for learning abstract technological concepts that are meaningful and engaging for the learner, such as exploring the ethics of the situation, or when examining the crumple zone (absorb energy on impact in a car accident) of the materials used in the manufacturing of cars. Developing understandings of abstract technological concepts means that design practices change, but also future student practices change, such as when purchasing a car later in life. Therefore, it is possible to see how "*conscious awareness enters through the gate opened up by the* [abstract technological] *scientific concept* (Vygotsky, 1987, p.191; original emphasis). The

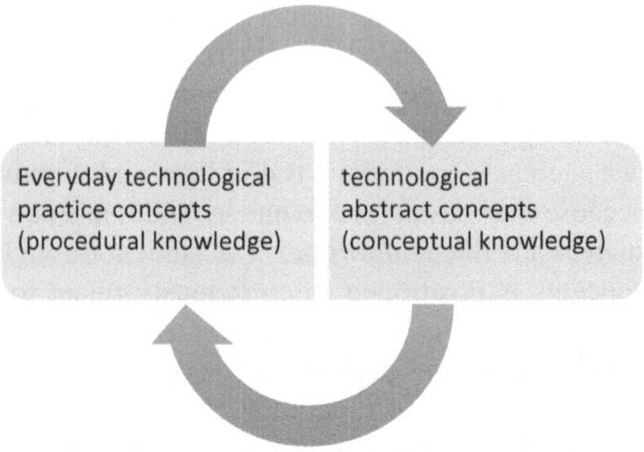

Fig. 1 The dialectical relations between everyday technological practices and abstract technological concepts

development of *conscious awareness of everyday practices* is an important outcome of technology education.

The relations between everyday practice and abstract concepts are central for the development of technological thought. A cultural-historical view of pedagogy foregrounds the importance of both forms of concepts – the practice concept and the abstract concept. Cultural-historical theory not only recognises the importance of each in the development of technological thought and action, but theorises the different pathways that together contribute to the development of conscious thought. That is, the "everyday [technological practice] concepts have blazed the trail for the continued downward growth of [abstract technological] scientific concepts" (p. 219). But also, the abstract technological concepts give new meaning to everyday practices. Figure 1 captures this dynamic relation.

Another key dimension of technology education alluded to by Williams (2016) is the social relations of people with practices. This is evident as learners work together in design teams, work with materials and project managers, engage with clients, etc. Cultural-historical theory focuses on the relations between the person and the social and material environment. They act in unity. Figure 2 captures the dynamic relations between people and practices in technology education. Here the agency of the learner, their motives and competencies determine how they read each technological situation. In the technological example of exploring the design and production of a vehicle, this social situation is likely to be read differently by different students.

4.2 Imagination and Creativity

In line with de Vries (2014) who has shown that technology education works with artefacts "not yet there", Vygotsky's (1987) dialectical conception of imagination and creativity also speaks to technology education. That is, secondary students can imagine practices, systems and artefacts in technology education. Imagining something that is not yet created (such as road systems for ethical autonomous vehicles) brings into reality something new, which changes the practice setting and interac-

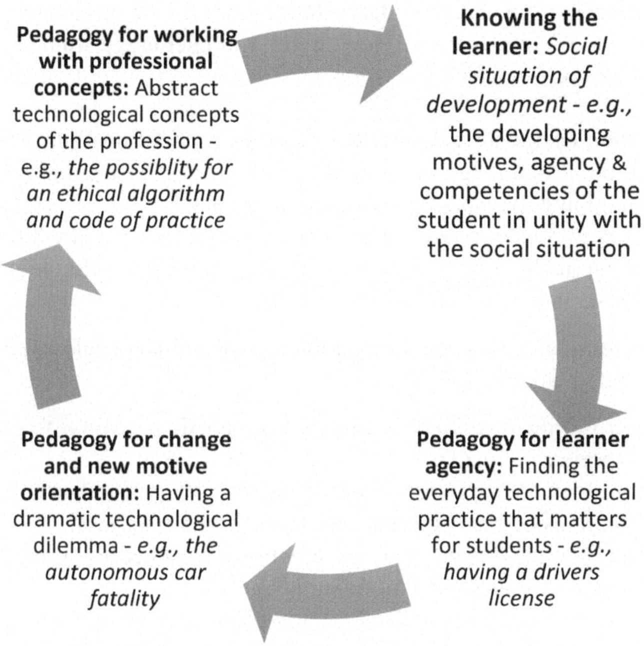

Fig. 2 The dynamic relation between people and practices in technology education – the drama of the ethical dilemma brings development in student thinking

Fig. 3 The dialectical relations between imagination and creativity in technology education

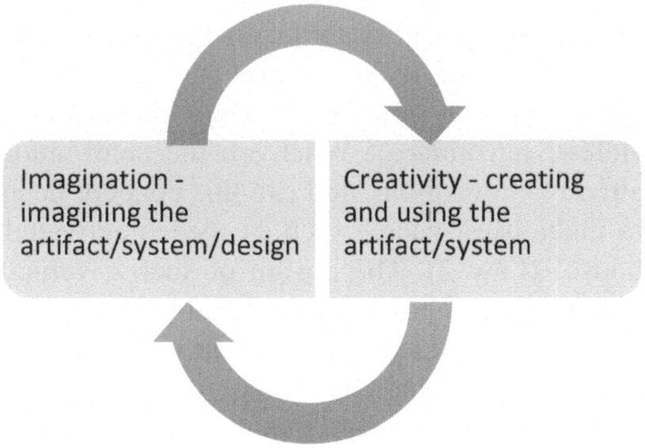

tions (of drivers, of accident statistics). To go beyond de Vries (2014) and in line with the logic of disruptive technologies (Barlex 2017), when something has already been created and exists in reality, it becomes possible to critique the artefact and the surrounding social practices.

Vygotsky (2004) offers technology education another way of considering the design and making process through introducing the dialectical relations between imagination and creativity. When something is imagined, it can then be created. When something is created, it gives new social possibilities, which in turn gives new ideas for imagining other design solutions. This dialectical relation is captured in Fig. 3.

Research reflection 3.8 In secondary classrooms, do we give enough time for developing students' imagination? How might you encourage students to imagine before creating? For instance:

1. Visualise their design ideas BEFORE drawing or modelling them.
2. Encourage them to engage in thought experiments – by testing out their ideas through imagining the working solutions in action.
3. Imagining new social relationships and interactions that are brought about because of the technology solution realised through the technological creation that now exists in the community.

There are four laws that are foundational to the dialectical relation between imagination and creativity. They are as follows:

1. Imagination is "always based on elements taken from reality, from a person's previous experience" (p.13).
2. Imagination does not reproduce what already exists, but rather brings together new combinations of past experiences to form something new.
3. Imagination includes images that are emotionally linked to an experience, but also experiences can provoke emotional images.
4. Imagination can also be crystallized as an object existing in the real world (creation), where it can affect other things (and even support the development of new images/imagining) (Vygotsky, 2004).

These four laws speak directly to technology education. For instance, imagining fully autonomous vehicles is based on the past experience of being in a car that needs a driver (Law 1). By bringing to the design process, different insights, as occurs when teams discuss and share their ideas, afford new imaginings that are collectively produced (group discussion based on the ethical dilemma that Helene introduces). What are the semi-autonomous features of cars that we already know about, and can this concept be applied to other features of the car, to make life easier – such as speech activated processes – 'Find my friend X's house' (Law 2). The design of such a vehicle is directly linked to something emotional and positive – to be able to visit a friend, or it could be linked to an ethical dilemma, such as "Would I buy a car that was pre-programmed to run me into a tree or to hit someone so that my life was saved" (Law 3). Once the design features exist in reality, this gives new possibilities for users of the new technology, such as being able to drive earlier than 17 years of age because the car is now programmed to drive more safely than is possible by humans – therefore giving the possibility of having a driver's licence earlier – something likely to be of great interest to secondary students (Law 4).

4.3 *The Social Situation and the Social Situation of Development*

The relation between the social situation (group discussion ecological footprints associated with design of autonomous cars) and the secondary students' motives and competences (concept of ethics, understandings of materials, values, beliefs) is captured through the concept of the *social situation of development* (Vygotsky, 1994). In technology education, students will always bring different experiences to the social situation, and this will mean problem scenarios introduced by the teacher are likely to make visible a diversity of social situations of development of the secondary students in the class. This concept supports the pedagogical practices of the technology teacher because it captures in unity the person, the practice, the problem scenario and the social relations between pupils – as the teaching lessons of Helene would illustrate in practice. The concept of the social situation of development is more than finding out what a student knows and can do; it is a concept that always socially positions the student in relation to others and to their environment as part of the teaching interaction.

Related to the concept of the social situation of development is the Vygotskian concept of *drama or dramatic moments*. This concept is also shown in Fig. 4. The scenario of the autonomous vehicle on face value is an example of a social situation, but not necessarily a social situation of development. This scenario becomes a learning or developmental opportunity in technology education, when the drama of the ethical dilemma is introduced. The dramatic situation brings together ethics and potentially the study of sustainable materials through exploring and researching the crumple zones in car crashes in a bid to save lives – and to work out how the pre-programmed response in an accident situation can deal with the possible crash situations – that is, both the car design and the materials used. The drama orients the secondary student to learning about materials and acts as a force in developing their motives and conceptual understandings of materials. The ethical dilemma is an example of how in secondary classrooms, a dramatic and emotionally charged social situation can orient the secondary students to develop new competencies and motives in technology education.

To realise a change in technological thinking, the concept of drama was introduced as the dynamic force that emotionally engages and activates the secondary student. These central cultural-historical concepts discussed in this section are brought together in Fig. 4.

This cultural-historical model draws attention to the social dimensions of technological learning which the literature has shown to be important, but which previous theories have neglected to fully explain, and therefore are not visible in the secondary teaching models they inform. Importantly, the word 'social' has been added on to 'constructivism' as one way to signal the need for movement from the lone learner to the social collective in technology education (as also noted by Barlex, 2012). Similarly, funds of knowledge, and communities of learners, as examples of cultural-historical theory, and references to activity theory through Engestom's

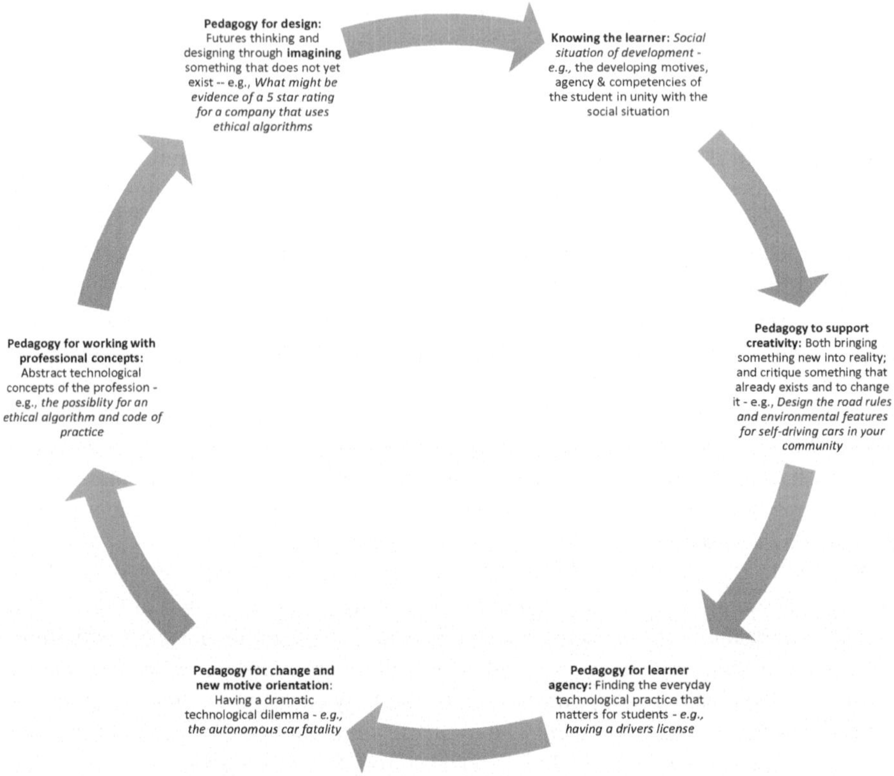

Fig. 4 A cultural-historical model of technology education

models, have also foregrounded the social dimensions of learning in technology education (Tables 1 and 2). Whilst the naming of theories in technology education could be misleading, as discussed by Liu and Matthews (2005), the directions being taken do align with Vygotsky's central concepts, therefore, do give eminence to a cultural-historical conception of technology education for secondary teachers. Therefore, Fig. 4 seeks to capture some of the key cultural-historical concepts discussed generally in the technology education literature, but does so through the original writings of Vygotsky. Together, these particular concepts speak directly to the social practices that are unique to secondary technology education.

Research reflection 3.9 Use the model in Fig. 4 to design a secondary lesson or program that draws on the cultural-historical concepts to guide pedagogical practices. Design challenges could be as follows:

1. Design a social media engagement plan that effectively deals with trolls.
2. Design, make and evaluate a *Dungeons and Dragons* event with a storyline that focuses on a favourite science fiction book.
3. Design a digitally infused party jacket which features sound and light capability.

5 Conclusion

This chapter sought to ask some fundamental questions about the relations between theory and practice in technology education. The theoretical muddle found in the technology education literature is to do with how theoretical orientations are named, rather than any particular models and theoretical framing being wrong, as some have claimed. What is clear is that the primary goal of scholars and the professionals who work in practice is to better name the social nature of their work but in a context of what de Vries (2014) has shown to be different to other professions.

Cultural-historical theory gives a holistic way of capturing and understanding the complexity of this important past-present-futures work, which other theories do not fully explain. In many respects, technology education has the potential to lead the education field in developing new models of practices captured through the lens of cultural-historical theory. Crystallising the theoretical lenses (Table 1), and clustering the models of practice (Table 2), is one step in supporting the shift from constructivism to cultural-historical theory as posited by Williams (2016) for developing technology education into the future.

After reading this chapter:

In developing your technology lesson or program, you will often think intuitively and devise lessons that are very successful in engaging your students and moving them forward in their technology learning. You won't necessarily have resorted to theory in order to devise the lesson. Devising a lesson from a theory as a starting point is challenging, and trying to do this may result in a 'deviser block'. Theory is important for helping you understand your lessons and pedagogical practices from different theoretical perspectives. This reflection on your practice in light of theory will give you both insight into and understanding of why your lessons are successful (or how they can be improved) and enable you to justify and improve your practice discussion with fellow teachers and senior leaders and in responding to the worth of new educational initiatives that may or may not be relevant to technology education. In light of this, consider the following questions:

Where am I now?
Where would I like to be in the future?
How can I get there?
What needs to change politically to enable technology education to be better understood?
What foundational theories and concepts can I bring to the discourses on STEM, twenty-first century learning and debunking of stereotypes?
What role can I play in articulating what is unique about technology education and how can this work productively contribute to general education?

References

AP images. (2018). http://www.apimages.com/metadata/Index/Self-Driving-Vehicle-Fatality/d7a9837e1275452d84aed6ea70a7f9ec/2/0). Accessed 12th July 2018.

Barlex, D. (2012). Curriculum development. In P. J. Williams (Ed.), *Technology education for teachers*. International Technology Education Studies. (pp. 196–230). Sense Publishers, Rotterdam.

Barlex, D. (2017). Disruptive technologies. In P. J. Williams & K. Stables (Eds.), *Critique in design and technology education. Contemporary issues in technology education*. Singapore: Springer.

De Vries, J. R. (2014). Technological knowledge and artifacts: An analytical view, In De Vries, J.R. (eds). Defining technological literacy. Towards an epistemological framework. Palgrave MacMillan, New York, US, (pp. 23–40).

Djordjevic, B., Spirtovic, O., & Acimovic, D. (2016). Social constructivism and technology. *International Journal of Business and Social Science, 7*(11), 178–189.

Eames, C. (2016). Exploring teacher pedagogical content knowledge (PCK) development using CoRes (content representations). In D. S. P. Gedera & P. J. Williams (Eds.), *Activity theory in education. Research and practice* (pp. 169–182). Rotterdam, The Netherlands: Sense Publishers.

Education Council. (2018). Optimising STEM industry-school partnerships: Inspiring Australia's next generation, final report, STEM Partnership Forum, Carlton South, Australia.

Engestrom, Y. (1999). Activity theory and individual and social transformation. In Y. Engestrom, R. Miettinen, & R.-L. Punamaki (Eds.), *Perspectives on activity theory* (pp. 19–38). Cambridge, UK: University of Cambridge.

Fox-Turnbull, W. H. (2015). Conversations to support learning in technology education. In P. J. Williams, A. Jones, & C. Buntting (Eds.), *The future of technology education* (pp. 99–120). Singapore: Springer.

Firat, M. (2017). Growing misconception of technology: Investigation of elementary students' recondition of and reasoning about technological artifacts. *International Journal of Technology and Design Education, 27*, 183–199. https://doi.org/10.1007/s10798-015-9351-y

Fleury, S., & Garrison, J. (2014). Toward a new philosophical anthropology of education: Fuller considerations of social constructivism. *Interchange, 45*, 19–41. https://doi.org/10.1007/s10780-014-9216-4

Gonzalez, N., Moll, L. C., & Amanti, C. (2005). Funds of knowledge. In *Theorizing practices in households, communities and classrooms*. Mahwah, NJ: Lawrence Erlbaum Associates Publishers.

Gumbo, M. T. (2015). Indigenous technology in technology education curricula and teaching. In P. J. Williams, A. Jones, & C. Buntting (Eds.), *The future of technology education* (pp. 57–76). Singapore: Springer.

Harvey, D. M., & Caro, R. (2017). Building TPACK in preservice teachers through explicit course design. *Technology Trends, 61*, 106–114. https://doi.org/10.1007/w11528-016-0120-x

Head, G., & Dakers, J. (2005). Verillon's trio and Wenger's community: Learning in technology education. *International Journal of Technology and Design Education, 14*, 33–46.

Hedegaard, M. (2012). Analyzing children's learning and development in everyday settings from a cultural-historical wholeness approach. *Mind Culture and Activity, 19*(2), 127–138.

Himmelreich, J. (2018). The everyday ethical challenges of self-driving cars Posted March 27, 2018 9.42pm AEDT, The Conversation, http://theconversation.com/the-everyday-ethical-challenges-of-self-driving-cars-92710. Accessed 12th July 2018.

Jones, A., Buntting, C., & de Vries, M. J. (2013). The developing field of technology education: A review to look forward. *International Journal of Technology and Design Education, 23*, 191–212. https://doi.org/10.1007/s10798-011-9174-4

Kotezee, B. (2010). Seven posers in the constructivist classroom. *London Review of Education, 8*(2), 177–187. https://doi.org/10.1080/14748460.2010.487340

Koycu, U., & de Vries, M. J. (2016). What preconceptions and attitudes about engineering are prevalent amongst upper secondary school pupils? An international study. *International Journal of Technology and Design Education, 26*, 243–258. https://doi.org/10.1007/s10798-015-9306-4

Krahenbuhl, K. S. (2016). Student-centered education and constructivism: Challenges, concerns and clarity for teachers. *The Clearing House: A Journal of Educational Strategies, Issues and Ideas, 89*(3), 97–105. https://doi.org/10.1080/00098655.2016.1191311

Lave, J. (1988). *Cognition in practice*. Cambridge, UK: Cambridge University Press.

Leontiev, A. N., & Luria, A. R. (2005). The problem of the development of the intellect and learning in human psychology. *Journal of Russian and East European Psychology, 43*(4), 34–47.

Liu, C. H., & Matthews, R. (2005). Vygotsky's philosophy: Constructivism and its criticisms examined. *International Education Journal, 6*(3), 386–399.

Lockley, J. (2016). Teachers designing classroom cultural through the lens of cultural-historical theory. In D. S. P. Gedera & P. J. Williams (Eds.), *Activity theory in education. Research and practice* (pp. 183–198). Rotterdam, The Netherlands: Sense Publishers.

Mapotse, T. A. (2017). Development of a technology education cascading theory through community engagement site-based support. *International Journal of Technology and Design Education*. https://doi.org/10.1007/s10798-017-9411-6

Papert, S., & Harel, I. (1991). Situating constructionism. In S. Papert & I. Harel (Eds.), Constructionism. New York: Ablex Publishing.

Pavlova. (2015). Design and technology education for sustainable futures: In preparation for global citizenship. In K. Stables & S. Keirl (Eds.), *Environment, ethics and cultures. Design and technology education's contribution to sustainable global futures* (pp. 87–100). Rotterdam Sense Publishers.

Petrina, S. (2000). The political ecology of design and technology education: An inquiry into methods. *International Journal of Technology and Design Education, 10*, 207–237. https://doi.org/10.1023/A:1008955016067

Scharff, R. C., & Dusek, V. (Eds.). (2003). *Philosophy of technology. The technological condition. An anthology*. Malden, MA: Blackwell Publishing.

Seeman, K. (2015). Designing for cultural groups and humanization: Two ideas from design anthropology. In K. Stables & S. Keirl (Eds.), *Environment, ethics and cultures. Design and technology education's contribution to sustainable global futures* (pp. 101–118). Rotterdam Sense Publishers.

Slatter, W., & France, B. (2011a). Taking part in the dance: Technology teachers interacting with communities of practice. *International Journal of Technology and Design Education, 21*, 217–233. https://doi.org/10.1007/s10798-010-9115-7

Slatter, W., & France, B. (2011b). The teacher-community of practice-student interaction in the New Zealand technology classroom. *International Journal of Technology and Design Education, 21*, 149–160. https://doi.org/10.1007/s10798-000-9111-y

Vygotsky, L. S. (1987). Problems of general psychology. In R. W. Rieber & A. S. Carton (Eds.), Trans by N. Minick. *The collected works of L.S. Vygotsky* (Vol. 1). New York: Plenum Press.

Vygotsky, L. S. (1994). The problem of the environment. In J. Valsiner & R. vander Veer (Eds.), *The Vygotsky reader* (pp. 347–348). Oxford: Blackwell.

Vygotsky, L. S. (2004). Imagination and creativity in childhood. *Journal of Russian and East European Psychology, 42*(1), 7–97.

Wajcman, J. (2004). *TechnoFeminism*. Cambridge, UK: Polity.

Wajcman, J. (2010). Feminist theories of technology. *Cambridge Journal of Economics, 34*(1), 143–152. https://doi.org/10.1093/cje/ben057

Williams, J. P. (2016). Research in technology education: Looking back to move forward … again. *International Journal Technological and Design Education, 26*, 149–157. https://doi.org/10.1007/s10798-015-9316-1

Williams, J., Eames, C., Hume, A., & Lockley, J. (2012). Promoting pedagogical content knowledge development for early career secondary teachers in science and technology using content representations. *Research in Science and Technological Education, 30*(3), 327–343. https://doi.org/10.1080/-2635143.2012.740005

The Case for Technology Habits of Mind

Janet Hanson and Bill Lucas

Abstract This chapter suggests how habits of mind for technology might be conceived and developed to offer technology teachers an alternative lens through which to explore their pedagogy. Habits of mind are ways of thinking or behaving intelligently when meeting new learning challenges. They include dispositions linked to academic success such as perseverance and curiosity and capabilities important for employability and long-term well-being such as creativity and sociability. Disciplinary habits of mind emphasise distinctive ways of thinking and behaving which support learning. In STEM subjects, they can help to bring learning to life for children by demonstrating links between the subject and its application in the world outside school. Technology habits of mind will complement those already developed for science, engineering and mathematics.

Keywords Habits of mind · Technology habits of mind · Learning dispositions

1 Introduction

This chapter offers a conceptual analysis of the ways in which understanding of technology education might be enriched if viewed through the lens of habits of mind. We discuss perceptions held about technology education and its place in the curriculum and propose that habits of mind offer a helpful way of framing its value to learning and its contribution to the educational outcomes of young people.

Given that technological literacy is increasingly seen as contributing to social and economic well-being, it is strange that the place of technology within the curriculum has not yet been firmly anchored in many education systems. Perhaps because technology itself is such a fast-moving field, it is challenging to design a curriculum that retains a sense of both currency and relevance. However, many tensions surrounding the positioning of technology education are relatively

J. Hanson (✉) · B. Lucas
Centre for Real-World Learning, University of Winchester, Winchester, UK
e-mail: Janet.Hanson@winchester.ac.uk; Bill.Lucas@winchester.ac.uk

P. J. Williams, D. Barlex (eds.), *Pedagogy for Technology Education in Secondary Schools*, Contemporary Issues in Technology Education,
https://doi.org/10.1007/978-3-030-41548-8_3

Table 1 Technology habits of mind (THoM)

Discipline	Habits of mind					
Technology	Pragmatic thinking	Critiquing	Imagining	Making	Human-centred designing	Maximising contexts

long-standing. They arise from factors such as definitions of the concept of technology itself, its origins as a school subject and its relationship with other subjects.

One approach to technology which has so far received little attention is to develop an understanding of its constituent habits of mind or those learning skills, dispositions and attitudes which are necessary for an individual to successfully engage with and apply technology knowledge to the solving of problems. The reframing of a discipline in terms of its habits of mind need not at all be at the expense of its foundational constructs, its content or its intellectual tools; these are still important. But the definition and understanding of habits of mind can add clarity to what learning is required to successfully engage with the discipline and to teach it. Disciplinary habits of mind also contribute to developing a sense of identity with the subject by those studying it, potentially leading to better student engagement. There has already been speculation that dispositions, or habits of mind, might form valuable learning goals for technology (Williams, 2011), and there is evidence from other fields close to technology that habits of mind have been worth articulating, including in engineering, science, mathematics, visual arts and creativity.

At the core of this chapter is our proposal that six habits of mind (Table 1) are important to technology and, although aligned with other disciplines, they are sufficiently distinct to constitute a unique differentiator for technology.

The chapter begins with a description of the research process we adopted for identifying the THoM, followed by an explanation of the concept of habits of mind and their value to a subject, with examples from science, mathematics, engineering, design/visual arts and creativity. This context underpins our review of a sample of the technology education literature from which we derive our argument that technology might benefit from identifying its habits of mind. After identifying the six THoM, we briefly illustrate how they might be cultivated in the classroom by teachers using specific thinking routines.

2 Our Research Process

As a conceptual paper which aims to explore constructs and propose new relationships to broaden the scope of thinking in a field (Gilson & Goldberg, 2015), in this case technology education, we gathered insights into the nature of technology from multiple sources. We began by reviewing definitions of technology in dictionaries and curriculum statements. We then undertook an integrative review (Torraco, 2005) of a sample of the literature about technology education in which dispositional issues could be identified and appeared to have an impact on pedagogy and learning.

We also identified some underlying tensions and ambiguities about the taught content and treatment of technology which might benefit from clarification through habits of mind.

Two main themes emerged which seemed to exercise significant impact on perceptions of what technology education is and how it should be taught: the relationship between technology and the other subjects in integrated science, technology, engineering and mathematics (STEM) programmes and the relationship between technology and design, including its association with making. These factors were then examined with reference to the research on HoM in other disciplines, and themes from this analysis were synthesised to support the development of our six THoM.

3 Why Habits of Mind Matter

In a world where knowledge rapidly becomes outdated or misrepresented as 'alternative facts', education should be equipping young people with the capabilities and dispositions to behave intelligently when faced with problems where the answer is not immediately known. Habits of mind provide this consistent way of approaching new learning (Costa & Kallick, 2002; Resnick, 1999). The deployment of habits of mind such as perseverance, sociability and curiosity has been linked to individuals' successful academic performance (Bjorklund-Young, 2016) and their increased long-term well-being (Heckman & Kautz, 2012).

Capabilities such as problem-solving, teamwork and communication are required by employers (World Economic Forum, 2016). Many national education systems acknowledge the importance of developing young people's habits of mind by incorporating them into curriculum goals together with knowledge and skills (McGuinness, 2018). In this case, they are often referred to by terms such as 'capabilities', 'personal competencies' or 'character'. Alternatively, psychologists may refer to them as 'non-cognitive skills'.

In this chapter, we choose to use the term habits of mind (HoM). This is to reinforce the idea that as habits, they can be cultivated concurrently with subject content through appropriate teaching and learning conditions (Claxton, Lucas, & Spencer, 2012) and that as powerful thinking tools they can be used habitually by individuals in many different contexts, both within and outside education.

4 The Value of Reframing Disciplines as Habits of Mind

The advantages of identifying discipline-specific HoM include the following:

- At the school/teacher level to clarify pupils' learning outcomes as thinking skills and personal dispositions required to study the subject and their relationship with

the subject knowledge; to increase teachers' confidence in recognising and developing habits of mind; and to enhance learning by providing more effective feedback to pupils (Lucas, Hanson, Bianchi, & Chippindall, 2017).
- At the discipline/subject organisation level to demonstrate the value of studying the subject (Leahy & Phelan, 2014; Lucas, Claxton, & Spencer, 2013); to make the subject more visible to stakeholders (Hetland, Winner, Veenema, & Sheridan, 2007) or to reframe the learning goals of a subject (Williams, 2011).
- At national government level to align national educational outcomes in response to international movements such as PISA or TIMSS, or employer demands for a more capable workforce (Lucas & Smith, 2018).
- At a global, transnational level often stimulated by individual thought leaders working with educators across the world (Fullan & Langworthy, 2014).

5 Disciplinary Habits of Mind

In this section, we briefly examine how studies at the first two levels of those above have produced HoM in a range of subjects, two related to engineering, engineering itself, visual arts and creativity (Table 2).

In mathematics, Cuoco, Goldenberg, and Mark (1996) articulated eight mathematical habits of mind (MHoM) thought to be important for every high school student to develop. These, they argued, would not only be useful in the real world but would also promote a sense of identity with the subject of maths. MHoM have since been expanded to include algebra (Matsuura, Sword, Piecham, Stevens, & Cuoco, 2013) and geometry (Erşen, Ezentaş, & Altun, 2018).

In science, Çalik and Coll (2012) developed a Scientific Habits of Mind Survey with seven SHoM with which they explored scientific literacy and the variation between public perceptions of socio-scientific issues and those of scientists.

Engineering habits of mind (EHoM) were first defined by the US National Academy of Engineering to be incorporated into the revised science curriculum (Katehi, Pearson, & Feder, 2009). In Britain, the Royal Academy of Engineering commissioned research aimed at reframing the desired outcomes of engineering education as EHoM (Lucas, Hanson, & Claxton, 2014) and six EHoM were identified, built around a central 'core engineering mind' of 'making things that work and making things work better'. They were initially validated through a small-scale pilot study where teachers incorporated EHoM into their teaching of engineering, science, mathematics and technology and evaluated the outcomes (Lucas et al., 2017).

Visual arts teachers were concerned about the narrowing appeal of their subject and sought a vehicle for articulating the broader learning gained through studying the subject, in addition to the more obvious craft skills of drawing and painting. To this end, Hetland, Winner, Veenema, and Sheridan (2007) observed videos of art teachers taking classes with students and discussed with the teachers what thinking habits they were aiming to instil in their students during the studio setting. The

Table 2 Disciplinary habits of mind

Discipline	Habits of mind							
Mathematics (MHoM)	Inventors	Tinkerers	Conjecturers	Experimenters	Pattern-sniffers	Visualisers	Describers	Guessers
Science (SHoM)	Scepticism	Objectivity	Curiosity	Open-mindedness	Rationality	Mistrust of arguments	Suspension of belief	
Engineering (EHoM)	Creative problem-solving	Improving	Problem-finding	Adapting	Systems-thinking	Visualising		
Visual arts (studio) (StHoM)	Develop craft	Engage & persist	Stretch & explore	Express	Observe	Envision	Reflect	Under-stand the art world
Creativity (CHoM)	Imaginative	Persistent	Inquisitive	Disciplined	Collaborative			

outcome was a set of studio habits of mind (StHoM) which raised the visibility of ways of thinking that had previously been implicit and unspoken in the art classroom.

Although cultivating creativity is now widely recognised as an important goal of education (Harris & De Bruin, 2018) and can be applied across the curriculum, there is still debate about what it is and how it can be developed, particularly at secondary school. The value of analysing creativity through habits of mind has been shown to be useful in this respect and for illustrating how progression in HoM might be tracked (Lucas et al., 2013). Following a small field trial of a five-dimensional model of creativity habits of mind (CHoM), this model is now the subject of an international study by the Organisation for Economic Cooperation and Development (OECD, 2018).

Each of these five fields faced an imperative to demonstrate value at a time when the knowledge content, outcomes and utility of the subject within education were open to debate and each has benefitted from a reframing as habits of mind. HoM demonstrated valuable links between the subject and the world outside school and offered teachers a more precise language to inform their teaching and feedback. This, in turn, supported young people to develop and display their capabilities and dispositions. In each case, the process involved a conceptual analysis of how the subject was taught, combined with experts' views on the how the subject is used in practice, to create a HoM framework. This was followed by proof-of-concept testing in schools. The first stage of this process, the conceptual analysis, is being applied here to technology.

Question for the Reader

Before moving on to the next section, you may like to reflect on the following:

Does your school promote specific 'habits of mind' that it aims to instil in its pupils? How might these align with the technology habits discussed in this chapter?

6 Understanding the Essence of Technology

We continue by exploring conceptions of technology in every-day and educational sources as the first step in identifying THoM.

At its narrowest, technology is the application of scientific knowledge in practice:

> Technology refers to methods, systems, and devices which are the result of scientific knowledge being used for practical purposes. (Collins Dictionary, 2019) (https://www.collinsdictionary.com/dictionary/english/technology)

At its broadest, it is a multidisciplinary subject that combines creativity with an understanding of technical knowledge drawn from a range of different disciplines:

> The branch of knowledge that deals with the creation and use of technical means and their interrelation with life, society, and the environment, drawing upon such subjects as indus-

> trial arts, engineering, applied science, and pure science. (Dictionary.com, 2019) (https://www.dictionary.com/browse/technology)

Technology is also used in the every-day sense to refer to examples of technology in the form of manufactured objects, everything from laptops and smartphones to satellites and spaceships. However, some technologists acknowledge that the term is misunderstood and that this narrow definition of technology should be challenged (Glover, 2016). Furthermore, jobs undertaken by technologist are often narrowly defined as being within information technology, although, as Eaton suggests "technologist is not a term we hear often in the technology industry" (2016), which is borne out by the job-titles listed among the "Top 10 tech industry careers", including software engineer, web developer and information security analyst (Crookes, 2014).

Turning to the educational treatment of technology expressed through definitions from curriculum bodies, the gap between the public understanding of technology and educational aspiration becomes more apparent. In England, the Design & Technology Association (D&TA) suggests that:

> Design and technology gives young people the skills and abilities to engage positively with the designed and made world and to harness the benefits of technology. They learn how products and systems are designed and manufactured, how to be innovative and to make creative use of a variety of resources including digital technologies, to improve the world around them. (D&TA, 2019)

D&TA positions digital technology as a resource to be harnessed in design and manufacture, rather than an end in itself, whereas the Australian Curriculum Assessment and Reporting Authority (ACARA) positions digital technologies as one of two distinct subjects in the learning area of technologies, design and technologies being the other area:

> Technologies draws together the distinct but related subjects of Design and Technologies, and Digital Technologies. It ensures that all students benefit from learning about, and working with, traditional, contemporary and emerging technologies that shape the world in which we live. In creating solutions, as well as responding to the designed world, students will contribute to sustainable patterns of living for themselves and others. (ACARA, 2016)

In America, the International Technology Educators Association (ITEA)[1] definition encapsulates both the craft origins of technology as well as its broader concerns:

> Broadly speaking, technology is how people modify the natural world to suit their own purposes. From the Greek word techne, meaning art or artifice or craft, technology literally means the act of making or crafting, but more generally it refers to the diverse collection of processes and knowledge that people use to extend human abilities and to satisfy human needs and wants. (ITEA, 2007:2)

The ITEA definition also presents an extended view of the concept which positions technology as an essential literacy necessary for all children to acquire in order to become technically literate citizens.

[1] Now known as International Technology and Engineering Educators Association (ITEEA)

The New Zealand curriculum statement reveals conceptions of technology which suggest technologists need to take a more critical view and balance human needs and wants with issues like equity and sustainability.

> Technology is intervention by design: the use of practical and intellectual resources to develop products and systems (technological outcomes) that expand human possibilities by addressing needs and realising opportunities. (Ministry of Education, 2007:32)

The scope of this chapter limits the extent to which we can develop this analysis, but through these definitions we see how potential misperceptions about the nature of technology might arise. Technology is not just about craft, or learning to make using manipulative skills, it is very much about designing as well as making. It possesses an interdisciplinary body of knowledge embedded in human activity, including digital activity, but it is challenging to categorise and codify this knowledge as the human focus of the activity is always changing, so there is no uniform pattern of thinking (Herschbach, 1995).

However, we can see some early candidate THoM emerging from this analysis; creativity is a key disposition, but so is empathy and relating to human needs. And needs must be balanced with understanding the contexts in which the technological solutions will be enacted. This all points to high-level thinking which requires a combination of subject knowledge (knowing what), skills (knowing how), combining these to being able to act (capability), and routinely knowing when and why to act (habit of mind) (Lucas & Smith, 2018).

Question for the Reader

Before moving on to the next section, you may like to reflect on the following:

What is your understanding of technology and how does this affect the way in which you teach it? How well does your curriculum balance knowledge, skills and capabilities?

7 Learning from Technology Pedagogy

In this section we explore themes from technology education pedagogic literature to inform our thinking about potential THoM.

The composition and focus of the modern technology curriculum vary according to the traditions from which it has emerged and the purposes ascribed to it. At least seven different treatments of technology education have been identified around the world: as craft, as industrial training, aligned with design, as applied science, integrated into a STEM programme, as a literacy or as individual technologies (e.g., manufacturing, food, textiles) (Jones, Buntting, & de Vries, 2013). These treatments reflect the drivers for its inclusion, ranging from economic to democratic arguments (Barlex & Steeg, 2017a), but they also influence the underlying knowledge structure of technology which, in turn, has an impact on content and pedagogy (Mioduser, 2015).

We selected two treatments to examine through the literature – technology in STEM (see also Chap. 13, John Wells & Didier Van de Veld) and technology with design (see also Chap. 7, Kay Stables) – because they are more likely to appear at secondary school level.

8 Technology and STEM

Many curriculum programmes adopt an integrated STEM approach as a vehicle for the application of science and mathematics to real-world problems. One reason given for this is to develop learners' ability to address complex problems, for example, climate change, which will require the use of multidisciplinary approaches and knowledge to generate technology solutions (Mayes, Gallant, & Fettes, 2018). Another reason is a desire to raise the status of technology knowledge by demonstrating its links with these subjects and to increase the flow of young people into engineering beyond school (Asunda & Quintana, 2018; Lewis, 2004). However, there is concern that STEM alignment may actually devalue technology. It has been suggested that the relative weakness of technology's identity tends to result in it being marginalized in comparison to the other subjects (Bell, Wooff, McLain, & Morrison-Love, 2017; Buntting & Jones, 2015), as it is framed as computing (Mayes et al.) or as a derivative of engineering, which emphasises technology knowledge as product rather than process (Lewis, 2004).

Furthermore, the aims and outcomes of the process of investigation in each subject are different; in science, inquiry tends to be driven by curiosity, in mathematics by proof-seeking and engineering problem solving is driven by need. But technology knowledge is perceived to be different, less objective and therefore lower in status than the knowledge derived through the other three subjects (Almutairi, Everatt, Snape, & Fox-Turnbull, 2014; Morrison-Love, 2017; Yasar, Maliekal, Veronesi, & Little, 2017).

So, integrating technology into STEM may not be helpful, particularly since teachers have a far weaker grasp of the epistemology of design and technology than they have of science and mathematics (Barlex & Steeg, 2013). However, if the differences and similarities between the four STEM subjects could be reframed through their habits of mind, or the dispositions and attitudes needed to successfully 'do' science, technology, engineering and mathematics, the legitimacy of the position of technology in relation to the other three might be better understood, since as we have seen earlier, these three already have defined HoM.

9 Technology and Design

Design has an interesting place in technology curricula. Design is viewed as a "methodology of technology" (Martin & Owen-Jackson, 2013: 34) or a process through which learners progress as they engage in designing solutions to technological problems, seeking information from multidisciplinary sources. However, the teaching of design is often the application of simplified design processes expressed though 'design-make-evaluate', which may have encouraged the development of a linear approach to teaching each stage of the process as a discrete activity rather than the iterative processes from which it was derived (Mawson, 2003). This does not give learners the space to find out what knowledge they need, for example, about the properties of materials (Martin & Owen-Jackson, 2013) and can create an artificial impression of how designers work in the real world (Esjeholm & Bungum, 2018).

Whereas, if the design process steps are understood as indicators of sources of information about the knowledge required for problem solving, the process becomes an iterative movement of "seeing, taking a step, and seeing again" (McRobbie, Stein, & Ginns, 2001). Therefore, much technology knowledge is provisional since its relevance is determined by its application to the problem, requiring learners to make high level decisions about which knowledge is relevant, where to find it and how to transform it into something useful, rather than just recalling and applying knowledge (Doyle, Seery, Gumaelius, Canty, & Hartell, 2018; Lawson, 2004). This rigour of thinking applied in technology is emphasised by Barlex and Steeg (2017a) who distinguish between knowledge of the problem, which differs according to the context, and knowledge for the solution, which is less context dependent. The latter is therefore more readily taught as content, for example, the principles underpinning gears, which remain constant, but the former requires learners to learn strategies for exploring the design context and know when and how to apply them.

Nevertheless, despite attempts to retain the complexity of design thinking (Stables & Kimbell, 2007), it has become apparent that much of its true nature has been lost in technology classrooms. A deeper understanding of the complexity of design thinking, as demonstrated by professional designers, might provide greater clarity about some of its key thinking process, which in turn might inform the development of THoM.

Professional designers apply abductive, or pragmatic, reasoning, using their creativity and analytical skills to develop possible solutions, continually testing and improving them, reframing the problem at the same time as seeking solutions (Dorst, 2010). For this reason, expert designers have been described as 'ill-behaved' problem solvers, compared with other STEM professionals, because they do not spend time defining the problem and then seeking solutions, but on scoping the problem, seeking information about it and generating possible solutions. They are not afraid to change the goal rather than stick to the problem as given (Cross, 2018) or redefine the rules of the game (Lawson, 2004).

The mental imagination required to produce novel designs has long been acknowledged (Ropohl, 1997) and the value of design thinking for generating creative and innovative solutions to problems has been recognized in many fields outside technology (Brown, 2009; Hassi & Laakso, 2011). The creative capabilities valued find expression in terms such as empathy, thinking by doing, challenging the given problem, finding a balance, multidisciplinary collaboration and being comfortable with ambiguity, all of which could all form the basis for candidate THoM.

Critiquing is another aspect of design practice that could inform THoM (Williams, 2017). Critique appears in technology in two senses, as practice and as disposition. As practice, it is the process of formative dialogue that takes place between student and teacher in class. It enhances students' powers of observation and reflection and in giving feedback to their peers. It encourages experimentation, risk taking and acceptance of failure as part of the design process. It operates in an inherently social environment (Crowther, 2013; Motley, 2017).

As a disposition, critiquing is "a frame of mind that imbues all aspects of designing in technology" (Williams, 2017: 145). In other words, it is not an activity that happens once in the application of a design cycle but should be seen as "as permeating all designerly and technological behaviours and circumstances" (Keirl, 2017: 3). Critiquing a design moves beyond evaluation, which asks whether or not the outcome meets the design criteria and does what it was intended to do, whereas critique asks "Is what it is supposed to do worth doing and what are its unintended consequences?" (Barlex & Steeg, 2017b: 10).

Despite the increased presence of design in technology leading to its move away from a product-focused, craft teaching tradition towards more holistic, process-driven pedagogical approaches (Leahy & Phelan, 2014; Mawson, 2003), many argue that both product and process approaches still belong in technology, since the learner's active engagement in making is essential for meaningful learning in technology. However, the technological artefact should not be regarded as the sole outcome of technology learning. The combination of action and reflection during the process of making also supports the construction and learning of conceptual knowledge (Mioduser, 2015). The educational implications of conceptualising technology learning as both process and product are exemplified by Barlex (2008) through his articulation of five interconnected design decision points, involving decisions about the conceptual, technical, aesthetic, constructional and marketing aspects of a design.

Furthermore, since manipulative skills are still essential for an individual to be considered technologically literate, they should now receive equal status as learning outcomes, together with problem-solving aptitudes. The inclusion of making in the articulation of THoM might reflect this status.

Question for the Reader

Before moving on to the next section, you may like to reflect on the following:

In your curriculum, what position does technology hold, and how might this affect your own thinking about its value to your pupils' education?

10 Potential Technology Habits of Mind

We can now begin to synthesise ideas to inform the identification of our candidate THoM. From curriculum statements we identified that technology is not just about learning to make or learning manipulative skills, it is very much about designing as well as making. Creativity is a key disposition, so is empathy, understanding human needs and being mindful of the contexts in which technological solutions will be enacted. However, approaches to delivering technology education, often influenced

Table 3 Technology habits of mind (THoM)

Technology habit of mind	Description
Pragmatic thinking	Moving from observations to possible explanations, generating the best possible explanation, challenging the norm
Critiquing	Observing, asking questions, testing ideas out on others, risk-taking, accepting feedback, evaluating, making judgements
Imagining	Flexible thinking, creative experimenting, developing multiple alternatives, accepting ambiguity, reflecting, transforming
Making	Sketching, modelling, prototyping, considering the properties of materials, generating solutions, accepting failure
Human-centred designing	Seeking to understand people, their culture and their interaction with technology, empathising, multidisciplinary collaboration
Maximising contexts	Appreciating the affordances and constraints of different situations, recognising the need to achieve balance, function and sustainability

by historic perceptions of the nature and place of technology in the curriculum, do not always result in these outcomes being fulfilled. Two treatments were explored to illustrate how viewing technology through a habits of mind lens might contribute to enhancing its position.

The delivery of technology through an integrated STEM programme can result in unequal treatment of the subject compared to the other three, particularly when science and mathematics are given pre-eminence, and technology is conflated with engineering or computing; therefore, the articulation of THoM concurrently with the existing statements of EHoM, SHoM and MHoM has the potential to clarify similarities and differences between the four fields and define a clearer space for technology.

The transformation of technology from product to process orientation has been encouraged through its closer association with design but its teaching can result in a restricted view of design thinking being applied to technology problems. If, however, a more fully articulated version of design thinking was to inform technology education, based on making and capabilities such as pragmatic thinking, critiquing, imagination and empathising, and drawing on existing statements for CHoM and STHoM, its transformational potential could be emphasised.

Technology education could therefore lay claim to the following habits of mind (Table 3).

11 Pedagogy for Cultivating THoM

From our experience of working with schools, we have identified four important factors that teachers should focus on when cultivating habits of mind (Lucas et al., 2014):

- When learners are introduced to a new way of thinking, they need to understand what it is, what it looks like when it is being used successfully and why it is important.

- The learning climate within the classroom should encourage and reward the new way of thinking.
- Teaching approaches should expect learners to use the new way of thinking.
- Learners should be encouraged to take ownership of the new way of thinking.

One approach that encapsulates these ideas is 'Visible Thinking', an initiative established by Project Zero at Harvard Graduate School of Education to develop a research-based approach to teaching thinking dispositions (Project Zero, 2016). Visible Thinking encourages teachers to regularly use patterns of action that promote new ways of thinking in learners until they become habitual behaviours that support independent learning. Many thinking routines have been developed, including those for design and making (Agency by Design, n.d.). Many of them consist of a set of questions or activities and are designed to be integrated by teachers with their existing content, not taught separately. We have adapted a few promising routines that teachers might use to cultivate THoM (Table 4) and expanded one of them into a classroom example (Vignette 1).

Vignette 1: Fostering Maker Capacities: On the Move (Adapted from The James Dyson Foundation, n.d.) Version 2 (29-05-19)
Lorna is introducing a lesson to her class of 11–12 year olds in year 7, the first year of secondary education in England. They are mid-way through a scheme of work for Key Stage 3 Design & Technology, exploring the power of wind and how this key natural resource might be harnessed by humans. One of the outcomes will be a simple prototype wind spinner designed and made by the students.

In previous lessons, the students have considered examples of biomimicry so that the creative adaption of designs from nature might inform their solutions. They have experimented with different seed types and observed how they are moved and dispersed through the air. They have sought out information about the physical properties of materials such as paper, card and various textiles, learnt how to use equipment and sketched possible designs.

Lorna explains the learning outcomes for this lesson:

- Master technical skills required to enable making and experimentation with a sufficient degree of precision.
- Understand sequences of actions so that the making of the planned prototype is not held up or interrupted.
- Consider, revise and amend plans to overcome problems or incorporate improved methods.

Then, using the approach known as 'split-screen teaching' (Lucas & Spencer, 2017) she also tells the students that they will be practising some specific THoM in this lesson. Pointing to a THoM poster on the classroom wall, she asks the students which ones they think are relevant. The children

(continued)

(continued)

quickly pick out Making, because they are about to create a prototype of their design, using skills such as folding, cutting and joining.

Lorna reminds them that they will also be Imagining because they should be reflecting on the notes they made when observing the seeds.

The children work in groups and having developed a production schedule, assign tasks to each member of the group.

Lorna notices that Rafael is hesitating about making the first folds in his cardboard and asks him why. He says he wants to get it right first time, or his group might blame him if they turn out to be in the wrong place. Lorna reminds him and the group that they are working quickly on this design, like professionals do, and that, like professionals, they should accept that they will sometimes make mistakes, but that's ok because design is an iterative process.

As they work on their designs, the children test them at each stage and record how well the design functions. They decide what changes are necessary to make it function more effectively. They make sketches or take photos so they can explain later how their design evolved.

A few minutes before the class is due to finish, Lorna asks each group to write on sticky notes: - what went well, what did not go so well and what they could do differently in the next lesson. She collects these from each group and sticks them on the wall next to the THoM poster, alongside other notes from previous lessons, to remind the children of the THoM they are developing.

Question for the Reader
Before moving on to the final section, you may like to reflect on the following:

How could you incorporate THoM into your technology teaching? Can you develop your own specific examples to embed within your lessons?

12 Conclusions and Next Steps

We conclude by reiterating the case for identifying habits of mind for technology. Technology presents a confused picture of its value as a curriculum subject to the wider world (Hardy, 2015) and its treatment within the curriculum often serves to compound that confusion. Reframing technology education through habits of mind offers an alternative perspective and may provide a sense of clarity of purpose and cohesion for the subject as these key THoM learning outcomes are made more visible to all. In particular, THoM can help to clarify the tacit knowledge valued in the successful 'doing' of technology.

Table 4 Thinking routines to develop technology habits of mind

Technology habit of mind	Thinking routine
Pragmatic thinking	*Parts, perspectives, me*
	Choose an object and ask:
	What are its various pieces or components?
	What perspectives can you look at it from? Users? Makers? Non-users?
	What assumptions or interests shape the way you see it?
	How are your views different to your classmates?
	Who is right?
Critiquing	*Taking a learning walk*
	With another classmate in your group, walk round the class quietly together, observing closely how other groups are working on their projects
	How have they organised their tasks?
	What materials are they using?
	How often do they use trial and error to move forward?
	How do they react if something does not work first time?
	Make notes and take a photo so you can share your thoughts with the class later
Imagining	*Imagine if*
	Choose an object and take it apart
	Consider its different parts, what are their purposes and how do they work together?
	Then ask:
	In what ways could this object be made more effective?
	In what ways could this object be made more efficient?
	In what ways could this object be made more ethical?
	In what ways could this object be made more beautiful?
Making	*Fostering maker capacities; see Vignette 1*
Human-centred designing	*Think, feel, care*
	Think of a technology artefact
	Who uses it? How do they use it?
	What emotions might they have as they use it?
	What is important to this person? Why do they want to use this artefact?
	How could you find out more about how they use it?
Maximising contexts	*Parts, people, interactions*
	Take an object and ask:
	What system might this object be connected to?
	Who are the people in the system who use the object?
	How do the people in the system interact with each other and with other parts of the system?
	How does a change in one element of the system affect the various parts and people connected to the system?

Technology works well when approached through interdisciplinary teaching, but it can be difficult to get secondary teachers to look outside their subjects. However, viewing subjects through the lens of habits of mind has proved to be a productive way of achieving collaboration and takes focus away from competing subject content, timetable space and resources.

THoM also offer an alternative way of framing the valuable contribution of technology, in place of the more generic term of technological literacy, which is frequently used to justify the need for technology education, but often criticised when presented in isolation from context (Herschbach, 1995).

Nevertheless, this is clearly not the end of the process. Our six THoM have been developed through a review of practitioner and research sources but they are a work in progress. They remain to be validated and further refined by technologists and technology educators and then trialled in classrooms. They are also currently described in terms that may not mean much to teachers and students, so in the process of validation, terms that are more school-friendly will be sought.

There are calls for a radical paradigm shift in education to re-imagine how technology might be facilitated through the curriculum (Buntting & Jones, 2015), but since this is unlikely in most scenarios (Barlex & Steeg, 2017a), THoM represent a more evolutionary process to bring about change by empowering teachers to make small steps of change to enhance their technology teaching and learning.

References

Agency by Design. (n.d.) Thinking routines. Available at: http://agencybydesign.org.s219538.gridserver.com/edresources/thinkingroutines/

Almutairi, A., Everatt, J., Snape, P., & Fox-Turnbull, W. (2014). Exploring the relationship between science and technology in the curriculum. *Australasian Journal of Technology and Education, 1*, 49–63.

Asunda, P. A., & Quintana, J. (2018). Positioning the T and E in STEM: A STL analytical content review of engineering and technology education research. *Journal of Technology Education, 30*(1), 2–29.

Australian Curriculum Assessment and Reporting Authority (ACARA). (2016). *Australian Curriculum: Technologies Foundation – Year 10*. Available at: https://www.australiancurriculum.edu.au/f-10-curriculum/technologies/structure/

Barlex, D. (2008). Assessing capability in design and technology: The case for a minimally invasive approach. *Design and Technology Education: An International Journal, 12*(2), 49–56.

Barlex, D., & Steeg, T. (2013). Student teachers' perceptions of 'enduring ideas' in design & technology. In M. J. de Vries (Ed.), *PATT conference proceedings* (pp. 2–9). Columbus, OH: International Technology Education & Engineering Association.

Barlex, D., & Steeg, T. (2017a). *Re-building design & technology in the secondary school curriculum. Version 2. A working paper.* D&TforD&T. Available: https://dandtfordandt.wordpress.com

Barlex, D., & Steeg, T. (2017b). *Big ideas for design & technology. A working paper.* D&TforD&T. Available: https://dandtfordandt.wordpress.com

Bell, D., Wooff, D., McLain, M., & Morrison-Love, D. (2017). Analysing design and technology as an educational construct: An investigation into its curriculum position and pedagogical identity. *The Curriculum Journal*. https://doi.org/10.1080/09585176.2017.1286995

Bjorklund-Young, A. (2016). *What do we know about developing students' non-cognitive skills?* Baltimore: Institute for Policy, Johns Hopkins School of Education.

Brown, T. (2009). *Change by design: How design thinking transforms organizations and inspires innovation*. New York: Harper Business.

Buntting, C., & Jones, A. (2015). The alignment of technology with other school subjects. In P. Williams, A. Jones, & C. Buntting (Eds.), *The future of technology education. Contemporary issues in technology education* (pp. 187–200). Singapore: Springer.

Çalik, M., & Coll, R. (2012). Investigating socioscientific issues via scientific habits of mind: Development and validation of the scientific habits of mind survey. *International Journal of Science Education, 34*(12), 1909–1930.

Claxton, G., Lucas, B., & Spencer, E. (2012). *Making it: Studio teaching and its impact on learners*. Winchester, UK: Centre for Real-World Learning.

Costa, A., & Kallick, B. (2002). *Discovering and exploring habits of mind*. Alexandria, VA: Association for Supervision and Curriculum Development.

Crookes, D. (2014). Careers in tech: Top 10 Tech industry careers. *The Independent*, Tuesday 18 November. Available at: https://www.independent.co.uk/student/career-planning/careers-in-tech-top-10-tech-industry-careers-9865807.html

Cross, N. (2018). Developing design as a discipline. *Journal of Engineering Design, 29*(12), 691–708.

Crowther, P. (2013). Understanding the signature pedagogy of the design studio and the opportunities for its technological enhancement. *Journal of Learning Design, 6*(3), 18–28.

Cuoco, A., Goldenberg, E. P., & Mark, J. (1996). Habits of mind: An organizing principle for mathematics curricula. *Journal of Mathematical Behaviour, 15*, 375–402.

Design & Technology Association. (2019). *What is design and technology and why choose to study it?* Available at: https://www.data.org.uk/campaigns/what-is-design-and-technology/

Dorst, K. (2010). *The nature of design thinking. Proceedings of the DTRS8 Interpreting Design Thinking: Design Thinking Research Symposium* (pp. 131–139). Sydney, Australia: DAB Documents. Available at: http://epress.lib.uts.edu.au/research/handle/10453/16590

Doyle, A., Seery, N., Gumaelius, L., Canty, D., & Hartell, E. (2018). Reconceptualising PCK research in D&T education: Proposing a methodological framework to investigate enacted practice. *International Journal of Technology and Design Education*. https://doi.org/10.1007/s10798-018-9456-1

Eaton, C. (2016). Who is a technologist? *Creating IT futures blog*. November 14. Available at: https://www.creatingitfutures.org/resources/blog/creating-it-futures/2016/11/14/who-is-a-technologist

Erşen, Z. B., Ezentaş, R., & Altun, M. (2018). Evaluation of the teaching environment for improve[ing] the geometric habits of mind in tenth grade students. *European Journal of Education Studies, 4*(6), 47–64. Available at: www.oapub.org.edu

Esjeholm, B.-T., & Bungum, B. (2018). Linking knowledge and activities: How can classroom activities in technology reflect professional technological knowledge and practices? In M. de Vries (Ed.), *Handbook of technology education* (pp. 567–578). Cham, Switzerland: Springer.

Fullan, M., & Langworthy, M. (2014). *A rich seam: How new pedagogies find deeper learning*. London, UK: Pearson.

Gilson, L. L., & Goldberg, C. B. (2015). Editor's comment: So what is a conceptual paper? *Group and Organization Management, 40*(2), 127–130.

Glover, D. (2016). What we talk about when we talk about technology. *ESRC STEPS (Social, Technological and Environmental Pathways to Sustainability) Centre*. September 08. Available at: https://steps-centre.org/blog/what-we-talk-about-when-we-talk-about-technology/

Hardy, A. (2015). What's D&T for? Gathering and comparing the values of design and technology academics and trainee teachers. *Design and Technology Education: An International Journal, 20*(2), 10–21.

Harris, A., & De Bruin, L. (2018). Creativity in education. Oxford Research Encyclopedias: Education. Online https://doi.org/10.1093/acrefore/9780190264093.013.383. [Accessed 17 January 2019].

Hassi, L., & Laakso, M. (2011). Conceptions of design thinking in the design and management discourses. In: N. F. M. Roozenberg, Chen, L. L., & Stappers, P. J. (Eds.) *Diversity and unity: Proceedings of IASDR2011, the 4th World Conference on Design Research*, 31 October – 4 November, Delft, the Netherlands.

Heckman, J., & Kautz, T. (2012). Hard evidence on soft skills. *Labour Economics, 19*(4), 451–464.

Herschbach, D. R. (1995). Technology as knowledge: Implications for instruction. *Journal of Technology Education, 7*(1), 31–42.

Hetland, L., Winner, E., Veenema, S., & Sheridan, K. M. (2007). *Studio thinking: The real benefits of visual arts education*. New York: Teachers College Press.

International Technology Educators Association (ITEA). (2007). *Standards for technological literacy: Content for the study of technology* (3rd ed.). Reston, VA: ITEA. Available at: https://www.iteea.org/42511.aspx

Jones, A., Buntting, C., & de Vries, M. J. (2013). The developing field of technology education: A review to look forward. *International Journal of Technology and Design Education, 23*, 191–212.

Katehi, L., Pearson, G., & Feder, M. (2009). *Engineering in K-12 education: understanding the status and improving the prospects. Committee on K-12 Engineering Education, National Academy of engineering and National Research Council*. Washington, DC: National Academies Press.

Keirl, S. (2017). Critiquing as design and technology curriculum journey: History, theory, politics and potential. In P. J. Williams & K. Stables (Eds.), *Critique in design and technology education. Contemporary issues in technology education* (pp. 109–133). Singapore: Springer.

Lawson, B. (2004). *What designers know*. Oxford, UK: Architectural Press.

Leahy, K., & Phelan, P. (2014). A review of technology education in Ireland: A changing technological environment promoting design activity. *International Journal of Technology and Design Education, 24*, 375–389.

Lewis, T. (2004). A turn to engineering: The continuing struggle of technology education for legitimization as a school subject. *Journal of Technology Education, 16*(1), 21–39.

Lucas, B., Claxton, G., & Spencer, E. (2013). *Progression in student creativity in school: First steps towards new forms of formative assessments*. Paris, Italy: OECD Education. https://doi.org/10.1787/5k4dp59msdwk-en

Lucas, B., Hanson, J., Bianchi, L., & Chippindall, J. (2017). *Learning to be an engineer: Implications for the education system*. London, UK: Royal Academy of Engineering.

Lucas, B., Hanson, J., & Claxton, G. (2014). *Thinking like an engineer: Implications for the education system*. London, UK: Royal Academy of Engineering.

Lucas, B., & Smith, C. (2018). *The capable country: Cultivating capabilities in Australian education*. Mitchell Institute policy report No. 03/2018. Mitchell Institute, Melbourne. Available at: www.mitchellinstitute.org.au

Lucas, B., & Spencer, E. (2017). *Teaching creative thinking: Developing learners who generate ideas and can think critically*. Carmarthen: Crown House.

Martin, M., & Owen-Jackson, G. (2013). Is design and technology about making or knowing? In G. Owen-Jackson (Ed.), *Debates in design and technology education* (pp. 64–73). Abingdon, UK: Routledge.

Matsuura, R., Sword, S., Piecham, M. B., Stevens, G., & Cuoco, A. (2013). Mathematical habits of mind for teaching: Using language in algebra classrooms. *The Mathematics Enthusiast, 120*(3), 735–776.

Mawson, B. (2003). Beyond the design process: An alternative pedagogy for technology education. *International Journal of Technology and Design Education, 13*, 117–128.

Mayes, R., Gallant, B., & Fettes, E. (2018). Interdisciplinary STEM through engineering design-based reasoning. *International Journal of Engineering Pedagogy, 8*(3), 60–68.

McGuinness, C. (2018). *Research informed analysis of 21st century competencies in a redeveloped primary curriculum*. Belfast, UK: Queens University. Final report.

McRobbie, C. J., Stein, S. J., & Ginns, I. (2001). Exploring designerly thinking of students as novice designers. *Research in Science Education, 31*, 91–116.

Ministry of Education. (2007). *The New Zealand National Curriculum for English medium teaching and learning in years 1–13*. Available at: http://nzcurriculum.tki.org.nz/The-New-Zealand-Curriculum

Mioduser, D. (2015). The pedagogical ecology of technology education: An agenda for future research and development. In P. J. Williams, A. Jones, & C. Buntting (Eds.), *The Future of Technology Education. Contemporary Issues in Technology Education* (pp. 77–98). Singapore: Springer.

Morrison-Love, D. (2017). Towards a transformative epistemology of technology education. *Journal of Philosophy of Education, 5*(1), 23–37.

Motley, P. (2017). Critique and process: Signature pedagogies in the graphic design classroom. *Arts & Humanities in Higher Education, 16*(3), 229–240.

OECD. (2018). Teaching, assessing and learning creative and critical thinking skills in education. http://www.oecd.org/education/ceri/assessingprogressionincreativeandcriticalthinkingskillsin-education.htm

Project Zero. (2016) .Visible thinking: Overview. Available at: http://www.pz.harvard.edu/projects/visible-thinking

Resnick, L. (1999). Making America smarter. *Education Week Century Series, 18*(40), 38–40.

Ropohl, G. (1997). Knowledge types in technology. In M. J. de Vries & A. Tamir (Eds.), *Shaping concepts of technology* (pp. 65–72). Dordrecht, Netherlands: Springer.

Stables, K., & Kimbell, R. (2007). Evidence through the Looking Glass: developing performance and assessing capability. In *The 13th International Conference on Thinking June 17–21, 2007 Norrköping, Sweden* (pp. 173–191).

The James Dyson Foundation. (n.d.). On the move: Design and technology scheme of work, Key Stage 3. Available at: https://www.jamesdysonfoundation.co.uk/resources/rethinking-design-and-technology-in-practice.html

Torraco, R. J. (2005). Writing integrative literature reviews: Guidelines and examples. *Human Resources Development Review, 4*(3), 356–367.

Williams, P. J. (2011). Dispositions as explicit learning goals for engineering and technology education. In M. Barak & M. Hacker (Eds.), *Fostering human development through engineering and technology education* (pp. 89–102). Rotterdam, Netherlands: Sense Publishers.

Williams, P. J. (2017). Critique as disposition. In P. J. Williams & K. Stables (Eds.), *Critique in design and technology education, contemporary issues in technology education* (pp. 135–152). Singapore: Springer.

World Economic Forum. (2016). *The future of jobs: Employment, skills and workforce strategy for the fourth industrial revolution*. Geneva, Switzerland: World Economic Forum. Available at: http://www3.weforum.org/docs/WEF_Future_of_Jobs.pdf.

Yasar, O., Maliekal, J., Veronesi, P., & Little, L. (2017). The essence of scientific and engineering thinking and tools to promote it. *Proceedings of the American Society of Engineering Education Annual Conference*, June 2017. Paper ID #18172. Available at: https://www.researchgate.net/publication/319952625_The_Essence_of_Scientific_and_Engineering_Thinking_and_Tools_to_Promote_It. Accessed 17 July 2018.

Making the Invisible Visible: Pedagogies Related to Teaching and Learning about Technological Systems

Jonas Hallström and Claes Klasander

Abstract Technological systems are interwoven into the very fabric of modern society to such an extent that we often take them for granted and they almost become invisible to us, because much of the infrastructure is hidden in the ground beneath us or behind walls. Many modern technological systems are also abstract in the sense that they include invisible connections and flows, for example, in cellular phone communications or GPS navigation. These systems also have societal components such as organizations, legislation and operators. Technological systems thus challenge traditional teaching and learning related to artefacts in technology education, since systems are much more difficult to grasp and also have some different characteristics and dynamics compared to single objects. The aim of this chapter is to address this challenge by presenting and discussing the characteristics of technological systems in relation to teaching and learning about systemic aspects of our lifeworld. We suggest four pedagogies to achieve this: interface pedagogy, holistic pedagogy, historical pedagogy and design pedagogy. Furthermore, we propose two ways of delimiting systems through two types of boundaries that are crucial in this regard: the systems horizon and the system border.

> It's Wednesday morning and Kiera, the technology teacher, is starting up a lesson on technological systems. Her class consists of 25 students aged 15 years old, and because of their interest in sustainable development they are pretty excited about the system example that Kiera has chosen: the deposit system for aluminium cans, glass bottles, plastic packaging, etc. Soon, however, it becomes obvious to Kiera that the students don't have a very clear idea of what a system is, nor do they know much about components of a system, their connections and flows, or where a system might end. She thinks she has prepared the students well by talking to them about systems and system concepts, but now she does not know what to do. An idea forms, based on what a colleague told her about having students draw their own systems, as a way of making the systems conceivable. She starts explaining how to draw a system map or model, in the form of a block diagram: "Listen, you know I showed you some pictures, or diagrams really, of systems last week. Do you remember? Well, now, that's what we're going to do this lesson, in order to ... "

J. Hallström (✉) · C. Klasander
TESER, Technology and Science Education Research, IBL, Linköping University, Norrköping, Sweden
e-mail: jonas.hallstrom@liu.se; claes.klasander@liu.se

P. J. Williams, D. Barlex (eds.), *Pedagogy for Technology Education in Secondary Schools*, Contemporary Issues in Technology Education,
https://doi.org/10.1007/978-3-030-41548-8_4

1 Introduction

Technological systems are part of our everyday lives; we use them, affect them and are affected by them. These systems are interwoven into the very fabric of modern society to such an extent that we often take them for granted and they almost become invisible to us, because much of the infrastructure is hidden in the ground beneath us or behind walls. Flows of matter like people, goods, water or food are less difficult to discern than flows of energy and information. These two often take on non-tangible forms such as electromagnetic waves or feedback loops. Many modern technological systems are, therefore, abstract and include invisible connections and flows, for example, cellular phone communications or GPS navigation as well as the stock exchange or a wind energy mill park. These systems also have societal components such as organizations, legislation and operators (Bijker, Hughes, & Pinch, 1987). Moreover, one counter-intuitive characteristic is that technological systems, in a sense, do not appear as systems until they are defined as such (Ingelstam, 2002). The boundaries of a system need to be decided, in the same way as, for instance, the boundaries of an ecosystem. Technological systems thus challenge traditional teaching and learning related to artefacts and objects in technology education, since systems are much more difficult to grasp and also have some different characteristics and dynamics compared to single objects.

The aim of this chapter is to address these challenges by presenting and discussing characteristics of technological systems in relation to teaching and learning about systemic aspects of our lifeworld.

2 Systems Theory, Systems Thinking and Technological Systems

This section relates to the questions of why and what to teach and learn about technological systems, that is, why systems and systems thinking are important in technology education and what system concepts are pivotal. Technology education curricula in many countries around the world promote a broad technological literacy that should encompass a variety of manifestations of technology, thus also a systems component (e.g. Standards for Technological Literacy: Content for the Study of Technology, 2000; Technology in the New Zealand Curriculum, 2007). One of the objectives of technology education is thus to provide all students with an understanding of how these systems work, their advantages and limitations (Keirl, 2006; Schooner, Klasander, & Hallström, 2018b). Technological systems are also important elements in engineering and STEM education, although we shall focus mainly on technology education and technological literacy in this chapter.

Systems thinking, in the context of technology, is the ability to think about, analyse and even design technology in terms of systems. We consider systems thinking at a higher level of generality, beginning with establishing some key concepts with regard to specific domains. A very basic definition of a technological system is that it fulfils a particular purpose; it consists of components, relations or connections between these components and has a system boundary. Beyond the system boundary there is the surrounding, which may interact with the system but is not part of it (Ingelstam, 2002). Most technological systems are sociotechnical systems in that they include not only technical but also societal components (Hughes, 1987; Vermaas, Kroes, van de Poel, Franssen, & Houkes, 2011), but it depends on the system in question. Below we will discuss various characteristics of technological systems in more detail.

What makes a system technological is an important distinction. Systems theory is concerned with an interdisciplinary study of systems, and systems theories have developed in different scientific fields such as mathematics, biology, economy and engineering. The most prominent technological aspect of a system is its physical-technical, human-made core, a difference made obvious in comparing the human digestion system with the national power grid. The first is a natural system, the second a technological system, but you can apply systems theory to both in order to understand them better. Systems theory has to do with scientific concepts, methods and knowledge claims, whereas when discussing the capability students should develop when it comes to understanding complex technological solutions and their impacts, the term systems thinking is better because it rather denotes a capacity on the part of the student (cf. Randle & Stroink, 2018).

The concept of technological system was derived from the broader systems theory and related concepts and theories (e.g. General Systems Theory and Large Technological Systems). Reviewing the systems literature led us to a broad synthesis of systems theories that could describe technological systems. We compiled them into a set of system significants (Barak, 2018; Bertalanffy, 1973; Bijker et al., 1987; Capra, 1996; Churchman, 1979; Ellul, 1980; Hughes, 1987; Ingelstam, 1996; Kroes, Franssen, van de Poel, & Ottens, 2006; Vermaas et al., 2011; Wiener, 1954). These significants characterize the most crucial aspects of technological systems. Eleven clustered groups of concepts make up the partly overlapping and mutually dependent system significants as follows:

1. *The technical core of a system* – the visible, tangible parts, e.g. the rails, water pipes, routers or chassis.
2. *Hierarchies, sub-systems, components* – e.g. the intended function (purpose) of a system, and the hierarchical structure of sub-functions distributed through sub-systems and components such as, in a car, the motor, the electrical system, the fuel system and the security system; or the regulatory level, the organizational level and the physical level of a national energy grid.
3. *Connections and wholeness* – the mental construction of the system and its representations in the form of e.g. models or diagrams, which show the connections between central sub-systems and components.
4. *System boundary and surrounding* – what is internal to or inside the system, and what is not; all that which the system can control is often regarded as being on the inside.

5. *Isolated, closed or open systems* – often used as methods for delimiting the analysis of the flows in and out of the system. Most technological systems are open and often then defined as sociotechnical systems because they include societal components and interact with the surrounding (other systems, society at large, etc.)
6. *Control, feedback, flow of information* – the flow of matter or energy through a system is what the system transforms in order to fulfil its intended function, but the flow of information between the different parts of the system is what keeps the system together.
7. *Systems' functions and behaviour, processes, models* – a system might be analysed as having different kinds of functions and related processes, e.g. the intended function of the road traffic system to deliver humans and goods, and the unwanted and unintended function of killing and maiming many people each year.
8. *Scale and complexity* – these are two dimensions that distinguish a system from a simpler artefact.
9. *Dynamics, development, change* – a system is dynamic and continuously changing, in the short term and historically over the years.
10. *Sociotechnological perspectives* – both the individual and society at large act in and interact with different systems. In many cases, people's engagement in and use of technological systems are crucial for understanding a system.
11. *Systems for innovation, conditions for production* – this represents the knowledge of how to develop and change technological systems.

One of the main features that systems theories – and the system significants – share, is that they are more holistic compared to the reductionistic perspectives common in certain natural sciences such as physics. Many system concepts, e.g. input – process – output, component, flow, connection, function, feedback, momentum, reverse salient, system boundary, surrounding and black boxing, are of great importance as tools for developing an understanding of the human-made world but could also be applied to many natural, social and economic systems. The above significants/concepts, derived from various systems theories, can help us understand, describe and analyse technological systems of different types and complexity, with different kinds of components and flows (Booth Sweeney & Sterman, 2007; Hallström, Klasander, & Svensson, 2015; Hughes, 1983, 1987; Klasander, 2010; Svensson, 2011, 2018). One important tool for such an analysis is the system model or diagram, see Figs. 1 and 2.

From a technological point of view, it is important to emphasize one other crucial difference between natural systems and human-made systems. This concerns the understanding of the concept of function. In a biological system, let us take a forest analysed as an ecosystem with flows, food webs or food chains, the components and sub-systems are attributed with functions. They play various parts in the system. In a technological system, e.g. a transport system, the rails, trains and switches also play parts. However, since it is human-made, in a technological system the functions are intentional, assigned by the designer, while the trees, grass, water and

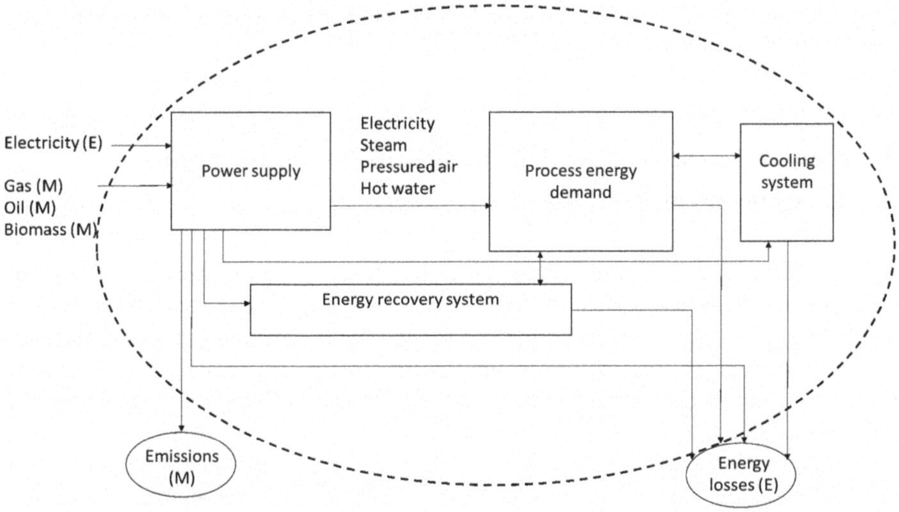

Fig. 1 System diagram of a typical industrial energy system. The diagram shows flows of matter and energy, but not flow of information, between components/sub-systems. Inputs and outputs over the system boundary are also shown. (Adapted from Handayani and Ariyanti (2012), p. 35)

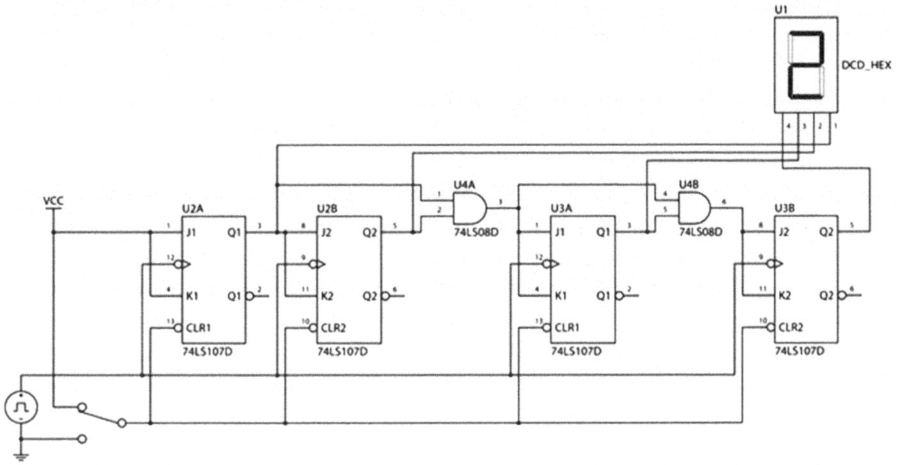

Fig. 2 Flow chart of an electronic circuit, in which the main flow consists of signals (information). (Source: Wikimedia Commons, https://commons.wikimedia.org/wiki/File:4_bit_counter.svg)

animals in the forest are not. This is one of the reasons why technological systems can be optimized and evaluated with respect to, for instance, ethical, sustainable and cultural perspectives of appropriateness. When it comes to controlling a human-made technological system, there are strategies and methods for measuring the level of optimization and feeding the information back into the system. This is not the case in natural systems, although they can be analysed in terms of their behaviour and feedback (e.g. Vermaas et al., 2011).

3 Current Research on Teaching and Learning About Technological Systems

This section continues the reasoning above and relates it to the question of what to teach and learn about systems, based on research on teacher and student conception and learning in technology education and related fields. The main strand of research concerns students' conception of technological systems, and teachers' and student teachers' conception, understanding and assessment of systems (Hallström & Klasander, 2017).

Mioduser, Venezky, and Gong (1996) investigated the mental models that American middle-school students produce of control systems such as automatic doors, heating/cooling systems and various household devices, before, during and after instruction. Furthermore, they studied the students' 'conceptions, missing conceptions and misconceptions' of these control systems on three main levels: device knowledge, perception of the control process and conception of the flow of information in the system. Their conclusions were that the students' understanding prior to instruction was very poor, but a little better after. Their device knowledge, for example, was poor, which means that they lacked an accurate understanding of common components and how these affect the system. Control features of the system were similarly poorly understood as well as the flow of information in the system, while system structure was well understood (Mioduser et al., 1996).

Ginns, Norton, and McRobbie (2005) carried out an intervention on technological systems in an Australian grade six class, and concluded that an 'improvement was observed in students' abilities to describe relationships between inputs, processes and outputs [...]' (Ginns et al., 2005, p. 47). Koski and de Vries (2013) designed an intervention study in which primary students and teachers did a pretest, the teachers thereafter taught lessons and then the students did a post-test, related to how they perceived various aspects of technological systems and how the teaching could be improved. They observed that the concept of input was clearer to the students than output, but that the latter conception improved somewhat during the intervention. Both before and after the intervention, however, the students had a linear conception of systems, and they found it difficult to separate between a process and a system. Flow of matter in a system was the easiest concept to understand, while information was the most difficult. Setting boundaries to systems was also a challenging task. Although their systems thinking was rather limited, the students were at least able to reach beyond basic descriptions of technological systems (Koski & de Vries, 2013).

Örtnäs (2007) studied how secondary school students perceived technological systems in their everyday life. Her conclusion was that with a little scaffolding they could understand how technological systems such as the mobile phone network and the washing machine work; at least, they understood the structure of the systems and how they related to sub-systems and humans. However, the older students showed a better understanding of single components than the younger ones (Örtnäs, 2007).

Reflection Point
Make a system diagram of a washing machine. Make sure you focus on the main components, and add flows of information (I), energy (E) and matter (M). See Fig. 1 and also Chap. 11 for examples of how to draw a block diagram.

Svensson, who studied 10- and 15-year-old students' experience of technological systems, concluded that they understood the structure quite well but were not so knowledgeable about how components interact and how humans fit in the systems (Svensson, 2011). Lind, Pelger, and Jakobsson (2018) investigated 13- to 14-year-old students' understanding of technological systems and their characteristics by studying the students' reasoning and collaboration in small groups. The researchers found situations where the students clearly demonstrated that they understood technological systems and the components and relationships between them. Some situations, on the other hand, indicated that students found it difficult to explain and understand systems concepts; students' understanding thus seems to be very dependent on the context. Most of the students had no difficulty describing a technological system as consisting of different components and that these work together to create a desirable function, but sociotechnical issues were increasingly difficult the more abstract and societal they became.

Klasander concluded that systems thinking among teachers is often hampered either by a focus on scientific, reductionist aspects of systems or a focus on single artefacts (Klasander, 2010). Svensson and Klasander studied how two groups of technology teachers plan their teaching about technological systems in lower secondary school. The study showed that the teachers require more knowledge about the similarities and differences between various technological systems. A better understanding of the system's components and different levels (physical, organizational, regulatory) could also contribute to a more developed understanding (Svensson & Klasander, 2012). Schooner et al. (2018b) investigated Swedish technology teachers' conceptions about technological systems. They found that the teachers focused on the technological core of the system, closely related to a philosophical conception of technology as objects, but also expressed views of systems similar to a sociotechnical understanding where humans play a significant role. There was one exception to this, namely how the systems are controlled, and here the teachers were ambivalent as to how much humans can intervene. The conception of technological systems as objects and the uncertainty about human control over these systems are two obstacles to well-designed systems teaching that can lead to technological literacy for students. Schooner, Klasander, and Hallström (2018a) investigated Swedish teachers' views of assessment of technological systems and found that they mainly focused on the systems' structure, relations outside the system boundary (e.g. consequences for the environment), and the historical change of the systems (Schooner et al., 2018a).

Hallström and Klasander (2017) studied student teachers' conceptions of technological systems. They concluded that the parts of the systems that the students understood and could describe what they do, were mostly the visible parts, either components, devices, or products, or the interface with the software inside a mobile phone. However, the 'invisible' or abstract aspects of the technological systems, such as flows of information, energy or matter or control operations, were difficult to understand for the majority of the students. The most important implication of this study is that students need to be trained in systems thinking, particularly regarding how components work and connect to each other, flows (especially of information), system dependency and the human role in technological systems (Hallström & Klasander, 2017).

Even though the previous research on technological systems in technology education is rather limited, one can draw a few conclusions about the teaching and learning of systems of relevance for this chapter:

1. Students understand systems better when they are scaffolded, either by an interviewer or by teaching interventions.
2. Unsurprisingly, students gain a deeper understanding of systems as they grow older, especially regarding the included components. However, more surprisingly, there is no significant difference between students and (student) teachers.
3. Both students and teachers are better at understanding structure, input and output of a system than its behaviour, and control mechanisms and flows of information are particularly difficult to grasp.
4. The role of humans in and around a technological system is difficult to understand, probably because humans fulfil so many different roles as designers, users and operators and thereby function as crucial but multifaceted components of the system (Vermaas et al., 2011).

Reflection Point

In the light of the above bullet point list, how would you design teaching about technological systems in order to facilitate student learning as much as possible? In your experience, are any of the system concepts mentioned in this chapter particularly difficult for students?

4 Pedagogies for Teaching and Learning About Technological Systems.

In this last section, we bring together the arguments from the two previous sections and feed them into the last topic of how to teach and learn about systems. Research on effective pedagogy for teaching and learning about systems is very limited, if not non-existent, although there are a few studies that will be used as a basis for a discussion on pedagogical approaches below. Underlying all of these approaches is the notion of active learning. A clear and active role is also assigned to the teacher since research suggests that teacher-led classroom leadership is effective for learning, both in education in general and in technology education in particular (e.g. Barak, 2018; Kirschner, Sweller, & Clark, 2006; Schooner, Nordlöf, Klasander, & Hallström, 2017). Good pedagogy thus requires both active students and teachers. The suggested pedagogies are not universal, however. The subject of technology takes on various forms around the globe, and so do classrooms in general; social, cultural, educational and political contexts vary a great deal. We thus point out under which conditions one or another pedagogy or teaching strategy might work (Hattie, 2012; Jones, Buntting, & de Vries, 2013; Wiliam, 2014).

Central to teaching and learning of technological systems is the use of appropriate concepts and models, adjusted to the students' age and prior knowledge (Martin, 1990; Svensson, 2018). We here suggest four pedagogies for system identification, coupled with two models – with associated concepts – to address two types of boundary in the identification and learning about technological systems: the "systems horizon" and the "system boundary".

4.1 Four Pedagogies for System Identification

When teaching about technological systems to secondary students research shows that teachers as well as textbooks and curricula can apply four different pedagogies or strategies for making systems conceivable to the students (e.g. Klasander, 2010). Underlying these pedagogies are two different aspects, one practical and one theoretical. The first aspect is that many complex technologies are hidden – under ground, behind walls, under a shell, or they constitute parts of such large networks that it is impossible to see them. They are thereby not directly accessible or visible, and this goes for a great deal of modern technology in, for example, cities where much of the infrastructure is hidden beneath the ground (e.g. Wallsten, Carlsson, Frändegård, Krook, & Svanström, 2013). The second aspect is that every system needs to be defined. Although, as was noted above, elements of the system often are visible objects, the system as a whole is a mental construction, not an ontological object, in the sense that its boundaries have to be pinned down. In this respect, it is not comparable to other technological artefacts. Even if the car or the computer as a system can appear to be a fairly well-defined object, you still need to define it as a system if that is what you want to study or understand, and system diagrams are important tools to achieve this kind of understanding.

So, the four pedagogies are as follows:

1. *Interface pedagogy*: Starting with the interface between the supposed system and the human beings using it. By starting, for instance, with the toilet seat you move the students towards the other important components and the wholeness of either the sewer system or the fresh water system, depending on the direction. Other

possible starting points are system dependent artefacts such as a smartphone or an ATM machine – both are examples of interfaces where students can meet a "hidden system" waiting to be explored.

2. *Holistic pedagogy*: Starting with a fairly agreed upon name of a known technological system (e.g. the railway system) you move from that wholeness and successively identify important sub-systems and components. The work to identify the characteristic sub-systems and significant components – without succumbing to an overwhelming level of detail – requires the use of "black boxing", that is, certain processes, flows, etc. need to be concealed for the students to be able to focus on the most crucial aspects of the system. Block diagrams might be useful here.
3. *Historical pedagogy*: Following the historical change – forwards from a prior point in time, or backwards from now – of a fairly well known and agreed upon technological system you can identify important structures, sub-systems and components, e.g. in the telephone system. With this method it is also possible to identify some of the most common patterns in technological change, such as small telephone systems a hundred years ago merging into larger systems, or the pattern of greater automation over time.
4. *Design pedagogy*: All the above pedagogies are about analysing existing systems, but many curricula refer to the notion that (design and) technology education is about designing products or systems, so designing would here include making, or prototyping, technological systems of appropriate complexity.

All four pedagogies have their strengths and shortcomings. What we want to show here is the problems associated with system delimitations, and how it is possible to overcome the shortcomings in pedagogical situations. Selecting a pedagogy in line with one or more of the above four can be a first step, because these pedagogies let teachers and students approach technological systems either by starting with well-known and well-defined technological systems, or conversely, beginning with well-known, mundane objects as interfaces. Furthermore, these pedagogies will cover a great deal of the technology curriculum since they include both historical and present-day systems, and both analysis of existing systems and design of students' own systems.

The four pedagogies can also be used to elaborate on the differences between natural and technological systems. Especially, the historical pedagogy opens up for such discussions since natural systems do not develop along the same lines as technological ones. An example with great potential is the two water systems that meet in a city: the natural one, with lakes, rivers, clouds and rain as components, and the technological one consisting of fresh water plants, pipes, watertowers and sinks/toilets.

With respect to the notion of active learning, all four pedagogies are promising for student engagement but they also require careful planning by the teacher. Since the interface pedagogy takes relevant everyday artefacts as its starting point, the students can begin by brainstorming the most relevant ones, before they start unravelling system components. In the holistic and historical pedagogies, the teacher needs to help students select relevant system exemples. The design pedagogy needs more teacher intervention at the planning stage to select a system with a reasonable level of complexity for designing.

4.2 Two Important Types of Boundary

To be able to employ one or more of the above pedagogies, one has to be able to identify two types of boundary related to technological systems. It is known from the previous research that teachers, student teachers and students find it diffcult to identify these boundaries (Hallström & Klasander, 2013; Koski & de Vries, 2013; Schooner et al., 2018b):

- The boundary between what can be defined as an artefact and what can be designated as a technological system (Fig. 3), and.
- The boundary between the technological system and its surrounding (Fig. 4).

The first type of boundary is not fixed but rather negotiable. Technologies can be categorized on the basis of their size – from the ones that we can hold in our hands (e.g. a stapler), via bigger technologies that we can only behold with our eyes (e.g. a hospital), to those that are so enormous in size and/or complexity that we are forced to abstract them (e.g. the Internet). All technology does not need to be described in terms of systems, but for the technologies with many components and complex connections both teaching and learning are facilitated when they are described as systems, and there are concepts, theories and models that help us do so (see Fig. 3).

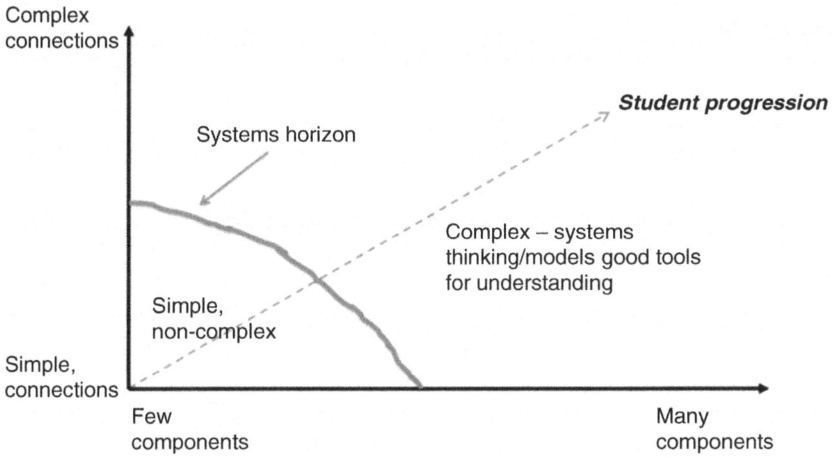

Fig. 3 Diagram of students' progression beyond the systems horizon between non-complex artefact and complex system. (Source: Hallström et al. (2015))

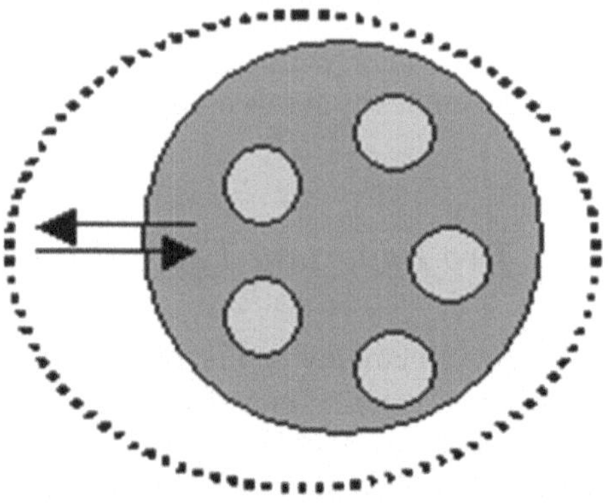

Fig. 4 Diagram of the system's relation to its surrounding. (Source: Klasander (2006), p. 23)

Close to origo, in the lower left-hand corner, are the most simple technological artefacts or objects which only consist of a few parts, for example, a butter knife or a stone axe. The latter has three components: a flint stone, a stick and some type of string to join everything together. Both these examples are so easy to understand that we do not need systems thinking or system models. However, as we move outward on the x and y axes the number of components will increase, and so will the complexity of the connections between the components. Somewhere we want to place a bicycle or a flashlight, somewhere a car or an electric grid. In the latter case we encounter a horizon, which constitutes the boundary to the complex technologies that we need to understand in terms of systems and begin to utilize system models and concepts. Some technological solutions will be border line cases depending on how we want to define them and what we want to do with our knowledge about them. There is also a progression over the school years in how the teacher can bring the students outward from origo toward increasingly complex technologies, across the systems horizon.

Reflection Point

Consider examples of technological systems from your own teaching that are borderline cases, and what needs to be done to move them one way or the other with regard to the systems horizon. What examples can you think of that are better described and understood as objects? What examples are better described and understood as technological systems?

The other boundary has to do with how we delimit the system itself. What is within the boundary and what belongs to the system's surrounding? What is controlled by the system is generally seen as a part of the system, while that which is not is part of the surrounding. Many systems are dependent on a stable surrounding that it can interact with in different ways. The surrounding can deliver flows of input into the system, or receive output from the system, depending on whether the system in question can be considered a closed or an open system (most technological systems should be seen as open systems). If we return to the vignette in the beginning of this chapter, one could put the system boundary of the deposit system around its main interfaces, the recycling centres and the garbage trucks. However, such a system boundary would delineate a system for recycling of household waste but exclude, for example, industrial, medical and hazardous waste, etc. By delimiting a system one thus includes certain components and excludes others. Another system boundary could encompass the complete deposit system of a whole city or nation.

Identifying a system boundary separates the technological system from the surrounding it interacts with, and the system thus becomes clearer and easier to discuss (see Figs. 1 and 4). One way is to use system models that convey a simplified, conceptual image of the reality that the technical system constitutes, for example, the input – process – output model (Hallström et al., 2015). This is also common in informatics and business analysis, where system models are often used to explain complex technical contexts as companies or different infrastructure systems for people of different backgrounds.

4.3 *The Importance of Progression and Relevant System Examples*

With regard to the conditions for teaching and learning about systems, therefore, one should start with the less complex technologies and as the students get older introduce more and more complex systems (see Fig. 3). This should come as no surprise given the results of the previous research presented above. Going from easy to difficult, and from simple to complex, is also a classic educational strategy and a fundamental notion of progression if we look at the history of education, pedagogy and developmental psychology (Herbart & De Garmo, 1913; Piaget, 1929/1971). Likewise, as the students grow older one can also, if needed, be more specific and detailed about the components in a system (Örtnäs, 2007). Conversely, the strategy of "black boxing", that is, just focusing on input into and output from a system, seems to work for students of all ages, provided that it is clear what the input and output are, what the parameters of the black boxed aspects are (see Fig. 5).

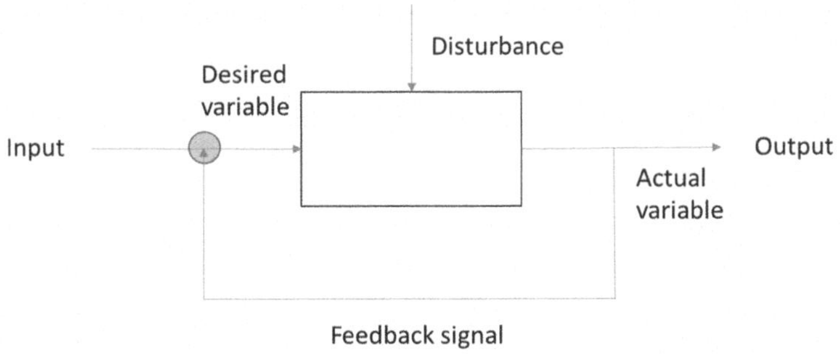

Fig. 5 System diagram of the flow of information in a feedback loop in a technological system, in which the square in the middle represents any black-boxed function, for example, automotive cruise control. (Source: Adapted from Grimvall (2014))

In a classroom context, where time is scarce, the question for the teacher is not "What specific system should I teach the students?", but rather "What system aspects can I teach the students, and what technological example can I best use to achieve that?" Therefore, it is pivotal that the example used is "larger than itself", that is, that it points to general principles or aspects of systems that could be applied even to other systems (Sjöberg, 2013). For example, if a dairy system is used it is not with the only aim of teaching the students about dairies but also about production systems in a more general sense, and to problematize systemic concepts like a system's border, input and output, sub-systems, feedback, flow of matter, energy and information, etc. Furthermore, questions concerning the human-system relationship, ethics, driving forces for change and consequences for the environment are important to address from technological perspectives (Ropohl, 1997).

For the teaching of systems to be most effective, it is imperative to find examples of technological systems relevant for teaching. Such systems could be electronic and/or control systems with feedback loops because of them being ubiquitous in much current technology. Barak (2018) gives some examples of how one can work with electronics to achieve not only knowledge of electronics per se but also systems thinking. Martin (1990) also describes a progression of working with school electronics as a path toward a more developed systems thinking in students. It is, as mentioned above, also important to recognize the importance of planning for progression. For young students, the examples and the questions about the systems' characteristics must be at a less complex level than for older students.

Large infrastructural systems such as water supply and electric grids are also relevant since they are commonplace particularly in urban environments. Such systems can also be easier to understand when presented at a general level with a degree of black boxing because of them sharing many structural similarities and having a fairly well-documented history (Ingelstam, 2002). Furthermore, there is also a well-developed set of concepts related to large technological systems, most notably by Thomas P. Hughes (e.g. Hughes, 1987) who introduced fairly intuitive, and therefore pedagogically powerful, concepts such as radical/conservative invention, system builder and bottle-neck/reverse salient. Even a less accessible concept such as momentum can be used as a basis for discussion with students about the way large technological systems tend to become difficult to change once they have been established (cf. Hallström, 2009).

5 Concluding Discussion

In this chapter, we have shown that technological systems must be seen as a central component of technological literacy. The tools that are available to study and understand technological systems are similar to those used in, for example, biology, physics, economics and social science and are derived from systems theory. The existing research on how students and teachers understand technological systems suggests that this is a very complex area of knowledge and that it is mostly artefacts and visible parts of the systems that they have knowledge about. Both teachers and students thus need to be trained in systems thinking, to get beyond the visual interface between human and system. It is imperative here to choose relevant examples of technological systems, and to learn about them in the right way. We have suggested four pedagogies to achieve this: interface pedagogy, holistic pedagogy, historical pedagogy and design pedagogy. Furthermore, we suggested two ways of delimiting systems through two types of boundaries that are crucial in this regard: the systems horizon and the system border.

To support a progression in systems thinking for students that go beyond the systems horizon, it is beneficial to choose the system examples along a progression line with consideration of the students' age and level of knowledge. Such lines of progression can be, for instance:

- From simple to complex systems.
- From small to large and widespread systems.
- From systems related to myself via us to others.
- From local systems via regional/national to global systems.

When planning teaching, teachers should also ask questions about how students should deal with technological systems. There are three questions that are particularly important, and they fall back upon the aim of the particular lesson (which should, of course, be based on the syllabus):

1. How can students describe a certain technological system?
2. How can students design, control or regulate a certain technological system?
3. How can students understand changes in a certain technological system over time?

One important aspect of especially points 2 and 3 is the sociotechnical perspective from which people themselves can be considered as part of the system and by their way of acting also contribute to the control and change of the system. On the other hand, design of sociotechnical systems is too complex in the educational context so the students should focus on small, less complex systems.

As in all other fields of knowledge, the students mature in their conceptual knowledge, and consequently the concept of "system" develops – broadens and deepens – over time. For younger students, describing a system and learning some basic system concepts might be sufficient. Continuing towards a more developed understanding of technological systems, students can both study and try out how to control systems – either in reality or in models/simulations. Likewise, the understanding of global technological systems demands a broad array of system concepts and might thus be better suited for older students. A student given a task to develop or design a technological system is put in a really challenging situation. Such a task needs to be adapted with respect to the two axes in Fig. 3 of the systems horizon, not too many components or complex connections – or given black-boxed components to use. An alternative is not to invent or design a new system but to use the same methods that have been significant for most technological change – you modify an existing system and try to fit it and optimize it against new demands.

References

Barak, M. (2018). Teaching electronics: From building circuits to systems thinking and programming. In M. J. De Vries (Ed.), *Handbook of technology education* (pp. 337–360). Dordrecht, the Netherlands: Springer.

Bertalanffy, L. v. (1973). *General system theory: Foundations, development, applications* (Rev. ed.). New York: Braziller.

Bijker, W. E., Hughes, T. P., & Pinch, T. J. (1987). *The social construction of technological systems: New directions in the sociology and history of technology*. Cambridge, MA: MIT Press.

Booth Sweeney, L., & Sterman, J. D. (2007). Thinking about systems: Student and teacher conceptions of natural and social systems. *System Dynamics Review, 23*(2/3), 285–312.

Capra, F. (1996). *The web of life: A new scientific understanding of living systems* (1st Anchor Books ed.). New York: Anchor Books.

Churchman, C. W. (1979). *The systems approach* (Rev. and updated. ed.). New York: Dell Pub.

Ellul, J. (1980). *The technological system*. New York: Continuum.

Ginns, I. S., Norton, S. J., & McRobbie, C. J. (2005). Adding value to the teaching and learning of design and technology. *International Journal of Technology and Design Education, 15*(1), 47–60.

Grimvall, G. (2014). *Teknikens metoder. Skolans teknikämne i senare skolår*. Lund, Sweden: Studentlitteratur.

Hallström, J. (2009). Systemteori och teknik. En introduktion till stora tekniska system. In P. Gyberg & J. Hallström (Eds.), *Världens gång – teknikens utveckling. Om samspelet mellan teknik, människa och samhälle* (pp. 99–117). Lund, Sweden: Studentlitteratur.

Hallström, J., & Klasander, C. (2013). Technology education for systems thinking and sustainability: What Swedish pre-service technology teacher students know about technological systems. In P. J. Williams & D. Gedera (Eds.), *PATT 27 technology education for the future: A play on sustainability: Christchurch New Zealand 2–6 December 2013*. Hamilton, OH: New Zealand University of Waikato.

Hallström, J., & Klasander, C. (2017). Visible parts, invisible whole: Swedish technology student teachers' conceptions about technological systems. *International Journal of Technology and Design Education, 27*(3), 387–405. https://doi.org/10.1007/s10798-016-9356-1

Hallström, J., Klasander, C., & Svensson, M. (2015). The black box and beyond: Introducing a conceptual model as a learning tool for developing knowledge about technological systems. In M. Chatoney (Ed.), *PATT 29 plurality and complementarity of approaches in Design & Technology Education, Marseille, France, April 2015*. Marseille, France: Presses Universitaires de Provence.

Handayani, N.A. & Ariyanti, D. (2012). Potency of Solar Energy Applications in Indonesia. International Journal of Renewable Energy Development, 1(2), pp. 33–38.

Hattie, J. (2012). *Visible learning for teachers: Maximizing impact on learning*. London: Routledge.

Herbart, J. F., & De Garmo, C. (1913). *Outlines of educational doctrine*. New York: Macmillan Company.

Hughes, T. P. (1983). *Networks of power: Electrification in Western society, 1880–1930*. Baltimore: Johns Hopkins University Press.

Hughes, T. P. (1987). The evolution of large technological systems. In W. E. Bijker, T. P. Hughes, & T. J. Pinch (Eds.), *The social construction of technological systems: New directions in the sociology and history of technology*. London: The MIT Press.

Ingelstam, L. (1996). *Complex technical systems*. Forskningsrådsnämnden (FRN): Stockholm.

Ingelstam, L. (2002). *System – att tänka över samhälle och teknik*. Stockholm: Energimyndigheten.

Jones, A., Buntting, C., & de Vries, M. J. (2013). The developing field of technology education: A review to look forward. *International Journal of Technology and Design Education, 23*(2), 191–212.

Keirl, S. (2006). Ethical technological literacy as democratic curriculum keystone. In J. R. Dakers (Ed.), *Defining technological literacy: Towards an epistemological framework* (pp. 81–102). New York: Palgrave Macmillan.

Kirschner, P. A., Sweller, J., & Clark, R. E. (2006). Why minimal guidance during instruction does not work: An analysis of the failure of constructivist, discovery, problem-based, experiential, and inquiry-based teaching. *Educational Psychologist, 41*(2), 75–86.

Klasander, C. (2006). *Innehållsaspekter av systemperspektiv i teknikundervisning. En studie av sex olika nationers styrdokument.* Paper presented at the Forskningskonferensen "Styrdokument och klassrumsverklighet i skolans teknikundervisning", Norrköping. http://www.ep.liu.se/ecp/017/004/ecp017004.pdf

Klasander, C. (2010). *Talet om tekniska system. Förväntningar, traditioner och skolverkligheter.* Norrköping, Sweden: Linköpings universitet.

Koski, M.-I., & de Vries, M. J. (2013). An exploratory study on how primary pupils approach systems. *International Journal of Technology and Design Education, 23*(4), 835–848. https://doi.org/10.1007/s10798-013-9234-z

Kroes, P., Franssen, M., van de Poel, I., & Ottens, M. (2006). Treating socio-technical systems as engineering systems: Some conceptual problems. *Systems Research and Behavioral Science, 23*, 803–814. https://doi.org/10.1002/sres.703

Lind, J., Pelger, S., & Jakobsson, A. (2018). Students' ideas about technological systems interacting with human needs. *International Journal of Technology and Design Education, 29*, 1–20. https://doi.org/10.1007/s10798-018-9449-0

Martin, D. J. (1990). Stages in systems thinking, modelling and realisation in school electronics. *International Journal of Technology and Design Education, 1*(1), 14–20.

Mioduser, D., Venezky, R. L., & Gong, B. (1996). Students' perceptions and designs of simple control systems. *Computers in Human Behavior, 12*(3), 363–388.

Örtnäs, A. (2007). *Elevers vardagsuppfattningar om tekniska system.* Linköping, Sweden: Bachelor's thesis/examensarbete. Linköping University.

Piaget, J. (1929/1971). *The Child's conception of the world.* London: Routledge & Kegan Paul.

Randle, J. M., & Stroink, M. L. (2018). The development and initial validation of the paradigm of systems thinking. *Systems Research and Behavioral Science*, 1–13. https://doi.org/10.1002/sres.2508

Ropohl, G. (1997). Knowledge types in technology. *International Journal of Technology and Design Education, 7*, 65–72.

Schooner, P., Klasander, C., & Hallström, J. (2018a). Swedish technology teachers' views on assessing student understandings of technological systems. *International Journal of Technology and Design Education, 28*(1), 169–188. https://doi.org/10.1007/s10798-016-9383-y

Schooner, P., Klasander, C., & Hallström, J. (2018b). Teknik, systemgräns och människa: Tekniklärares uppfattningar om vad tekniska system är. *NorDiNa, Nordic Studies in Science Education, 14*(4), 427–442.

Schooner, P., Nordlöf, C., Klasander, C., & Hallström, J. (2017). Design, system, value: The role of problem-solving and critical thinking capabilities in technology education, as perceived by teachers. *Design and Technology Education: An International Journal, 22*(3), 60–75.

Sjöberg, S. (2013). Grundläggande tekniska förmågor. In J. Hallström & C. Klasander (Eds.), *The Ginner handbook of technology education: Some theses about technology, school and society* (pp. 49–60). Linköping, Sweden: Linköping University Electronic Press.

Standards for Technological Literacy: Content for the Study of Technology. (2000). Reston, VA: International Technology Education Association.

Svensson, M. (2011). *Att urskilja tekniska system. Didaktiska dimensioner i grundskolan.* Norrköping, Sweden: Linköpings universitet.

Svensson, M. (2018). Learning about systems. In M. J. de Vries (Ed.), *Handbook of technology education* (pp. 447–462). Cham, Switzerland: Springer International Publishing.

Svensson, M., & Klasander, C. (2012). *Teacher's professional growth in planning and teaching technological systems.* Paper presented at the Technology Education Research Conference, Surfers Paradise, Australia.

Technology in the New Zealand Curriculum. (2007). Retrieved from The New Zealand Curriculum Online website: http://nzcurriculum.tki.org.nz/

Vermaas, P., Kroes, P., van de Poel, I., Franssen, M., & Houkes, W. (2011). *A philosophy of technology: From technical artefacts to sociotechnical systems*. San Rafael, CA: Morgan & Claypool Publishers.

Wallsten, B., Carlsson, A., Frändegård, P., Krook, J., & Svanström, S. (2013). To prospect an urban mine – Assessing the metal recovery potential of infrastructure "cold spots" in Norrköping, Sweden. *Journal of Cleaner Production, 55*, 103–111.

Wiener, N. (1954). *The human use of human beings; cybernetics and society*. Boston: Houghton Mifflin.

Wiliam, D. (2014). Why teaching will never be a research-based profession and why that's a Good Thing. *Dylan Wiliam's website.* Retrieved from https://www.dylanwiliam.org/Dylan_Wiliams_website/Presentations_files/2014-09-06%20ResearchED.pptx

Maker Education: Opportunities and Threats for Engineering and Technology Education

Gerald van Dijk, Arjan van der Meij, and Elwin Savelsbergh

Abstract Over the past decade, the maker movement and in its slipstream maker education have attained worldwide popularity among educators, politicians, and the media. Makers' enthusiasm for creative design and construction, using old and new tools has proven contagious, and is worth exploration and critical reflection by the community of engineering and technology education (ETE). This chapter describes what has been said about "making" by philosophers and educators; what maker education is, and what is new and not so new about it; why it has gained momentum; what the evidence is about its effectiveness and its possible weaknesses; and how mainstream technology education may benefit from maker education. This chapter concludes with ideas for a research agenda.

1 The Maker Movement and Making in Education

1.1 The Maker Identity

Making has been a defining trait of humanity since the first tools for carving stone and wood were used. The ancient Greeks' attitude towards making was ambivalent: at the one hand, for free men, logic (episteme) was regarded as the highest form of knowledge. On the other hand, the value of the knowledge involved in making was also acknowledged, certainly for the lower classes. A Homeric hymn to Hephaestus, the god of craftsmen, testifies of this.

G. van Dijk (✉) · A. van der Meij
University of Applied Sciences Utrecht, Utrecht, The Netherlands
e-mail: gerald.vandijk@hu.nl; ajvdmeij@gmail.com

E. Savelsbergh
Christelijk College De Populier, The Hague, The Netherlands
e-mail: elwin.savelsbergh@hu.nl

P. J. Williams, D. Barlex (eds.), *Pedagogy for Technology Education in Secondary Schools*, Contemporary Issues in Technology Education,
https://doi.org/10.1007/978-3-030-41548-8_5

> Sing clear voiced Muse, of Hephaestus famed for skill.
> With bright-eyed Athena he taught men glorious crafts throughout the world.
> Men who before used to dwell in caves in the mountains like wild beasts.
> But now that they have learned crafts through Hephaestus famous for his art
> they live a peaceful life in their own houses the whole year round
> (Anonymous)

Hephaestus possessed a combination of techné (knowledge to produce) and metis, which is a form of cunning intelligence, opportunism and experience that was held in high regard. The Greeks' ambivalent attitude has parallels in our present-day educational system, as well as in present-day philosophy. Modern day philosophers who regard making as a highly valuable element of humanity, include Hannah Arendt and Richard Sennett. Arendt acknowledges the liberating power of making by contrasting the "animal laborans," who are occupied with daily chores without any progress, to "homo faber," the making human, who is free to make and destroy things and who, therefore, "*conducts himself as lord and master of the whole earth*" (Arendt, 1958, p. 139). Sennett (2008) admires the advanced skill of the craftsman, which he sees as an embodied engagement that keeps humanity rooted in materiality and which changes an individual's view of the world. This, he argues, is important as a balance in a world that tends to be dominated by mental activities. The liberating power of "making" is also emphasized in the present-day maker movement.

In this chapter, our focus will be on making and the maker identity as they have emerged from the maker movement of the past two decades. It is hard to pinpoint a definitive start of this movement, because people have always made things and there is no sharp boundary between for instance radio electronics hobbyists in the previous century and present-day makers. Moreover, the maker movement is not a single movement, but rather a diverse coalition of (disruptive) entrepreneurs, ethical (and perhaps unethical) hackers, sustainability activists, crafts hobbyists, and so on. Nevertheless, Blikstein (2013) and others do see unique characteristics of the maker movement since 2000. The availability of relatively cheap equipment such as laser cutters and 3D printers enabled individuals to prototype artefacts that could only be made in specialized facilities before. Moreover, the maker movement has been linked to the rising popularity of entrepreneurship. Fablabs, Maker Spaces, and Maker Faires facilitate kick-starter projects for entrepreneurs in previously unthinkable ways. Mark Hatch (2014) characterized the movement in a 9 word summary: Make, Share, Give, Learn, Tool up, Play, Participate, Support, and Change. Among the first features that stand out if one visits a Maker Space or Maker Faire, are the innovative use of tools (Tool up) and the playful creativity. Playfulness can be seen in the type of products that are made, but it is also regarded as an important contribution to a problem solving mindset. The words "share, give, and participate" denote the movements propensity to create networks of makers in which knowledge, skills, and half-products are generously shared. Makers share their ideas in magazines such as "Wired" and "Make," on blogs and forums such as instructables.com and hackable.com. Not only do they share ideas but also they submit complete product designs and software libraries to Github, as part of the Creative Commons.

The ease of sharing knowledge and experiences through the internet has thus been instrumental in the rise of the maker movement, as well as being a fundamental driving force for those many makers who are inspired by techno anarchistic ideals. At a societal level, democratic and entrepreneurial values are attributed to the maker movement, because it gives individuals the capacity to make sophisticated products that can change the world, in a way that was previously unthinkable (Halverson & Sheridan, 2014; Vossoughi & Bevan, 2014). The word "Change," finally, refers to the idea that the act of making is fundamentally human and that "*you will become a more complete version of you as you make*" (p. 31).

> Reflective question: Who could be the "maker heroes" of our students? A grandparent who makes beautiful textile products? An artist who makes fascinating installations out of scrap materials? A metal worker who modifies a car in a cool way?

1.2 Making in Education

As early as the Greeks, "making" has been taught as a skill, as illustrated by the Homeric hymn above. In our days, making obviously is an important part of vocational and general technology education. Construction workers and electricians learn to make and repair products in vocational education, and so on. These vocations can also be of an industrial kind, whereby artefacts are made in mass production and where students are prepared to contribute to this process by learning to make standardized objects at school. In general engineering and technology education (ETE), where objectives focus on technological literacy, making also has its place. Even though "designing without making" can have a legitimate place in such curricula (Barlex & Trebell, 2008), more often making is an inherent part of the iterative process from conceptualizing an idea based on some human need to realizing and testing the product. Experiencing and reflecting on this process teaches children how the designed world comes into existence and is, therefore, an inherent part of a curriculum that targets technological literacy (ITEA, 2007).

In crafts oriented education, students obviously learn knowledge and skills needed for making. Scandinavian Sloyd is an example of crafts oriented education since 1865, in which the slow and attentive process of manually constructing artefacts is highly valued (Whittaker, 2014) as contributing to personal development. Reformist educational approaches as instigated by Fröbel, Montessori, Dewey, and Malaguzzi (in Reggio Emilia) acknowledged that making is not only valuable in itself but that it can also contribute to development of conceptual knowledge. Making thus has had its well-established place in ETE objectives and pedagogies for many years, so how can be explained that Maker Education has been embraced as a novel phenomenon so eagerly?

2 The Rise of Maker Education

The maker education movement has its roots in universities such as Stanford and MIT, where the first fablabs were established. From there, Gershenfeld, Papert, and Blikstein and others began to explore the educational potential of digital fabrication for extracurricular activities in education, as early as 2003 or the 1980s of the previous century in case of learning to code (Martinez & Stager, 2013). This, in turn, led to the development of new tools for programming and digital fabrication, with varying degrees of openness, that were suitable for use in schools, such as Scratch, MaKey MaKey, Little Bits, and so on. In the meantime, the cost of tools such as 3D printers, laser cutters, and simple programmable computer chips dropped dramatically. In 2019, the cost of a computer chip that can be programmed by secondary school students was below one euro. The availability of digital fabrication tools may have boosted the maker movement and maker education, but a quick glance in any maker space, fablab or makers' blog reveals that low-tech tools and materials are also used in maker classes, often in combination (see Fig. 1).

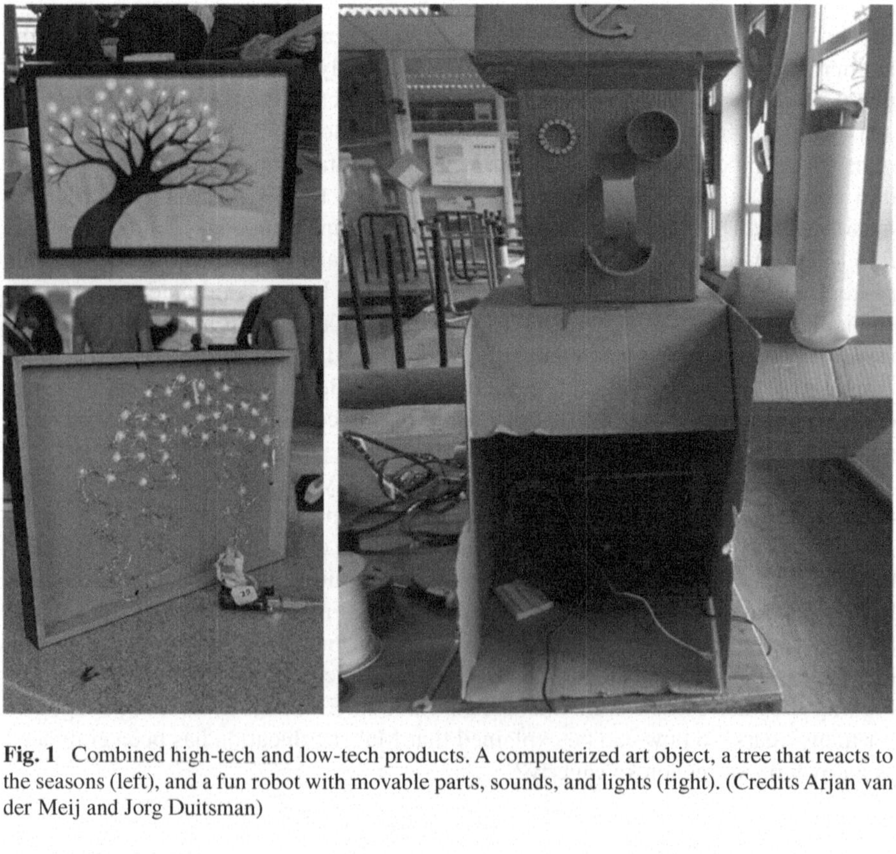

Fig. 1 Combined high-tech and low-tech products. A computerized art object, a tree that reacts to the seasons (left), and a fun robot with movable parts, sounds, and lights (right). (Credits Arjan van der Meij and Jorg Duitsman)

In the meantime, educationalists started using Papert's (1987) learning theory of constructionism to shape a maker education pedagogy and describe its foundations. Constructionism asserts that embodied experiences and the production of artefacts are central to the ways people learn and that pedagogies should be interest-driven rather than content-driven. The constructionists' idea that education should not stifle children's natural propensity to learn can be traced back to Rousseau and is also prevalent in visions of the early reformists mentioned above.

Papert (1987) illustrates constructionist teaching and learning with a contest in which students make a vehicle of their own idea with Lego bricks. The vehicle runs down a slope freely and it is then supposed to continue across a horizontal floor. Students are challenged to adapt the vehicle so that it runs further than their classmates' vehicles. Papert suggests that this activity, in itself, develops students' understanding of concepts such as mass and friction. It is worth noting that Papert, like Sennett, extends his conception of making to activities such as drawing with the aid of a computer (Barak, chapter "Pedagogical Approaches to Vocational Education").

Martinez, Stager, and Blikstein were among the first to use and extend Papert's ideas to shape a maker pedagogy for primary and secondary education. Martinez and Stager (2013) criticize what they call "design models" as commonly used in schools for containing too much instruction, which would lead to too many interruptions of the learning process (p. 52). The vignette in Table 1 illustrates how such interruptions are typically avoided in maker education. For the sake of contrast, the left column displays a more conventional approach, even though in reality there is no sharp line between such approaches.

Instead of regular design models, Martinez and Stager (2013) promote the use of the Think, Make, Improve model. The importance of collaboration and tinkering are foregrounded in this model. Tinkering is described as "*a playful way to approach and solve problems through direct experience, experimentation, and discovery*" (p. 32) that is much more open than recipe-like modes of assembling (Vossoughi & Bevan, 2014). Resnick and Rosenbaum (2013) describe three characteristics of construction kits such as Scratch and Makey Makey that make them suitable for tinkering: Immediate feedback, fluid experimentation, and open exploration. Table 2 lists these characteristics and illustrates how they are part of the programming language Scratch (Resnick & Rosenbaum, 2013).

It is worth mentioning that the use of Scratch, for instance as a programming language for the Arduino, has been criticized on the grounds that it does not resemble professional programming languages and that it is, therefore, unsuitable to teach about (software) engineering. However, text-based and numerical languages used by robotics engineers also evolve in the direction of more user friendliness, for instance by using a more Scratch-like visual architecture (Essers, 2016).

"Think, Make, Improve" (Table 3) is both a design method and an approach to teach designing. A selection of elements within this model as listed by Martinez and Stager (2013, p. 52) reflects common characteristics of maker education.

These key words and examples of maker education assignments that are found in research literature and on blogs illustrate the somewhat anarchistic and playful nature of maker education.

Examples of assignments typically found on maker education blogs and in literature

1. Hack a toy: bring a toy from home and make it move/blink or do other cool things with it.
2. Build a marble track that takes exactly 60 s to complete.
3. Make something you really want to make.
4. Build an object that shows how the liver works.
5. Build a musical instrument. You have to play a song on it that the teacher is able to recognize.
6. Hack your school: build something that improves your or the teacher school life.
7. Build a cardboard box with something locked inside that takes considerable effort to open it up (no force!).

Table 1 Avoidance of interruptions in maker education

Engineering and technology education	Maker education
Teacher: This week you'll design an alarm for the small statue that you've made in your art class. Statues need protection because people might want to steal them, right? You will use a buzzer and a LED as an alarm. It will be set off by a tilt switch in the pedestal, which is made of plywood. We'll first construct the pedestal. The electronics will come later. Carl: What about the size and shape? Teacher: Oh, I forgot. You'll get the design brief listing some specifications and an instruction to construct the pedestal. Other things are for you to choose and work out. Later Teacher: Class, we'll now study the issue of connecting the tilt switch, buzzer, and LED to the battery. Study the handout about series and parallel circuits and then draw a circuit diagram in your design portfolio. Carl: I'm ready, teacher. Can I start soldering after you've checked?	Teacher: What do you want to make this week, Jasmin? Jasmin: I want to make a light box for my aunt. She'll come over after a long time. Teacher: A light box? Jasmin: Yes. With my image engraved and a LED behind it. And it has one of those switches that flips when you turn the box, so my image appears. Later Teacher: Great! I see you made it easy to change the battery. Jasmin: Yes, but it's not bright, so I need more LEDs. Do they have to be in series or in parallel? Teacher: In this case both could be fine but, but if you put them in parallel, you'd somehow have to tinker with the voltage. Check the theory behind series and parallel circuits, and then you'll see that you'd burn them with this 9 V battery. If you need it, I'll explain.

Table 2 Tinkerability in scratch

Tinkerability characteristics	Sub category	Example how this characteristic is built into Scratch
Immediate feedback	See the result	Code is visible through blocks. Clicking on a block immediately shows what the block does to the object that is being programmed
	See the process	Chunks of code are being highlighted as the program is running
Fluid experimentation	Easy to get started	There is a default object that can be tinkered with right away
	Easy to connect	As with LEGO, in scratch it is easy to see which coding blocks can be connected, as a result of the color and shape of the blocks
Open exploration	Variety of materials	Sounds and music, images, backgrounds are available in a library that is easy to access from within the coding environment. Users continually extend this library
	Variety of genres	Scratch can be used to make computer games, animations, stories, music, interactive art, and as input/output computation for a robot or Arduino

Table 3 Think, make, improve according to Martinez and Stager (2013)

Think	Make	Improve
Brainstorming, talking it out, predicting, gathering materials, identifying expertise, deciding who to work with, setting goals, sketching, outlining, flowcharting, researching, planning	Play, build, tinker, create, program, experiment, construct, deconstruct, test strategies/materials, observe others, borrow and share code, document process, look for vulnerabilities, ask questions, repair	Conduct research, talk it out, discuss, look at it from a different perspective, use different materials, change one variable at a time, think how you solved similar problems in the past, play with it, find a similar project to analyze, ask a peer or expert, be cool, get fresh air, sleep on it

Such examples illustrate what is perhaps the most alluring quality of maker education: making with a combination of traditional and digital tools is fun. Mitch Resnick uses the metaphor "low floor, high ceiling, wide walls" to explain that maker education should be accessible for novices, challenging for learners and provide a great variety of pathways and projects for learning (Resnick, 2018). Maker education practitioners have also contributed to practical knowledge that helps to establish "low floor, high ceiling, wide walls." With regard to tools they for instance experienced that a laser cutter is more accessible than a 3D printer, whereas it still provides "wide walls" and a "high ceiling" (Van der Meij, Kloen, Hazelaar, & Van Oven, 2018).

Maker education is often considered as contributing to STEM learning objectives, and its openness and emphasis on student-centered learning resonates with inquiry-based STEM approaches. Knowledge and skills from a variety of disciplines such as physics, chemistry, mathematics, and art are used to make products, but generally it is not explicated what the relation is with these disciplines. Maker education is often located outside schools, for instance in libraries and museums. Where it is school-based, it is often part of extracurricular activities, whereby students voluntarily participate and without high-stakes formal assessment. The learning environment is often called a "maker space." The space and the tools reflect the characteristics of maker education as described above. However, regular technology workshops in schools vary widely and some can doubtlessly be used as maker spaces.

Reflective question: How do technology workshops at schools often look like, and how could they be improved to cater for:
A "low floor, high ceiling, wide walls" pedagogy
Creativity
Tinkering
A culture of sharing of ideas and (half) products

3 Research Findings: Pedagogy and Outcomes of Maker Education

Empirical research on learning outcomes in maker education is sparse and often scattered across journals from different disciplines (Troxler, 2016; Bevan, 2017). Empirical studies have largely been qualitative and descriptive and focused on out-of-school and after-school settings such as activities in museums and school-based maker clubs.

A review by Vossoughi and Bevan (2014) summarizes research findings in three categories. A first category is about how maker education gives young people a chance to develop identities as participants within STEM practices, specifically as makers. Second, the review gives examples of how learning and development can be structured. Third, the review shows how maker education promotes collaboration and fluid roles of novices/experts, rather than fixed roles of teacher and learner. We will use these three categories to elaborate on research findings.

3.1 Maker Identity

An outcome of maker education that is foregrounded in much research is what is called a "maker identity," or self-efficacy as "maker," a change agent in the material world. Participants take on new roles as makers, using computational media and craft technologies and they connect these roles to long-term interests. Blikstein (2013) agrees and illustrates this from his extensive experiences. He emphasizes that this identity is multidisciplinary and in some cases as strongly associated with engineering as with other domains. As an example, Blikstein describes the case of

Max, who made a robotic flute. Blikstein argues that Max learned a lot about engineering, but the main learning outcomes were about music interpretation. In another example, students made a historical model using digital fabrication equipment and learned about history, engineering, and mathematics in the process.

A critical note with regard to this "maker identity" is concerned with maker education's power to bolster this identity across a diverse student population. Scholars from within the maker education community have observed that maker education presents a picture of "making" as a white, middle class and male enterprise (Vossoughi, Hooper, & Escudé, 2016). Blikstein and Worsley (2016) warn that maker education in schools may increase inequality when middle class boys who have had access to maker culture get to do more demanding work in schools than girls and boys from lower-income groups. If no precautions are taken, this can be a result of maker education's tendency to leave decisions about what is being learned and who takes up tasks in a project group to students themselves. In addition, Vossoughi et al. (2016) argue that maker education generally fails to appreciate valuable accounts of making from different cultures. Moreover, they assert that maker education finds its legitimacy one-sidedly in economic value of commercial innovations, which will only strengthen existing economic structures and imbalances of power.

3.2 *Structuring Learning*

With regard to pedagogy, Vossoughi and Bevan's review (2014) highlights how maker education provides meaningful contexts for learning STEM concepts, particularly the STEM that is grounded in a socio cultural approach to action learning. Although the argument seems plausible, there is little concrete evidence about how the STEM-learning could be implemented from such context. Details with regard to sequencing of activities, the choice for specific making assignments, sources for students, teaching materials etc., have also not been provided thus far. Blikstein describes possible drawbacks of a maker education pedagogy, an important one being the "*keychain syndrome*" (p. 8). He shows how students' success in making keychains with the aid of a laser cutter came at the cost of their willingness to "invent." Students valued product over process, a problem that is exacerbated due to the very nature of the machine that facilitates the production of flashy products in a relatively simple way. The machine becomes "*a Trojan horse*" unless the educational designer intervenes (Blikstein & Worsley, 2016, p. 9).

3.3 *Roles*

With regard to the roles that teachers and students can take on in maker education, much of the research describes these roles as fluid. Teachers are role models as "learning makers," but at times they can also adopt roles as coaches or instructors.

In this latter role, they also make knowledge explicit. Words that are also used for teacher's roles in Vossoughi and Bevan's (2014) review are "sparking" (interest), "sustaining" (participation), and "deepening" (participation). The latter role resonates strongly with Martinez and Stager's (2013) emphasis on improvement of students' products, as far as possible. But also there, students' agency in the way their product could be improved is given priority.

In addition to what has been said thus far, Smith, Iverson, and Veerasawmy (2016) described the following challenges for teachers to adopt a role that fits the ideas behind maker education: bridging the gap with formal curricula, emphasizing the learning process over the product, and nourishing a design language, the latter being a general challenge in technology education (Van Dijk & Hajer, 2017).

4 Some Doubts from the Learning Sciences

In this section, we provide a few links with insights from the learning sciences that could help to identify opportunities and threats for technology education, as we try to learn from maker education.

Nowadays, school-based maker education is often part of extracurricular activities. In this case, students follow compulsory science and technology classes and voluntarily come to "maker activities" after regular hours. In this context, the fun factor of maker education can be fully exploited and regardless of pedagogy, some students will certainly learn many things. However, where maker education is adopted as a replacement for more traditional science and technology lessons, a discussion about pedagogy is paramount. In this section, we raise two issues that are of concern when maker education becomes part of regular curricula: the role of teacher guidance and problems with transferability of skills. The issue of content will be dealt with in Sect. 6.

Martinez and Stager's (2013) argument for less teacher instruction, fewer interventions, and fewer interruptions resonates with radical constructivists' view on learning as well as with Papert's constructionism (in particular its child-centeredness). Publications by other proponents of maker education follow the same lines and it is hard to find any publication about maker education where guidance and explicit instruction are being advocated (Stockard, Wood, Coughlin, & Caitlin, 2018). Constructivism and constructionism, however, are certainly not uncontested, in particular for learning new content (Andersen & Andersen, 2017; Klahr & Nigam, 2004; Kuhn, 2007). Mayer (2004) reviewed a body of research on constructionist' methods of teaching to code in the LOGO language. He comes to the conclusion that "*author after author noted the role of guidance in learning to program*" (p. 17), which explains the failure of "discovery oriented" environments for learning to code. Furthermore, evidence is inconclusive with regard to the equity effects of student-centered and constructivist pedagogies: while some authors found that open and child centered pedagogies tend to widen the gap in learning outcomes between high and low SES students (e.g., Andersen & Andersen, 2017), others report beneficial effects particularly for low SES students (Mehalik, Doppelt, & Schuun, 2008). This issue deserves careful attention if we consider implementing Martinez and Stager's advice for less instructions in formal curricula.

For learning to design and make, whole task approaches that are prevalent in constructivist pedagogies have been explored by Van Breukelen (2017). He asserts that many design-based learning approaches involve too many objects of integration (skills, practices, attitudes and content) for students, "*that often remain underexposed in the case of unexperienced practitioners.*" Content is often overlooked as a result (Van Breukelen, 2017: p. 102). This too raises questions about claims that "making" in itself is a powerful vehicle for learning conceptual and procedural knowledge for all children. This does not mean that maker education as a whole can be disqualified on research grounds. In some cases, publications that criticize constructivist teaching approaches can also be used to highlight strengths of maker education. Clark, Kirschner, and Sweller (2012) for instance emphasize the effectiveness of learning with the aid of "worked examples." Following the "share" principle, the maker education community has found many ways to make "worked examples" easily accessible for students. The Scratch platform, for instance, provides students with many examples of chunks of code, whereby it is easy to see how the code works (Table 2).

The learning sciences are also useful to shed light on the issue of transferability of skills. The rhetoric in maker education is abound with claims that students learn skills such as creative thinking and problem solving that they will need in life, perhaps for different things than making. A quote from Mitch Resnick, LEGO Papert Professor of Learning Research at MIT and long standing and influential developer of maker education tools for learning (Scratch, MaKey MaKey) illustrates this idea.

> As people learn to code, they think systematically. They start to identify bugs and problems and fix them in ways that carry over to other activities. You learn basic strategies for solving problems, designing projects and communicating ideas. That will be useful to you even if you don't grow up to be a programmer, but a journalist or a marketing manager or a community organizer. We sometimes say, it's not so much about learning to code, but coding to learn. As you code, it's helping you learn other things (Resnick, in Barshay, 2013).

However, research has shown that "thinking systematically," creativity, and skills generally develop in interaction with domain knowledge (Bransford, Brown, & Cocking, 2003). In case of creativity, Lucas, Claxton, and Spencer (2013) have described this in general and Christiaans and Venselaar (2005) have done so for ETE.

5 The Potential of Maker Education for Engineering and Technology Education

In the previous sections, we have described features, strengths, and weaknesses. Now, we will consider how these features can be used to strengthen our ETE agenda. Maker education has been able to exploit the "fun factor" of making, in particular

Table 4 ETE concepts according to Rossouw, Hacker, and de Vries (2011)

Main concept	Sub-concepts
Designing ("design as a verb")	Optimising, trade-offs, specification, invention, product life cycle
Systems	Artefacts "design as a noun", structure, function
Modelling	
Resources	Materials, energy, information
Values	Sustainability, innovation, risk/failure, social interaction, technology assessment

with the aid of accessible high-tech tools in combination with traditional tools and materials. ETE can benefit from maker education's potential to develop "maker identities" by attracting more students and keeping them engaged. In order to utilize this potential, we need a clear understanding of similarities and differences. This will not amount to a simple addition – some elements can strengthen each other, but other elements may be at odds.

Making has always had a prominent place in ETE. However, there is not only overlap between maker education and ETE (Bell & Quinn, 2013), but there are also important differences. Rossouw, Hacker, and De Vries (2011) used a Delphi study to identify five key concepts in ETE, each with associated sub-concepts (Table 4). The concepts were chosen by experts from technology education (secondary education, technology teacher education, and associated research), engineering education (tertiary education and engineering associations), philosophy of technology, design methodology, and science and technology communication.

Of these five, designing is often used as an overarching concept. Design in ETE is goal directed, meant to arrive at the best possible solution for a problem, within technological and other constraints. The goal is usually some customer need or other human desire that guides an iterative process of setting requirements, specifications, construction, testing and evaluation of different solutions (Björnberg, 2013). Maker education does not necessarily regard making as part of a design process that is meant to identify and solve authentic problems in the world outside schools (Bevan, 2017). Many of the examples of maker education tasks in this chapter testify of this difference, for instance "*make something that you really want to make*" and "*hack a toy*." This reflects the emphasis on the personalized fun factor of making in the maker movement,[1] rather than an emphasis on design in an ETE sense.

Differences are also easy to find if we zoom into the ways learning to design is scaffolded with the aid of functional or structural design models (Mioduser & Dagan, 2007), as stated earlier. Such models, however, serve as scaffolds for novices to learn to work systematically during designing, for instance to regularly check whether the process is still on track to meet user requirements and technical requirements. Thinking is terms of systems is another important aspect of learning about technology (Hallström & Klasander, this volume). This helps students to gain

[1] A YouTube search on '*useless machine*' will result in many examples.

insight in working principles, input and output, and the way systems work together to realize a function, a type of transferable knowledge that is particularly important because technologies evolve very rapidly (Hallström & Klasander, this volume). Systems thinking is not only of importance to be able to design, but also to maintain and repair consumer products as well as industrial assets. For these reasons, systems thinking is usually explicitly addressed in ETE. Systems thinking can perhaps be developed through maker education activities, but there are few examples that demonstrate how this is done explicitly, with guidance by the teacher. This is the case for the remaining core concepts (modelling, values, resources) too. The relation with values, for instance in case of sustainability, is particularly interesting. Maker education enthusiasts claim that they can make a contribution to sustainability, because they can help reduce waste by teaching how to reuse materials, to repair broken products, and to become active participants in the search for technical solutions to sustainability. Also, learning to design and make in local communities can decrease carbon footprints because less transportation is needed, it is claimed (Kohtala & Hyysalo, 2015). However, one could also argue that making rather useless fun products at school adds waste unnecessarily and it does not raise awareness of the limits that we face in our use of energy and materials. Furthermore, the maker education literature does not yet demonstrate how values such as sustainability can be addressed explicitly. In contrast, the literature about ETE is rich with regard to values and sustainability (Keirl, this volume). A more overarching concern with regard to sustainability is that participants in the maker movement as a whole are often unaware of major environmental implications of their work, such as toxicity of materials, or they do not even regard sustainability as key issue for future maker spaces (Kohtala & Hyysalo, 2015). Whether this problem with knowledge, beliefs and attitudes transcends from the maker movement into maker education is unknown.

As a conclusion, we assert that replacement of a regular technology, engineering and science curriculum with a radical form of maker education is likely to result in a misfit with generally accepted learning objectives. This has been acknowledged from within the maker education community (Smith et al., 2016; Christensen, Hjorth, Iversen & Blikstein, 2016), and we have attempted to specify this risk a little further. Nevertheless, maker education can potentially strengthen more traditional forms of ETE. As a starting point for further discussion, Table 5 identifies three principles to achieve that.

However, each of these principles is likely to result in trade-offs for both maker education and ETE. Depending on national and local contexts, decisions can be made to arrive at a responsible and feasible balance in the applications of the principles.

Reflective question: What possibilities do you see, to use the three strengthening principles in one technology education practice that you know well?

Table 5 Principles for strengthening ETE curricula

Strengthening principle	Examples of interventions in ETE curricula
Maker education projects as curricular add-ons for increasing students' ownership Allocate time for free, and collaborative making and tinkering, whereby students set their own agenda. Include "making challenges" that fit well with different cultures and gender identities.	Include "making without designing" challenges. This means that a user's needs and other constraints (e.g., technical) do not always have to be specified. "Step-by-step design methods" are not necessarily explicated before or after the process
Frontloading conceptualization and procedural understanding in "making challenges" Set "making challenges" that can only be met if scientific and technical conceptual and procedural knowledge is applied	Start a project with a brief such as: *You are going to make a cool "class promotion board" that interacts with sounds and movements in the environment. It has pulsating LED's and moving "old school" meters with arrows that indicate current and voltage as the Led's pulsate. Next year's class must be able to adapt your product and they can only do so if you provide a written explanation of working principles of your system, which includes calculations of power, voltage, current, and resistance* Give instruction and guidance to enable students to meet the challenge
Backtracking for conceptual and procedural understanding in making challenges Use a product from a maker class or maker faire as an artefact to be understood in terms of regular ETE and science concepts	Ask questions such as: *How do the subsystems in the fun robot work together? How can we visualize that with a diagram to arrive at a better understanding about why it worked so well in the end? What are potential environmental problems with the production and disposal of this fun robot? Which procedures were useful to follow when you needed to find out why a subsystem was not working? How does that relate to formal procedures that engineers use?* Give instruction and guidance to enable students to answer such questions

6 Towards a Research Agenda

Learning by making is alluring, but what is being learned in terms of regular and credible ETE objectives? Van Breukelen's study (2017) shows that conceptual development through designing and making is possible, if students are provided with teacher-led scaffolds that put science content central, such as direct instruction. This was accomplished in a class were students all worked on the same design task, which made it possible for the teacher to give instruction and guidance that was relevant for all students. Moreover, the design task was set in such a way that science content was indeed needed. In a student-centered maker pedagogy this is harder to achieve. The issue of hybridization of maker education and ETE curricula, i.e., explored in Table 5, is worth further research. Such research should also take diversity in student populations into account.

And perhaps we should not just search for hybridization. Extracurricular maker activities deserve a place in their own right, unhindered by demands that would undermine their successes, as long as they do not jeopardize legitimate ETE curricula. In any case, it will be worth the effort to learn from maker education practitioners how they manage to "teach" so beautifully in the spirit of Hephaestus, who rightfully acknowledged that *mankind is a making kind.*

References

Andersen, I. G., & Andersen, S. C. (2017). Student-centered instruction and academic achievement: Linking mechanisms of educational inequality to schools' instructional strategy. *British Journal of Sociology of Education, 38*(4), 533–550.

Arendt, H. (1958). *The human condition* (2nd ed.). Chicago: University of Chicago Press.

Barlex, D. M., & Trebell, D. (2008). Design-without-make: Challenging the conventional approach to teaching and learning in a design and technology classroom. *International Journal of Technology and Design Education, 18*(2), 119–138.

Barshay, J. (2013). *MIT technology trailblazer is a critic of computerized learning (interview with Mitch Resnick).* http://hechingerreport.org/mit-technology-trailblazer-is-a-critic-of-computerized-learning/

Bell, P., & Quinn, H. (2013). How designing, making, and playing relate to the learning goals of K-12 science education. In M. Honey & D. Kanter (Eds.), *Design, make, play : Growing the next generation of STEM innovators* (pp. 17–32). New York: Routledge.

Bevan, B. (2017). The promise and the promises of making in science education. *Studies in Science Education, 53*(1), 75–103.

Björnberg, K. E. (2013). Rational goals in engineering design: The Venice dams. In M. De Vries, S. O. Hansson, & A. W. M. Meijers (Eds.), *Norms in technology* (pp. 83–102). Dordrecht, the Netherlands: Springer.

Blikstein, P. (2013). Digital fabrication and 'making' in education: The democratization of invention. In J. Walter-Herrmann & C. Büching (Eds.), *FabLabs: Of machines, makers and inventors* (pp. 203–223). Stanford, CA: Stanford University.

Blikstein, P., & Worsley, M. (2016). Children are not hackers: Building a culture of powerful ideas, deep learning, and equity in the maker movement. In K. A. Peppler, E. Halverson, & Y. B. Kafai (Eds.), *Makeology: Makerspaces as learning environments* (pp. 64–79). New York: Routledge.

Bransford, J. D., Brown, A. L., & Cocking, R. R. (Eds.). (2003). *How people learn: Brain, mind, experience and school. Expanded edition.* Washington, DC: National Academy Press.

Christensen, K. S., Hjorth, M., Iversen, O. S., & Blikstein, P. (2016). Towards a formal assessment of design literacy: Analyzing K-12 students' stance towards inquiry. *Design Studies, 46*, 125–151.

Christiaans, H., & Venselaar, K. (2005). Creativity in design engineering and the role of knowledge: Modelling the expert. *International Journal of Technology and Design Education, 15*(3), 217–236.

Clark, R. E., Kirschner, P., & Sweller, J. (2012). Putting students on the path to learning: The case for fully guided instruction. *American Educator, 36*(1), 6–11.

Essers, M. S. (2016). *Design of a novel, hybrid decentralized, distributed, modular architecture for manufacturing systems.* Enschede, Netherlands: Universiteit Twente.

Halverson, E., & Sheridan, K. (2014). The maker movement in education. *Harvard Educational Review, 84*(4), 495–504.

Hatch, M. (2014). *The maker movement manifesto.* New York: McGraw-Hill.

ITEA. (2007). *Standards for technological literacy: Content for the study of technology* (3rd ed.). Reston, VA: International Technology Education Association.

Klahr, D., & Nigam, M. (2004). The equivalence of learning paths in early science instruction: Effects of direct instruction and discovery learning. *Psychological Science, 15*(10), 661–667.
Kohtala, C., & Hyysalo, S. (2015). Anticipated environmental sustainability of personal fabrication. *Journal of Cleaner Production, 99*, 333–344.
Kuhn, D. (2007). Is direct instruction an answer to the right question? *Educational Psychologist, 42*(2), 109–113.
Lucas, B., Claxton, G., & Spencer, E. (2013). *Progression in student creativity in school: First steps towards new forms of formative assessments*. OECD Education working paper 86, OECD Publishing.
Martinez, S. L., & Stager, G. (2013). *Invent to learn*. Torrance, CA: Constructing Modern Knowledge Press.
Mayer, R. E. (2004). Should there be a three-strikes rule against pure discovery learning? *American Psychologist, 59*(1), 14–19.
Mehalik, M. M., Doppelt, Y., & Schuun, C. D. (2008). Middle-school science through design-based learning versus scripted inquiry: Better overall science concept learning and equity gap reduction. *Journal of Engineering Education, 97*(1), 71–85.
Mioduser, D., & Dagan, O. (2007). The effect of alternative approaches to design instruction (structural or functional) on students' mental models of technological design processes. *International Journal of Technology and Design Education, 17*(2), 135–148.
Papert, S. (1987). *A critique of technocentrism in thinking about the school of the future*.http://www.papert.org/articles/ACritiqueofTechnocentrism.html ed.
Resnick, M. (2018). *Designing for wide walls*.https://design.blog/2016/08/25/mitchel-resnick-designing-for-wide-walls/
Resnick, M., & Rosenbaum, E. (2013). Designing for tinkerability. In M. Honey & D. Kanter (Eds.), *Design, make, play : Growing the next generation of STEM innovators* (pp. 163–181). New York: Routledge.
Rossouw, A., Hacker, M., & De Vries, M. J. (2011). Concepts and contexts in engineering and technology education: An international and interdisciplinary Delphi study. *International Journal of Technology and Design Education, 21*(4), 409–424.
Sennett, R. (2008). *The craftsman*. New Haven, CT: Yale University Press.
Smith, R. C., Iversen, O. S., & Veerasawmy, R. (2016). Impediments to digital fabrication in education: A study of teachers' role in digital fabrication. *International Journal of Digital Literacy and Digital Competence, 7*(7), 33–49.
Stockard, J., Wood, T. W., Coughlin, C., & Caitlin, R. K. (2018). The effectiveness of direct instruction curricula: A meta-analysis of a half century of research. *Review of Educational Research, 88*(4), 479–507.
Troxler, P. (2016). *Niet alleen "omdat het kan": Een onderzoek naar bestaande kennis over maker education (translation: Not just because "we can": A study into research about maker education.)*. Amsterdam: Platform Maker Education/De Waag.
Van Breukelen, D. (2017). Teaching and learning science through design activities. (PhD, Delft University of Technology).
Van der Meij A., Kloen, P. I., Hazelaar, M. & van Oven, R. (2018). *Basisscholen, koop geen 3D printer*.http://makered.nl/basisscholen-koop-geen-3d-printer/
Van Dijk, G., & Hajer, M. (2017). Teaching the language of technology: Towards a research agenda. In M. J. De Vries (Ed.), *Handbook of technology education* (pp. 537–549). Dordrecht, the Netherlands: Springer.
Vossoughi, S., Hooper, P. K., & Escudé, M. (2016). Making through the lens of culture and power: Toward transformative visions for educational equity. *Harvard Educational Review, 86*(2), 206–232.
Vossoughi, S., & Bevan, B. (2014). Making and tinkering: A review of the literature. Washington DC: The National Academies Press.
Whittaker, D. J. (2014). *The impact and legacy of educational Sloyd: Head and hands in harness*. New York: Routledge.

Signature Pedagogies for Designing: A Speculative Framework for Supporting Learning and Teaching in Design and Technology Education

Kay Stables

Abstract In this chapter, I focus on the challenge of learning and teaching designing and offer an approach to overcome this challenge by proposing a pedagogic framework that enables teachers to focus on the why, what and how of teaching and learning designing in a flexible and creative way. Exploring the importance of design as a human capability, I unpack this capability in terms of its complexity, ongoing discussions about the nature of design knowledge and the concept of designerly ways of knowing. I then turn to explore a concept less well developed in design and technology education, that of signature pedagogies. Drawing on this concept and ways in which it has been developed in higher education design, parallels are explored with mainstream school education. Collected ideas are drawn together to create a framework for signature pedagogies of learning and teaching designing in schools. This framework is structured on design pedagogic purposes of speculation, imaging and modelling, materiality, need-to-know, critiquing and collaboration linked to pedagogic actions that can be taken and pedagogic tools that facilitate such actions. Finally, I provide two vignettes that illustrate how the framework could be used in classrooms.

1 Why Focus on Pedagogies for Designing?

Across national and provincial borders, many versions of curricula for Technology Education, Design and Technology Education, Science and Technology Education and Technology and Engineering Education exist. Each has its own local focus, philosophy, structure and content. But there are also common, ubiquitous

K. Stables (✉)
Goldsmiths, University of London, London, UK
e-mail: K.Stables@gold.ac.uk

P. J. Williams, D. Barlex (eds.), *Pedagogy for Technology Education in Secondary Schools*, Contemporary Issues in Technology Education,
https://doi.org/10.1007/978-3-030-41548-8_6

curriculum elements, one of which is the centrality of processes of designing. Despite its centrality, there is general recognition that teaching and learning designing is a challenge, not least because of confusion over definitions and models of designing (Kimbell and Stables, 2007; Mawson, 2003; McGimpsey, 2011; McLain, 2012; Ofsted, 2002b, 2008). The aim of this chapter is to provide support for developing approaches to overcome this challenge by proposing a pedagogic framework that crosses borders and enables teachers to focus on the why, what and how of teaching and learning designing in a flexible and creative way.

It is important to first consider what designing is and why people might want to learn or teach it. What is it about designing that makes it something that is valuable to learn? Why do human beings need to be able to do it? A considerable amount has been written about the extent to which design capability is one of the defining characteristics of being human. Bronowski (1973) writes of how human creativity and imagination allows us to "visualise the future" (p. 56) "not to accept the environment but to change it" (p. 19). Archer links designing with an "envisaging what" capacity that he sees as "the third great defining characteristic of humankind" (Archer & Roberts, 1992, p. 9). Nelson and Stolterman (2003) remind us that "Humans did not discover fire – they designed it. The wheel was not something our ancestors merely stumbled over in a stroke of good luck; it, too, was designed." (Nelson & Stolterman, 2003, p. 9). Baynes draws directly on cognitive science "to show conclusively that designerly thinking and action are features of the mental activities of all humans. …. The highly complex skills of the professional engineer, fashion designer or CGI artist are simply the specialist development of abilities and understandings we all have" (Baynes, 2006, p. 7).

2 Building Capability in Designing

Although we all have potential design capability, for the potential to be realised it needs to be developed through learning experiences. Baynes (1992) highlights how design learning starts as soon as a child engages with their material world and that when a child starts formal schooling they bring their early capability with them. His concern is that these early school experiences need to go beyond conventional 'making' projects to those that "bring a deeper understanding of what we might call 'design intelligence' – that is, the particular ways in which children and adults think and act when they are designing." (Baynes, 1992, p. 1).

Cross (2008) also makes a case for design being a particular kind of intelligence, a cognitive function that humans have whether they are professional designers or not. Like all cognitive functions, it needs to be nurtured. He makes a case for design in general education, developing "innate abilities in solving real-world, ill-defined problems … cognitive development in the concrete/iconic modes of cognition … development of a wide range of abilities in non-verbal thought and communication" (Cross, 1982, p 226). This case is based on his identification of what he labels 'designerly ways of knowing', particular ways of thinking and acting.

3 Designing: A Complex Activity

What Cross is describing is not a simple activity and, by definition, not an easy thing to learn or teach. Others have contributed to this view, identifying problem solving within design as being of 'wicked problems' (Buchannan, 1995; Rittell & Webber, 1973) for which there is no correct answer, just a range of different solutions that could be seen as better or worse, depending on the lens they are viewed through. Designing is characterised as solution focused (Cross, 2006). It involves cognitive and external modelling of speculative ideas, the mind and hand working together, to make models visible (to self and others) through words, drawings, physical models and prototypes (Archer & Baynes, 1992). There is no doubt that designing is a complex process and this complexity is what McGimpsey (2011) identifies as being at the heart of the challenge of teaching it.

Central is the procedural nature of designing and the ways in which, in developing ideas, humans iterate between thinking and doing. A focus on process in design has been increasingly present in school curricula since the 1970s as a shift in emphasis has taken place between the product made and the process this was achieved by. This shift echoed a quest among professional designers to create a modernist, universal approach to designing, which was made manifest through a clinical, linear set of steps to be followed – a model that was adopted in schools and reinforced through assessment structures. Drawing on more recent design research, the linear model that persists as something often called 'the' design process has been challenged, questioning whether it is either true or helpful in developing design capability. (see, e.g. Flowers, 2010; Kimbell & Stables, 2007; Mawson, 2003; Williams, 2000). This more recent analysis has provided a less simplistic, more authentic perspective, drawing together the speculative, wicked, uncertain, iterative nature of designing, negating an idea that there is one standard design process. In giving an illustration of the complexity of designing and the uncertainty inherent in such activity, Lawson (2004) exemplifies what chess might be like if it was a design activity.

> Designing then, in terms of chess, is rather like playing with a board that has no divisions into cells, has pieces that can be invented and redefined as the game proceeds and rules that can change their effects as moves are made. Even the object of the game is not defined at the outset and may change as the game wears on. Put like this it seems a ridiculous enterprise to contemplate the design process at all. To try to understand how it proceeds and what knowledge is used and develop some structure for that may seem foolhardy. (Lawson, 2004, p. 20)

Lawson suggests that it would be foolhardy to create a procedural structure or knowledge base for design and, for educators, it could also be seen as foolhardy to imagine that it is possible to work out what to teach and how to teach people to develop their design capability. But if alternatives to linear models are to be embedded in school designing, then fresh and refreshed frameworks need to be established. It is critical that teachers have access to a repertoire of pedagogies that they can utilise, modify and exploit in their endeavours to develop designerly ways of knowing and doing in young people.

4 Design Knowledge?

Focusing on what could be the specifics of knowledge for designing has received considerable attention and debate. Pinning down exactly what knowledge is 'design knowledge' is an attractive idea. But such specificity seems akin to looking for a holy grail – and Lawson's description of design chess gives some clues to the challenge. A concept of 'knowledge' as a fixed resource is unhelpful as, in any design situation, you can't know everything in advance. There may be a repertoire of skills – for imaging, modelling, reflecting, investigating, prototyping and so on. But the context in which designing takes place will inevitably be full of unknowns, suggesting a need for a more fluid concept of design knowledge. Take, for example, designing a learning aid for a child with cerebral palsy. A successful solution will require contextual knowledge, which in this example would include knowledge of both learning and of cerebral palsy. As a design solution begins to develop, there will also be a need for technical knowledge, for example of specific materials or mechanisms and tools that need to be used, or procedural knowledge of how to go about creating a prototype. So, in this example, the contextual, technical and procedural knowledge all become design knowledge, required in response to needs in the task. Kimbell and Perry (2001, p19) highlight the interdisciplinary nature of such knowledge and referred to this as knowledge and see it as characterising design and technology as a "restive, itinerant, non-discipline".

Cross coined the phrase 'designerly ways of knowing' and this, in itself, requires a stance that takes 'ways of knowing' beyond what could be seen as a traditional Eurocentric concept of knowledge. In exploring differences between a Eurocentric concept and one commonly found in indigenous societies, Akinhead and Elliot (2010) suggest that

> the word knowledge is embedded in a Eurocentric epistemology and should be replaced by other expressions that more authentically capture an Indigenous worldview, such as Indigenous ways of knowing, living or being. Concomitantly, the Eurocentric meaning of to learn becomes coming to know in most Indigenous contexts, a meaning that signifies a personal, participatory, holistic journey toward gaining wisdom-in-action. (p. 3)

Designerly ways of knowing and doing has resonance with this broader view. Designing is an activity that constantly provokes a need for new learning, whether it is a specific skill or particular understanding, such as those suggested above in designing in the context of cerebral palsy. Gaining wisdom-in-action equates well as an underlying principle for pedagogies of designing. This broader view of designerly ways of knowing, thinking and acting is both more authentic in terms of how humans act as designers and more holistic in considering what needs to be taught and learnt.

Norman and Baynes (2017) suggest that, in the context of the English national curriculum, there has been a lack of understanding of the significance of developing designerly ways of knowing in schools. They make the point that a 'designerly way of knowing' is not an imaginary academic construct, but an everyday reality that requires appropriate consideration and weight in curriculum planning (Norman & Baynes, 2017, p. 6). This, in turn, highlights its importance when considering pedagogic approaches to designing.

5 Signature Pedagogies for Teaching and Learning Designing

In tertiary design education, attention has been given to Shulman's concept of signature pedagogies (Shulman, 2005a) – pedagogies that he identifies as being those that prepare people for their profession, in ways of thinking, performing and acting with integrity that have their own 'signature' methods of teaching and learning, depending on the profession. Exploring this concept, Tovey (2015) suggests that, in design, there are conventional signature pedagogic elements or arenas; the studio, the design tutorial, the library, and the 'crit' (Tovey, 2015). Shreeve (2015) and Orr and Shreeve (2018) present a similar list, again highlighting the studio and the crit, but also identifying the project and brief, materiality, dialogue and research. Each of these can be readily seen to match to Shulman's concept. He identifies signature pedagogies as having pervasive and ritualistic or routine methods of learning and teaching which, in the context of the profession being prepared for, have a surface structure of "concrete, operational acts of teaching and learning, of showing and demonstrating, of questioning and answering, of interacting and withholding, of approaching and withdrawing"; a deep structure of "assumptions about how best to impart a certain body of knowledge and know how" and an implicit structure that provides "a moral dimension that comprises a set of beliefs about professional attitudes, values and dispositions. (Shulman, 2005a, pp. 54–55). In addition, he identifies that signature pedagogies

> form the habits of the mind, habits of the heart, and habits of the hand ... [and] prefigure the culture of professional work and provide the early socialisation into the practices and values in the field. (p. 59)

Shulman goes further to draw attention to the potential similarities between educating for a profession and a general, liberal, education, such as school education, in relation to a further set of pedagogies of uncertainty, (Shulman, 2005b). He asks (and answers) the following question.

> How then does a professional adapt to new and uncertain circumstances? She exercises judgment. One might therefore say that professional education is about developing pedagogies to link ideas, practices, and values under conditions of inherent uncertainty that necessitate not only judgment in order to act, but also cognizance of the consequences of one's action. In the presence of uncertainty, one is obligated to learn from experience.
>
> Are there connections between these ideas and the goals of liberal education? I would say that learning ideas, practices, and values, and developing the capacity to act with integrity on the basis of responsible judgments under uncertainty, and to learn from experience, is a reasonable description of what liberal learning should be about, as well. (p. 19)

Shreeve's signature pedagogies for designing (Shreeve, 2015) link directly to the nature of uncertainty in designing. She identifies *The project* and *The brief* as signature pedagogies of project-based, experiential learning. With open-ended outcomes, largely unknown at the outset by teacher or learner, these support learner autonomy. Shreeve (2015) also highlights pedagogies of critique, of the studio and of dialogic exchange as signature pedagogies of design. *The studio* is portrayed both as the site of learning and as a signature pedagogy that removes a teacher from the centre of learning, supports a student-centred approach, focuses on dialogue, peer engagement and peer learning and creates a community of practice and design culture.

6 Signature Pedagogies for Learning and Teaching Designing in Schools

Each of the pedagogies highlighted above also have relevance for mainstream school designing, suggesting value in identifying 'signature pedagogies for design' as a part of general education. But what might these signature pedagogies be? What are the designerly ways of knowing, thinking and acting that all learners should have an entitlement to in order for their design capability to flourish? What are the pedagogies that will support their development?

The ubiquitous linear notion of 'the' design process has been seen by some as creating a systematic process that learners can learn and then apply to design problems. Despite this not being borne out "either in reality or in the classroom" (Williams, 2000) its existence has, by default, created prescriptive approaches that could be viewed as current 'signature pedagogies' of Design and Technology Education:

- Pedagogies of identifying a problem (e.g. brainstorm possible problems or 'needs', write a specification);
- Pedagogies of conducting research (e.g. internet or magazine search for objects that have solved similar problems);
- Pedagogies of generating an idea (draw six possible ideas then choose one);
- Pedagogies of making (e.g. draw on a set of skills related to specific materials and tools that are on the syllabus for a particular age group that have been taught in advance);
- Pedagogies of evaluating (e.g. write an evaluation).

A key driver in both the creation and continuation of these approaches has been the ways in which they have been linked to assessment systems that award marks for each stage. This has exacerbated a narrowness of pedagogic approaches and a consequent fixed, rather than fluid concept of knowledge as examination criteria have required evidence of each 'stage'.

7 An Overarching Pedagogic Ethos?

Focusing pedagogies on producing assessment evidence creates a danger of assessment leading the pedagogic ethos of designing. It creates an atomised approach that detracts from establishing a more holistic, overarching, pedagogy for the context in which learning takes place. Shreeve (2015) proposed the *studio* as both the site for, and culture of, learning. Focusing on school education, Claxton, Lucas and Spencer (2012) explored the studio from a similar perspective. They asked

> If you were trying to create an ideal learning environment of the kind that the very best craft apprentices or artists or technologists or designers would thrive in, what would it look and feel like? How would it be different from a typical school classroom? What would the teacher do and not do? How would learner roles be different? How would the physical space be organized? (Claxton et al., 2012, p7)

Their conclusion was that this environment would be a studio, identifying seven key dimensions that would contribute to this environment.

1. The role of the teacher – facilitative or didactic?
2. The nature of activities – authentic or contrived?
3. The organization of time – extended or bell bound?
4. The organization of space – workshop or classroom?
5. Levels of interaction – group or individual?
6. Visibility of processes – high or low?
7. The role of the learner – self-managed or directed?

(Claxton et al., 2012, p. 7)

Each dimension exists on a continuum and the closer the learning environment is to being facilitative, having authentic activities, extended time, a workshop organization, involving group interactions, high visibility of processes and self-managed learners, the more the environment would match their concept of studio teaching. The parallels with Shreeve are clear. Both approaches highlight not just a place, but the pedagogic approach that is embedded in that place. Such environments can be found in Design and Technology education learning spaces, but there is typically less emphasis on the focus of creating the pedagogy of the studio than on its physical attributes. In a highly detailed, small scale, research project exploring a studio teaching approach with Design and Technology teachers in English schools, Claxton et al. found that focusing on their seven dimensions of studio teaching brought new and challenging pedagogic approaches to a regular Design and Technology workshop but that, over time, and with the support of an intervention to develop a studio pedagogy, the majority of teachers changed their practice. Learners became more engaged, resilient, resourceful, reflective and collaborative. Almost as a by-product (rather than as the priority focus) a third exceeded predicted grades in external assessments (Claxton et al., 2012).

Kimbell and Perry (2001), in their report to the UK Engineering Council on *Design and Technology in a Knowledge Economy,* outline what they term a 'distinctive pedagogy' for Design and Technology – enriching further a perspective on pedagogic ethos. At the heart, their distinctive pedagogy is project-based and 'wicked' task centred. It has a methodology that involves unpacking the wickedness of tasks, identifying values, engaging in creative exploration, modelling futures and managing complexity and uncertainty. Rather than proposing a set of pre-defined technological knowledge and skills, they focus on "the skill of acquiring task-related knowledge" and creating "new, task-related knowledge." (p. 8). In a similar vein to Lawson, they don't deny that there is specialist design and technology knowledge. But they prioritise creating a climate of enquiry where learners identify what they need to know as they progress through a task, working from tacit to explicit knowledge as needs become clearer. Their vision is of an autonomous learner and of design capability as

> that combination of understanding, skill, insight, imagination and motivation that enables creative development. It provides the bridge between what is and what might be. Specifically in technological terms it mediates between human desires and dissatisfactions on one hand and technical constraints and possibilities on the other. (Kimbell and Perry, 2001, p. 7)

This highlights again the range of knowledge that might be needed in any design and technological project. And it is this potential range that provides one of the biggest challenges to learning and teaching in design and technology – how to manage and balance what a teacher plans to teach and what a learner needs to know. I will return to this later in the chapter.

8 Pedagogies that Enable Designing within an Overarching Ethos

A pedagogical ethos of designing, as indicated by the collected ideas above, is akin to a philosophy, a stance, a particular learning 'soup' in which young designers can flourish. It lays down perspective and principles and doesn't shy away from what might make the task of enacting the ethos complex and challenging. This challenge was recognised by George Hicks, a leading innovator in the pioneer days of the nascent English Design and Technology curriculum. He also acknowledged the learning potential, when he stated

> Teaching facts is one thing: teaching pupils in such a way that they can apply facts is another, but providing learning opportunities which encourage pupils to use information naturally when handling uncertainty, in a manner which results in capability, is a challenge of a different kind. (Hicks, 1983, p. 1)

Providing tools to support teachers meeting this challenge is critical. Many such tools (strategies, methods, activities and interventions) have been created over the years to develop designing skills. Some tools have been designed for a single purpose. For example, a tool such as *user profiling* helps learners to create a tangible persona that makes it easier to think deeply about the particular needs of a client group. Like many 'single purpose' tools alternative uses might be identified. In much the same way that a screwdriver might be used to open a can of paint, user profiling might also be used as a tool to critique a developing solution. Other tools are intentionally multipurpose, for example, the *design tutorial* (Stables et al., 2016), which might be used to explore early ideas, or equally to provide information about a needed skill or material, or to discuss a tricky problem that has arisen.

Exploiting a tool pedagogically is a valuable skill for a teacher, but it requires an open mind and creative thinking to see below the surface level of a tool's function. An example of this is how (and when) annotated sketching might be used. In the 1980s, our research team at Goldsmiths in the Technology Education Research Unit (TERU) had the particular challenge of assessing the design and technological capability of 10,000 15-year-olds from UK through a 90-minute design task (Kimbell, Stables, Wheeler, Wozniak & Kelly, 1991). A major challenge was finding ways of evidencing capability of having, developing and critiquing ideas in this short time frame. The activity was structured so that, early on, the learners were asked to use annotated sketching to move initial design ideas from inside of their heads out onto paper so they could develop them in a more tangible way. This is a very common use of annotated sketching in learning and teaching designing. But the short time frame for the whole activity focused our mind on how we could extend the use of annotated sketching to gain insight throughout the activity. To do this, we turned to annotated sketching for three distinct purposes; early in the activity in initiating and develop their ideas, midway to critically annotate the strengths and weaknesses of developing ideas and finally to 'fast forward' to imagine what their outcome would look and be like as a finished product.

Not only is the purpose of the tool a consideration, but also the point in an activity when a tool might be used. Deciding when and where to use a pedagogic tool is illustrated in McLain's (2018) research exploring *demonstration* as a signature pedagogy for school teaching. In Shulman's terms, demonstration is a constant and routine approach but the term 'routine' masks a hidden depth of pedagogic value. McLain provides a detailed and deep discussion that provides insight into demonstration's underpinning educational theory and how the term encompasses teacher as expert; modelling and explaining (both physically and linguistically), gently moving learning into Vygotsky's Zone of Proximal Development (Vygotsky, 1978) and raising the need for a teacher to make judgments, consider the impact of when, if and how to demonstrate. Should it be "front-loaded", "just-in-time" or "after-failure" (McLain, 2018, p. 987). His discussion emphasises the importance of going underneath the surface to expose the purpose of the pedagogic approach taken.

9 Speculating on a Pedagogic Framework for Designing

9.1 The Choreography of an Iterative Approach

Based on research conducted in the APU D&T project, (Kimbell et al., 1991), an alternative to a linear model of designing was proposed, an iterative model based on to-ing and fro-ing between thought and action as a hazy design idea develops to

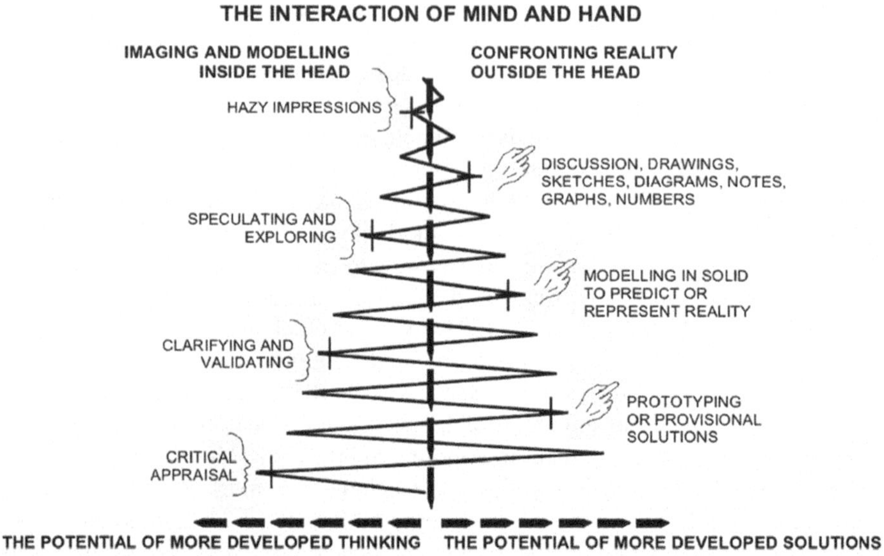

Fig. 1 The APU design and technology model (Kimbell et al., 1991, p. 16)

successful resolution (see Fig. 1). Progress is driven by responding to the needs in developing the idea, rather than adhering to a set of steps prescribed in advance.

This model sits comfortably with the concept of designerly ways of knowing, doing and acting, through a recognition of the 'wicked' nature of designing, of uncertainty and responding to the needs in a design task. Pedagogic purpose guides what needs to be taught and learned at a particular time to support the developing ideas. From a perspective of managing teaching learning and assessment this may be challenging. But from a perspective of designerly ways of knowing, thinking and acting it will be more authentic. The purpose forms the basis of a pedagogical choreography which may provide a pre-planned framework for an activity, but also allows for flexibility to amend a structure as need arise, either on the basis of a whole class, or on the needs of an individual project. The choreography allows a teacher to deal with the chess-like nature of a learner's designing as they "adapt to new and uncertain circumstances … [and] exercise judgment" (Shulman, 2005b pp. 18–19) in pedagogical approach.

9.2 Constants and Phases

The notion of an iterative model that starts with a spark of an idea and makes a meandering design journey from that hazy starting place to a point of resolution is, in itself, a challenge of pedagogical uncertainty. How can learning and teaching be focused and structured when there are so many unknowns? From the APU project

(Kimbell et al., 1991) and further research (Kimbell and Stables, 2007), certain dimensions have emerged that, pedagogically, begin to provide some form of structure. Preset, linear locksteps are unhelpful, but there are other structural constants. First is a studio-based pedagogical ethos of learning and teaching designing. A second is the design journey of an undeveloped idea to becoming an effective solution, from early sketchy ideas, through modelling and prototyping to a final solution. Third is the design context that a project is both embedded in and responds to. A design context provides the background to the people, places and purposes at the heart of a design challenge and the drive for Kimbell and Perry's 'task-related' knowledge requirements. It provides the impetus for a project and so is often seen as important in the early stages but then fades into the background. However, losing sight of the context can result in tokenistic and ill-developed projects, so the context needs to be a constant.

While the choreography will vary in relation to the needs of the developing idea, these 'constants' create their own structure of the rhythm of the activity. TERU projects have shown the value of recognising phases in a project which speak to the reality of the notion of a journey from hazy ideas to well developed, articulated prototypes. It can be helpful in supporting learners to recognise, explore and manage the complexity in their projects by seeing the design journey in three phases. An initial phase of *setting the scene* is important, capturing the learners' imagination, provoking initial ideas and providing insights and perspectives into a situation rich with design issues. A middle phase, when ideas are being modelled towards some visible reality, provides a heightened focus on *understanding the needs in the task* allowing learners to step back and think about the people and places they are designing for and to develop a more rounded understanding of what they have embarked on. As their project approaches resolution, *checking the effectiveness of a prototype* allows them to consider the overall success and effectiveness of their project identify its strengths and getting user feedback. There is an inevitable logic to these stages that provides a helpful framework for structuring a project but one that recognises that, in each phase, any or all pedagogic purposes may be present (Fig. 2).

9.3 A Pedagogic Framework

In is not the intention that this chapter should provide a catalogue of pedagogic tools, but to propose a framework to help decision-making about why, pedagogically, one would choose a particular tool at any given time. Looking at this challenge in a structural way, three levels are apparent; the *purpose* of the pedagogy, the subsequent *pedagogic actions* that could be taken and, following this, the *pedagogic tools* that might be used (Fig. 3).

Reflecting back on higher education signature pedagogies (Shreeve, Orr, & Tovey), studio learning and teaching (Claxton et al.) and distinctive pedagogies (Kimbell & Perry) I propose, an overarching pedagogical ethos of studio teaching that has the following *pedagogic purposes*.

Fig. 2 Three phases in a project

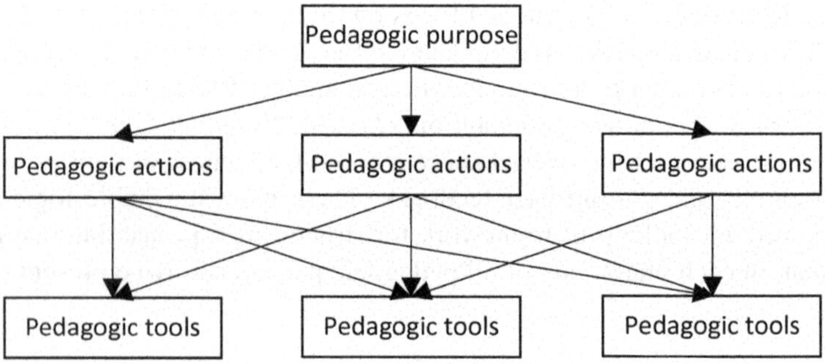

Fig. 3 An outline framework

- Pedagogies of *speculation*: that support learners to consider 'what if', 'what might be', 'how could'.
- Pedagogies of *imaging and modelling*: that support learners to test their speculative ideas by bringing them into some form of reality.
- Pedagogies of *materiality*: that enable learners to understand and develop knowledge and skills in bringing ideas into physical being.
- Pedagogies of *need-to-know*: that enable learners to have the confidence and competence to acquire knowledge, skill and understanding as the needs in their design tasks arise.

- Pedagogies of *critiquing*: that allow learners to make thoughtful decisions and judgements, based on values and ethics.
- Pedagogies of *collaboration*: that support learners to develop skills in working with and for others.

Each of these potential 'signature' pedagogies of designing enable a teacher, and potentially a learner, to make a decision about the facet of designing that needs to be focused at any given time. Would the learner and their project benefit from speculating, finding something out, collaborating with others? Having a focus on purpose moves attention away from a linear set of steps towards a more responsive approach, led by the learning and teaching needs, to move a learner forward with their designing.

With a background of the 'constants', identifying the pedagogic purpose at any stage is a first step. This can then be linked to the nature of an appropriate *pedagogic action* to be taken. Does the learner, for example, need to engage in creative exploration, identify values, be helped to avoid fixation, assisted in handling uncertainty, modelling futures? Once these options have been explored, a decision needs to be made about the particular *pedagogic tool, or tools* that could achieve the purpose. Consideration needs also to be given to who is taking the decision – the teacher or the learner?

Figure 4 represents how the structure of decision-making might look. At the top are the signature pedagogies – the purposes of pedagogic intervention. The middle level provides examples of the focus of actions that could address a pedagogic purpose. Following this are examples of tools that could be used to achieve the purpose and action (see the Appendix for glossary of tools included). Taken together, the three levels provide elements within a pedagogic choreography.

The framework provides a way of considering options for making decisions. But making pedagogic decisions can be complex, both in terms of how a decision supports progression in both the learning and the learner's project and also in terms of who is making the decision – the teacher or the learner. With less experienced learners it is likely that the teacher will be the key decision maker. From a management perspective, this would likely be the same for a teacher less confident in supporting more open-ended design projects. But as confidence and expertise grows, building towards greater autonomy and the pedagogic voice of the learner (Baroutsis, McGregor & Mills, 2016), the balance of decision making can shift.

The following two 'vignettes' illustrate how the framework could be employed.

The first vignette (Fig. 5) is based on the early stages of a project with a class of 14 year olds. The topic for their project is 'design for disability'. The teacher wants to encourage the learners to be innovative so has decided that the outcomes will be working prototypes, not completely finished products. The project is due to last for 8 lessons, each 75 min long. In the first lesson, the teacher introduced the context for the project and learners explored different potential problems and scenarios. The vignette is drawn from the second lesson. It illustrates a structure where the teacher's initial pedagogic purpose is to support learners to engage in speculation about their developing ideas. She chooses to start this by encouraging creative exploration

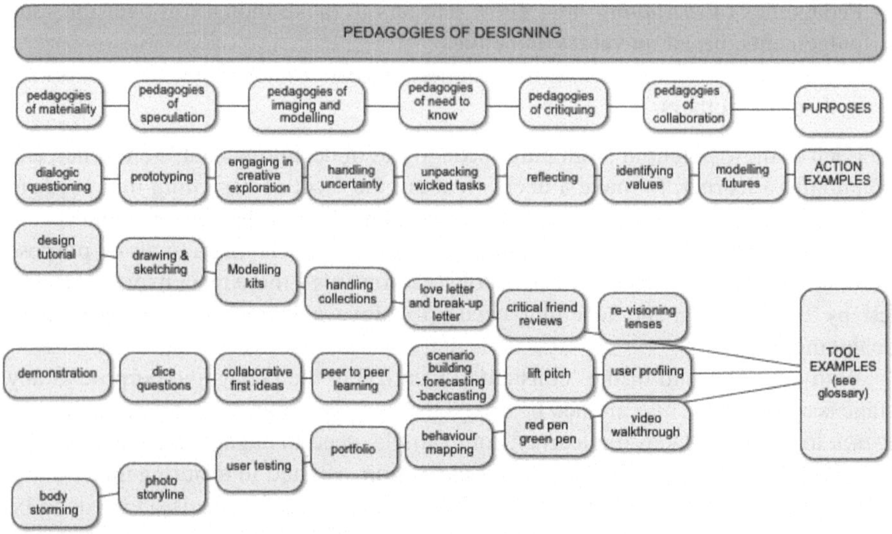

Fig. 4 The pedagogic framework exemplified

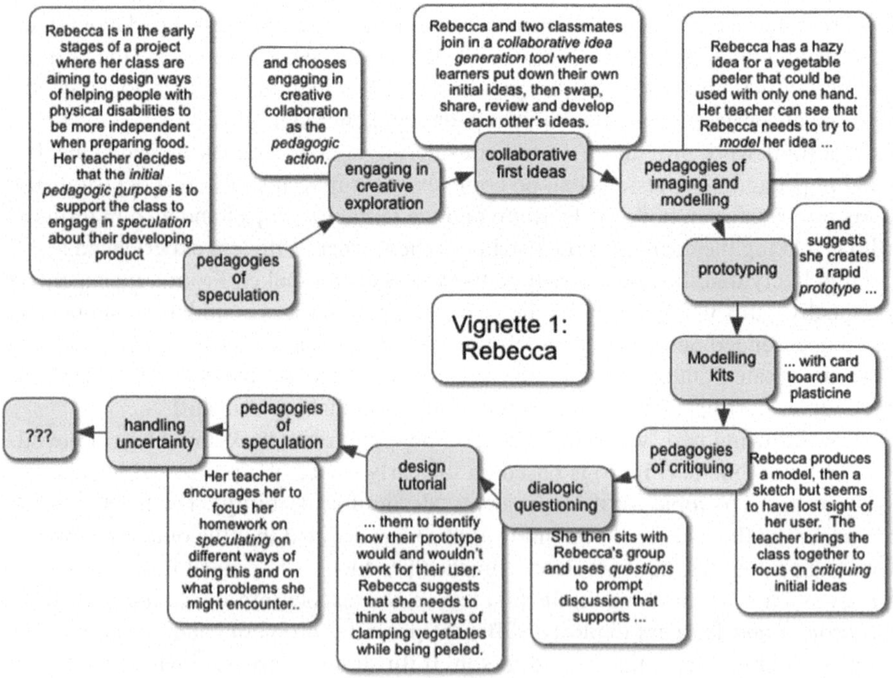

Fig. 5 Vignette 1 – a potential route through the early stages of a project

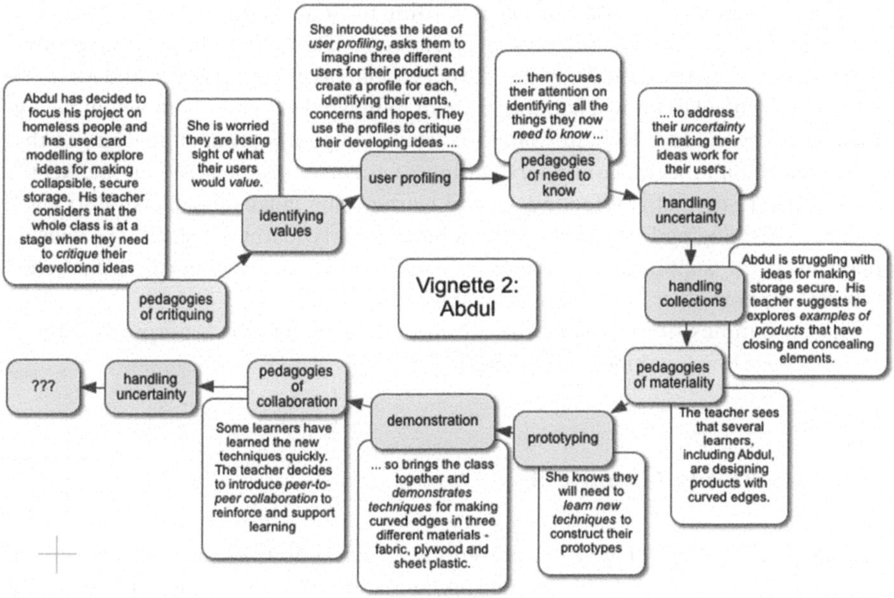

Fig. 6 Vignette 2 – a potential route through the middle stages of a project

using a collaborative idea generation strategy before each learner uses modelling resources to begin to visualise their ideas in three dimensions. She then focuses on dialogic questioning, through small group design tutorials, initially to encourage learners to critique their developing ideas in terms of how effectively they meet the needs of their proposed user group and then to engage in further speculation of how their ideas could be developed to be more effective. The vignette illustrates the actions of the teacher and one learner, Rebecca, as she moves through her project.

In this vignette, the majority of the pedagogic decisions are made by the teacher as the learners are less experienced in prototyping ideas, and she wants to maintain control of the overarching lesson structure. But through collaborative work and dialogic questioning, she is shifting some responsibility to the learners and by the end of the lesson is prompting Rebecca to make her own procedural decisions on how her design might function and what technical knowledge she might need to learn to achieve this.

The second vignette (Fig. 6) is set in the middle stage of a project with 16 year olds who are working on projects related to displaced people. It illustrates a structure where the teacher's pedagogic purpose is to support learners to critique their developing ideas by creating user profiles to check out their current ideas. The purpose then shifts to identifying what they need to know in order to address issues raised through their critique. The pedagogic purpose then shifts again to supporting material skills needed to create a prototype, both through demonstration and peer-to-peer collaborative learning. Again, the vignette illustrates the approach by focusing on one learner, Abdul.

In this second vignette, the teacher is shifting most decision-making to the learners. Abdul has decided on his own project focus. Through creating user profiles further decisions will be made by the learners about task-related issues and identifying what they need to know encourages them to make decisions about technical and procedural knowledge. To help Abdul his teacher suggests product analysis so that he can build some technical understanding of security systems. When she sees a construction issue that is affecting several learners she makes the decision to teach specific skills. Seeing how some learners have quickly learnt the skills, she sees an opportunity for supporting 'pedagogic voice' by having the newly established 'experts' support others in learning the skills they need.

These examples illustrate just two of multiple ways in which purposes, actions and tools could be combined in any project.

10 Concluding Comments

This chapter has set out with an ambitious target, ambitious in both the aim of speculating on a fresh way of considering the nature of a pedagogic framework for learning and teaching designing and with an ambition for teachers to implement the approach proposed. Design capability is at the heart of technological activity but teaching and learning designing in the context of mainstream schooling is challenging. By moving from a prescribed approach, following a linear set of steps, to a responsive approach, where there is recognition that, as a project and its needs become clearer, the ground will shift, a teacher's pedagogic practices will inevitably be unsettled. An alternative structure is needed, and one that has the potential to scaffold both a teacher and their learners. By identifying key distinctive, potential signature pedagogies, and proposing a framework of pedagogic purpose, actions and tools, this chapter offers an alternative. In considering the proposed framework, teachers will want to reflect on their existing practices and the successful tools they already employ. Equally, more formulaic practices may need to be critiqued. This chapter aims to provide approaches to refresh, energise and provide opportunities to create new pedagogic approaches that, in time, will be seen as signature pedagogies for designing that result in a flourishing of design capability in learners.

Initial points for a reader reflecting on current practice could include the following

- What is your view of design capability? How do you develop this through your current perspective on design process? What drives your current approach?
- What would you identify as your current 'signature pedagogies'? How do these relate to those proposed in this chapter? How and why might you modify your curriculum stance to refresh your pedagogic approach?
- What would you currently see as the range of pedagogic purposes, actions and tools in your learning and teaching repertoire? How might the proposed framework extend your repertoire?

- How would you critique the speculative framework presented in this chapter? What do you see as the major challenges and benefits? How could you overcome the challenges and take advantage of the benefits?

Appendix 1: Pedagogic Tool Glossary

Below is a brief overview of pedagogic design tools mentioned in this chapter. Some have been developed through research undertaken in the technology education research unit (TERU) at Goldsmiths, University of London, some have been 'borrowed' from elsewhere, some are common, ubiquitous approaches.

Where possible and helpful, links to further information are provided.

Annotated sketching	*Annotated sketching* is an approach that enables the thoughts behind a developing idea to be made visible. The annotations can be both descriptive and evaluative and should be made 'in the moment' to capture the iterative relationship between thought and action. In encouraging learners to verbalise their thoughts in an informal way, they make their thinking more visible to both themselves and others
Behaviour mapping	Behaviour mapping is a tool that allows learners to focus in detail on how a user interacts with and uses a product, system or environment. It involves then in closely observing a user (or a series of different users) for example, peeling vegetables, using a self-service supermarket checkout. Using words, drawings, photos and/or videos, learners capture the detailed actions of the user in ways that they can then analyse when designing
	See also
	Martin, B., & Hanington, B. (2012). *Universal Methods of Design: 100 ways to research complex problems, develop innovative ideas, and design effective solutions*. Beverly, MA: Rockport Publishers
Body storming	*Bodystorming* is a user-centred tool that is a subset of user testing, that is, a bit like physical brainstorming. It involves creating a physical situation through which a learner can experience specific needs, such as poor vision and lack of mobility, such that they can gain insights and inspiration to design for people who have these needs. It allows learners to develop empathy with a user group and then simulate user testing of models and prototypes as their design ideas evolve
	See also
	Martin, B., & Hanington, B. (2012). *Universal Methods of Design: 100 ways to research complex problems, develop innovative ideas, and design effective solutions*. Beverly, MA: Rockport Publishers
	Stanford D school https://dschool-old.stanford.edu/groups/k12/wiki/48c54/Bodystorming.html

Collaborative first ideas	Collaborative first ideas is an approach where, early in a project, learners work in groups of three as 'critical friends' to review and develop each other's ideas. The activity starts with each spending a short time (e.g. 5 minutes) quickly putting down (drawing and/or words) any initial ideas they have. The ideas are then swapped and each learner reviews and develops the ideas in front of them (more drawings and/or words). After the same time frame, the swap is repeated. Finally, the ideas return to their owner, who reflects/acts on the ideas and comments as they continue with their designing
Critical friend reviews	In a similar way to *collaborative first ideas*, the 'critical friends' come together at appropriate times during the length of a project to review each other's ongoing work. This can usefully be structured by each learner providing three comments on what they think is working really well and three comments where they consider that more work is needed – Effectively giving a 'thumbs up' to the good bits and a 'thumbs down' to bits needing attention
	See also:
	Iterative design in action https://www.data.org.uk/resource-shop/iterative-design-in-action/
Demonstration	*Demonstration* is effective when a teacher decides that an individual, small group, or the whole class need to 'see' and understand how something works. This might be a physical piece of equipment, such as a pillar drill, or equally an action or strategy, such as providing feedback on a learner's work. It can be the teacher demonstrating and/or learners taking the demonstration role. Critical in the endeavour is observing the impact of the demonstration as learners then practice for themselves and, where necessary, correcting any mistakes or misunderstandings before continuing.
	See also: McLain, M. (2018). Emerging perspectives on the demonstration as a signature pedagogy in design and technology education. *International Journal of Technology and Design Education, 28*(4), 985–1000
Design tutorial	In a *design tutorial*, a teacher discusses project work with an individual or small group of learners. The teacher encourages learners to explain their work and critique it, and follows this by encouraging them to speculate on how they could improve it. Finally a discusses takes place on what the next steps could be and how to proceed. Small group tutorials have the added advantage of involving peer-to-peer discussion and feedback
	See also: Ward, M. (2013). Design tutorials: The basics*2015*. Retrieved from http://sb129.com/2013/11/08/design-tutorials-the-basics/
Dice questions	*Dice questions* or *Left field questions* is a tool developed by TERU as a way of disrupting learners' thinking by asking random questions about their designing that encourages them to think differently. The first iteration of the tool involved rolling a dice that was linked to a set of questions. The idea was developed further using an on-screen avatar to ask random questions. Other ways could be used to provide questions randomly, e.g. a set of cards. Central to the tool are the questions themselves – And teachers can devise these. Examples include "would your product work under water? "… in the dark?" "… be made from custard?"
	See also: Stables, K. (2017). *Talking with avatars: the potential and impact of design dialogue with an on-screen avatar on the development of a learner's design and technology project work.* Paper presented at the PATT 34 technology and engineering education: Fostering the creativity of youth around the globe, Philadelphia

Handling collections	A *handling collection* is a set of carefully chosen objects that learners can pick up, examine and fiddle around with in order to stimulate design thinking. They can provide creative inspiration throughout a task. They can familiarise learners with ideas, concepts and issues within a task; help learners understand how features of a product function; inspire learners to explore unexpected, novel and provocative ideas; and 'unstitch' a concept from one product to use in a new way in their own designing
	See also:
	Iterative design in action https://www.data.org.uk/resource-shop/iterative-design-in-action/
Lift pitch	Creating a *Lift pitch* is a way of presenting all of the positives of a designed outcome in a short, snappy way. The idea is that the designer steps into a lift in a very tall building and realises that they have a potential manufacturer or retail manager who they could 'sell' the idea of their design outcome to. They have less than a minute (when the list reaches the top of the building) to present a 'pitch' for their idea. The tool can be used both as a final evaluation of a learner's project or earlier on when they are speculating on the qualities of the product they are designing
Love letter and break up letter	The *Love letter and break up letter* is a novel design tool for helping learners to analyse and evaluate an object (or service/system/environment) – Typically an existing one but possibly the one they are designing. The *Love letter* allows for everything good about a 'relationship' with an object to be describe – Physical and emotional, providing an analyse of all the positives. The *Break up letter* does the opposite, including providing insights into why the learner has fallen out of love with the object. Originally created as a design strategy by smart design, they advocate writing the letters individually and then sharing them with a group to initiate discussion. Teachers will want to customise this approach for use in classrooms
	See also
	Smart Design's video at http:// www.vimeo.com/ smartdesign/ breakupletter for an example of love and breakup letters
	Martin, B., & Hanington, B. (2012). *Universal Methods of Design: 100 ways to research complex problems, develop innovative ideas, and design effective solutions*. Beverly, MA: Rockport publishers
Modelling kits	A modelling kit is a diverse collection of plentiful, cheap, easy to use materials and a range of basic tools that enable learners to rapidly mock-up whole or parts of design ideas – What are sometimes called 'sketch' models. The kits allow learners to explore ideas at speed and without wasting expensive materials. They need to include tools and materials that allow for 'box', 'skeletal' and 'organic' shapes to be modelled
	See also:
	Iterative design in action https://www.data.org.uk/resource-shop/iterative-design-in-action/

Peer-to-peer learning	*Peer-to-peer learning* involves less experienced learners being, for example, taught a skill, given advice or gaining needed understanding by working with a more experienced learner. It can be formally structured – Such as when a learner is given a 'mini expert' label for something they have mastered, or used informally, for example by a teacher encouraging learners to ask a peer for help before they ask the teacher. It requires a teacher having a well-developed understanding of the abilities of learners in the class. It has the potential to support metacognitive development – As the 'expert' has to externalise their knowledge or understanding; increase self-esteem of the 'expert'; increase collaboration; and help a teacher confirm the level of understanding of those on both sides of the exchange
Photo storyline	Creating a *photo storyline* as a project progresses involves periodically taking photos of work as it progresses in order to quickly capture each stage, particularly when ideas are being modelled and prototyped, including things that didn't work and were discarded or remodelled. The photos can be annotated to provide further reflection and insight into how the designing is developed. Creating the storyline has benefits for both formative and summative assessment, as both thought and action can be captured throughout a whole project or as a detailed 'cameo' of a particular stage in a project.
	See also
	Iterative design in action https://www.data.org.uk/resource-shop/iterative-design-in-action/
Red pen, green pen Review and development	Red pen, green pen is a tool developed by TERU to be used for onging evaluation throughout a project. Learners are asked to pause and review their work to identify what is and is not working well. They are asked to directly annotate their portfolio with a red pen or pencil to identify what isn't working and a green pen or pencil to identify successes. The use of coloured annotation directly onto work aims to break the mould of neat, tidy, after-the-event annotation, thus promoting portfolios as working documents, not presentation pieces
Revisioning lenses User focus	*Revisioning lenses* is a tool created by pi-studio at goldsmiths UoL to provide a physical artefact that focuses a designer on different facets of their design. Each lens is a card that has images relating to the focus, such as materials, environment, culture and disposal. It has a circular hole cut in the card that the designer looks through to focus on their design. Used with learners, it raises their awareness of a breadth of issues that affect the success of a design, whilst also allowing them to focus on a specific perspective at any one moment. Cards can be made that are generic, or specific to a particular context
	See also:
	Iterative design in action https://www.data.org.uk/resource-shop/iterative-design-in-action/

Scenario building, forecasting, backcasting	Scenario building is a tool that allows learners to speculate on a future design situation in a way that is manageable and realistic without becoming pure fantasy. It allows learners to think about what the near future might be like and how this can create a focus for their designing. *Forecasting* allows them to think about positive and negative future scenarios, for example a world where, by 2025, there are no plastic bags or a world that has been overtaken with plastic bags. *Backcasting* allows them to think about how design can work incrementally towards the positive scenario – To hit the 2025 deadline, where do we need to be by 2021, 2023
	See also
	Mathilda Tham getting people to speculate on future clothing habits by asking simple 'what if' questions http://www.wowtalks.tv/mathilda-tham/.
	Martin, B., & Hanington, B. (2012). *Universal Methods of Design: 100 ways to research complex problems, develop innovative ideas, and design effective solutions*. Beverly, MA: Rockport Publishers
User profiling, personas	User profiling involves creating personas for the people that are being designed for to help learners focus on the needs their designs should address. They can be created by teachers or the learners can create them in groups or on their own, possibly using a template created by their teacher. Details of the personas include lifestyle, behaviour patterns, likes and dislikes, special interests, special needs etc
	See also:
	Personas in Martin, B., & Hanington, B. (2012). *Universal Methods of Design: 100 ways to research complex problems, develop innovative ideas, and design effective solutions*. Beverly, MA: Rockport publishers
Video walkthrough	Towards the end of a project, before the design outcome is finalised, ask the learners to make a video walkthrough of their design that explains all of its features, how it works etc. partner them with another learner who can ask questions, critique and provide feedback to identify any final changes or developments that can be made

References

Aikenhead, G., & Elliot, D. (2010). An emerging decolonizing science education in Canada. *Canadian Journal of Science, Mathematics and Technology Education, 10*, 321–338.

Archer, B., & Roberts, P. (1992). Design and technological awareness in education. In P. Roberts, B. Archer, & K. Baynes (Eds.), *Design occasional paper no. 1, modelling: The language of designing* (pp. 3–4). Loughborough, UK: Logughborough University of Technology.

Baroutsis, A., McGregor, G., & Mills, M. (2016). Pedagogic voice: Student voice in teaching and engagement pedagogies. *Pedagogy, Culture & Society, 24*(1), 123–140.

Baynes, K. (1992). Children Designing. In *Learning design: Occasional paper no 1*. Loughborough, UK: Loughborough University of Technology.

Baynes, K. (2006). Design education: what's the point? *Design and Technology Education: An International Journal, 11*(3), 7–10.

Bronowski, J. (1973). *The ascent of man*. London: British Broadcasting Corporation.

Buchanan, R. (1995). Wicked problems in design thinking. In V. Margolin & R. Buchanan (Eds.), *The idea of design* (pp. 3–20). Cambridge, MA: MIT Press.

Claxton, G., Lucas, B., & Spencer, E. (2012). *Making it: Studio teaching and its impact on teachers and learners*. Winchester, UK: Centre for Real-World Learning.
Cross, N. (1982). Designerly ways of knowing. *Design Studies, 3*(4), 221–227.
Cross, N. (2006). *Designerly ways of knowing*. Berlin, Germany: Springer.
Flowers, J. (2010). The problem in technology education (a definite article). *Journal of Technology Education, 21*(2), 10–20.
Hicks, G. (1983). Another step forward for design and technology. *APU Newsletter, 4*.
Kimbell, R., & Perry, D. (2001). *Design and technology in a knowledge economy*. London: Engineering Council.
Kimbell, R., & Stables, K. (2007). *Researching design learning: Issues and findings from two decades of research and development* (Hardback ed.). Berlin, Germany: Springer.
Kimbell, R., Stables, K., Wheeler, T., Wozniak, A., & Kelly, A. V. (1991). *The assessment of performance in design and technology*. London: SEAC / HMSO.
Lawson, B. (2004). *What designers know*. Oxford: Elsevier.
Mawson, B. (2003). Beyond "the design process": An alternative pedagogy for technology education. *International Journal of Technology and Design Education, 13*, 117–128.
McGimpsey, I. (2011). *A review of literature on design education in the national curriculum* (p. 23). London, RSA.
McLain, M. (2012). An (auto)ethnographiic narrative of the teaching of designing within design and Technology in the English curriculum. *Procedia Social and Behavioural Sciences, 45*, 318–330.
McLain, M. (2018). Emerging perspectives on the demonstration as a signature pedagogy in design and technology education. *International Journal of Technology and Design Education, 28*(4), 985–1000.
Nelson, H. G., & Stolterman, E. (2003). *The design way: Intentional change in an unpredictable world*. Englewood Cliffs, NJ: Educational Technology Publications.
Norman, E., & Baynes, K. (Eds.). (2017). Design education and curriculum planning. Rothley, UK.: Loughborough Design Press Ltd.
Ofsted. (2002a). *A report of the OFSTED conference for secondary teachers of design and technology 'teaching designing'*. Birmingham, UK: Jaguar Cars Limited.
Ofsted. (2002b). *Secondary subject reports 2000/01: Design and technology*. Retrieved from London.
Ofsted. (2008). *Education for a technologically advanced nation Design and technology in schools 2004–07*. Retrieved from London.
Orr, S., & Shreeve, A. (2018). *Art and design pedagogy in higher education*. Abingdon, NY: Routledge.
Rittel, H., & Webber, M. (1973). Dilemmas in a general theory of planning. *Policy Sciences, 4*, 155–169.
Roberts, P., Archer, B., & Baynes, K. (1992). *Modelling: The language of designing. Design: Occasional paper no. 1*. Loughborough, UK: University of Loughborough.
Shreeve, A. (2015). Signature pedagogies in design. In M. Tovey (Ed.), *Design pedagogy: Developments in art and design*. London: Routledge.
Shulman, L. (2005a). Pedagogies of uncertainty. *Liberal Education, 91*, 18–25.
Shulman, L. (2005b). Signature pedagogies in the professions. *Daedalus, 134*(3), 52–59.
Stables, K., Kimbell, R., Wheeler, T., & Derrick, K. (2016). *Lighting the blue touch paper: Design talk that provokes learners to think more deeply and broadly about their project work*. Paper presented at the PATT 32: Technology Education for 21st Century skills, Utrecht, Netherlands.
Tovey, M. (Ed.). (2015). *Design pedagogy: Developments in art and design education*. London: Routledge.
Vygotsky, L. S. (1978). *Mind in society; the development of higher psychological processes*. Cambridge, MA: Harvard University Press.
Williams, P. J. (2000). Design: The only methodology of technology? *Journal of Technology Education, 11*(2), 48–60.

Pedagogies for Enabling the Use of Digital Technology

Deborah Winn

Abstract The variety, volume and rapidly evolving digital technologies that are required to be taught in Design Technology lessons to prepare students for an unknown future are vast. This often leads to anxiety in the teachers and students as they battle through these complex technologies alongside their other teaching and learning load. This chapter seeks to break down what the students and teachers actually need to know and teach about these technologies rather than attempt to teach everything about all of them. In addition, the chapter also suggests some strategies that may be useful in the classroom regardless of the level of teacher expertise to inspire a future generation to become confident users of the technologies.

Digital technologies used in Design & Technology lessons are numerous and vary greatly across the facets of the subject area, ranging from regular administrative tasks, animation, image manipulation and coding through to design and make tasks using CAD and CAM.

From a teacher's perspective, the digital technologies are taught alongside traditional design and make skills so therefore are only a part of the everyday teaching load. As the technologies evolve, keeping knowledge up to date with such a wide variety of software in addition to the traditional skills can be problematic. The Fujitsu report (2017) reveals that digital literacy is quite low among teachers generally, and education faces challenges in 'teaching the teachers'. It goes on to state that 51% of IT departments in education feel they can't keep up with technological advances; however, 84% of respondents say they have a duty to prepare their students for a digital future. Teachers clearly understand the importance of teaching students to use digital technologies but many struggle to keep up with the demands.

D. Winn (✉)
Neale-Wade Academy, March, UK
e-mail: DWinn@neale-wade.org

P. J. Williams, D. Barlex (eds.), *Pedagogy for Technology Education in Secondary Schools*, Contemporary Issues in Technology Education,
https://doi.org/10.1007/978-3-030-41548-8_7

It, therefore, becomes ever more important in the classroom to know what exactly needs to be taught and the most effective ways in which to teach it. This chapter seeks to consider the following:

- What do students really need to know?
- How can we foster resilience in the students?
- What approaches can be used to teach digital technologies?

Inevitably, teaching digital technologies involves a significant number and range of different types of software. Some of the software is arguably easier to use and teach than others. Software used for administrative tasks, image manipulation software and vector-based CAD programs do not rely on precise parameters, and it is therefore difficult to 'get wrong'. Coding has specific difficulties, mainly involving sequencing rather than difficulties with the procedure. At a beginner level in all of these, the students are able to use trial and error to explore and create designs in a way that is far harder in parametric-based CAD programs. Whilst the pressures of teaching the more complex software are undeniable, it is vital that certain aspects of CAD and other technologies are taught. The Fujitsu report (2017) states that CAD prepares students more fully as a future information worker and informed consumer. Cox (2012) observes that 'CAD design is used all over the place; you have to design something to build it. From that standpoint, it is important to give them the tools they need, and one of those tools is design software'. He believes there are significant benefits to using CAD to teach math and science principles. 'CAD gives students a way to focus on working through a design process to understand what must be done to solve a problem'.

As many of the digital technologies are relatively easy to learn through trial and error and CAD is such a difficult area to teach, yet such an important one, the majority of this chapter is therefore devoted to the teaching and learning of CAD.

From a student perspective, the ease of use across the technologies varies, as do attitudes to using digital technologies, and this appears to have little connection to their ability. Reactions from students when told they are learning some technologies are diverse and can be extreme. It is an area of the curriculum where students (and teachers) either love or hate it (Winn 2014).

Taking this into consideration and to help ensure a positive teacher attitude, resources to support both the student and the teacher can be drawn upon to encourage confidence. How to achieve this requires consideration.

Musta'amal, Norman, and Hodgson, (2009, p. 54) write that the perceptions that users have of CAD systems and their expertise can significantly influence their performance. Similarly, Bransford, Brown and Cocking (2000) extend this by stating that students come to the classroom with preconceptions about how the world works. If their initial understanding is not engaged, they may fail to grasp any new concepts or information that is taught, or they may remember them for the purposes of a test but then revert to their preconceptions when outside the classroom. This applies to both sides of the spectrum of beliefs. Therefore, success or failure when using digital technologies often boils down to attitude, resilience and how we as teachers respond to this in the initial part of the lesson.

1 What Do Students Need to Know?

Given that there are many different digital technologies available to schools and that updates are made regularly, it seems sensible to focus on the commonalities of the various systems and versions of the technologies. These aspects are mainly vocabulary, what the programs can do and be used for and what makes an outcome fail.

1.1 Vocabulary

The first stumbling block to teaching and learning digital technologies is likely to be vocabulary. The programs are littered with complex commands or terms that students, particularly younger ones, find confusing. Words such as 'workplane', 'orientation' and 'extrude' are just a few. Other words they may recognise but are unable to explain or place in context. To illustrate the effect of this, I recently asked a class to highlight all of the words they did not understand on a screenshot from a 3D CAD program. Around 48% listed between 20 and 30 words, and 52% simply highlighted the whole page saying they didn't understand any of it even though there were words present such as front, draw and sketch, which they would be familiar with in some form. In addition to the language issue, it clearly illustrates that the mindset for some students is to see what they don't know and make the assumption they can't do it rather than see what they do know and try, which will be considered in more depth later. The task also further highlighted how language could be an ongoing issue when the argument previously stated by Bransford, Brown and Cocking (2000) is considered. If the student already has the idea they are not good at the task and then is faced with language they can't access, they fail to be engaged and their preconception is reinforced.

Obviously, it is not necessary for the students to understand all of the terms, and in a limited time frame, it is also not practical to teach all of them. In the software I use regularly, I count ten unfamiliar terms the students would need to know by the end of the topic. Realistically focussing on five terms at the beginning of a topic and adding or reinforcing an additional two or three each session is more than enough and makes the task more achievable from the students' viewpoint. Setting this expectation from the outset could help to alleviate students' concerns and negativity around the subject. The key is not necessarily the vocabulary itself but the students' attitude to its importance and how likely it is going to be in preventing them being able to use the program.

In order to make the students more comfortable with the vocabulary, it may be tempting to simplify the language but I do not believe it is beneficial in the long term. It would almost require relearning the correct vocabulary in later stages of using the software, thereby potentially putting the students at a disadvantage. Lane and Allen (2010) write:

> One of the biggest barriers to vocabulary growth in school is the simplistic way many teachers talk to children. We have all been guilty at one time or another of using words beneath our students' level of understanding. Simplistic vocabulary may be appropriate for initial instruction, as a support for students' understanding of a new concept. But once students develop a basic understanding, it is time to elevate our instructional language to enhance our students' vocabularies. (p.367)

Once the core vocabulary is identified, how to teach the commands and make students comfortable is the next challenge. A common way of teaching new vocabulary, especially in a foreign language, is to use pictures to reinforce the words but this is not always possible as simple pictures do not always illustrate the words effectively. Similarly asking students to copy definitions into their books without a link to context is also unlikely to help many students. Lane and Allen (2010) writes that 'many of the traditional techniques teachers and students use to learn vocabulary do not work because most students, not just those with learning problems, rarely remember the meanings of new terms beyond the test' (cited in Jones, 2018).

Modelling the vocabulary by showing it on the screen is a successful method. However, if the student already has a lot of negative feelings about the program, it may not be sufficient to explain the command. The use of everyday 3D objects that students may be familiar with that demonstrate the meaning of the commands can be a helpful strategy. For example, a simple Playdoh shape set or piping nozzle and Playdoh for 'extrude' demonstrates how the shape stays the same but is pulled out to length. A paper lantern decoration can be used to show 'revolve' (Fig. 1). It is also something tangible the student can do themselves to reinforce the process before seeing it on screen.

1.2 What the Programs Can Do: Declarative, Procedural and Strategic Knowledge

Rynne, Gaughran and Seery (2010) outline a framework for developing cognition and expertise in 3D part modelling which gives distinct categories for the types of knowledge used in CAD practice. They define awareness of what the tools do and

Fig. 1 Demonstrating revolve feature

modelling techniques such as extrusion and revolve as declarative knowledge, specific knowledge of how to use the tools as procedural knowledge and knowing the best way to create the model as strategic knowledge. Declarative knowledge is important because digital technologies are widely used to design and make products, images, apps and websites, and knowing how these are made at the very least leads to an informed consumer able to make reasoned choices in their purchases. Ideally, students would be able to remember what the programs can do and if not taught how to complete the task at a later date may be able to find out for themselves from the wealth of online videos and tutorials. This is increasingly pertinent as CAM equipment such as 3D printers become more affordable and reach a home market. It allows for tinkering as encouraged by the maker movement. Realistically, the software for this equipment will not be the same as what they are taught to use in school and changes so rapidly that the specifics of the program will be of little long-term use. Knowledge of the procedure is of little use but declarative knowledge becomes highly beneficial.

The range of digital technologies used within Design & Technology is wide with some commonalities within them. Basic tools such as crop, fill, trim, transparency, extrude, round, etc. tend to remain similar regardless of the software used. Table 1 summarises the main areas that students should be aware of over their education.

Table 1 Digital technology functions

Areas which use digital technology in education	Tasks which use digital technologies
Image manipulation	Graphical representation and enhancement
Animation	Advertising
Robotics	Creating for the television and film industry
Coding	Sound editing
Product designing	Developing systems
Manufacture	Software and web design
Simulation	Designing a 2D and/or 3D product
Administration	Photorealistic rendering
	Converting a 3D design into a technical drawing
	Annotating the ideas
	Rapid prototyping
	Sending the ideas and feedback to third parties
	Sending the idea to be made via CAM
	Simulating fluid flow, functional aspects of a product or the mechanical interaction between components
	Identifying component clashes
	Estimating potential weight and cost of a product and comparing possible options
	Being able to make changes quickly
	Save multiple variations of ideas easily
	Problem solving
	Collaborative working
	Data tracking and analysis

A younger student would not be able to perform all of these tasks but the knowledge of what is possible and what transfers to 'real life' outside of school is powerful. Used effectively it will promote the subject matter by suggesting possibilities they may not of thought of before, creating an interest in the subject or developing confidence in the student to want to try. Perhaps changing a nervous student who believes they can't do it to one that has a can-do attitude may also sow the seeds of thought for a potential career path. I recently attended a construction careers event in which the teams were showing how they used digital technologies to map out the roads, tunnels, cabling and bridges to aid planning, visualisation and problem-solving, which had saved considerable amounts of time and money by completing this in the virtual world before construction teams and materials were involved. I asked the person demonstrating what skills he would like to see students leave school with and he replied that he hadn't known much about the systems he was using when he had left school but felt a real interest in the subject and a positive attitude to the subject would be the most useful attributes teachers could encourage. This suggests that the intention of the teacher should be less about developing skills and more about developing a positive attitude and an awareness of the possibilities. As digital systems become more complex and are able to accomplish more tasks, an enthusiastic workforce and an inquisitive mind willing to try will be helpful both in industry and the home. Sparking an interest or passion for digital technologies, especially the more complex types, is certainly something that could be achieved in schools but this may need to start with the teacher delivering the scheme of work as just the thought of it can strike fear into the hearts of some teachers who see it as a hurdle (Winn 2014).

This viewpoint is shared by a representative of Autodesk who stated that they had researched what industry is looking for in their future workforce and strategic knowledge featured highly with little interest in procedural knowledge. Essentially, the need is for design thinking and problem-solving with a view to flexibility in the design. Encouraging students into a Design & Technology-related career may seem a little premature at this early stage. However, with recent curriculum changes in England and a potential skills shortage looming (Design Council 2017), survival of the subject may depend on getting students confident and excited across a breadth of possibilities enabled by D&T.

1.3 *What Makes an Outcome Fail?*

Strategic knowledge can be understood regardless of whether the model works or not and often more is learnt when the outcome fails. The ability to be able to identify and correct the error is extremely important and can be far more valuable than merely rote learning the process, as sooner or later a model will fail and as the models become more complex, it may be harder to rectify. As teachers, an emphasis may be on progression within a lesson, and over time, this may be seen as completing something, getting it right. Therefore, we reward the completed outcomes and see failed ones as not there yet, rather than learning opportunities with value. Gershon (2016) writes in *How to Develop Growth Mindsets in the Classroom* that mistakes should be redefined as good mistakes, and that from the outset, it should be explained to the class that mistakes are expected rather than a failure. Explaining to a class that you expect them to make mistakes and that they are a good thing may seem a little difficult to start with but when a person starts to learn anything new and complex, it is inevitable that mistakes will happen; it is the reaction to these mistakes that is relevant. Imagine the students starting computerised work with little confidence; the first thing they may say or think is, 'I can't do it, I'll get it wrong'. If the teacher has already explained or has said 'mistakes are OK, as it helps you learn', the student cannot help but be more positive or at the very least is unable to respond negatively. Seeing 'failure' as progress is vital for students who believe they can't do it and removes or alleviates the failure stigma. This is also true for teachers of the subject. It is often expected that the teacher should know more than the students but with digital technologies, this cannot always be the case. It is not possible to know everything about all of the facets and software we are required to teach and stay up to date. The most successful lessons can be where the teacher works through the problem or learns from the student. It sends a clear message to the student that it's OK to get it wrong but you need to work through it to find out why rather than be given the answer.

One method to explore, as suggested by Gershon (2016), would be to use a 'mistakes log' where the students write down the mistake they have made and how they solved it. By doing this, they can see the connection between mistakes and learning, and the students and teachers can monitor how the mistakes have helped them learn.

In conjunction, a points system could be used where a completed model is rewarded with points but a failed model, where the fault has been identified in the log, is also rewarded. If the student is also able to solve the fault, more points are awarded. Progress is then identified and proven with the points to both the student who may have been disappointed at first and to education hierarchy who may be observing.

To aid the learning gained from making mistakes, the early learning for fault finding abilities can begin before the computer is turned on with sheets showing models with the common errors and asking the students to identify them. These can start by being very obvious and continue to more discreet. Common errors across most CAD software, for example, include crossing lines, gaps and double lines. These can be more easily identified if the student zooms in sufficiently although surprisingly students often don't do this. Failure to zoom in adequately is also a common error in image manipulation software and on 2D vector-based CAD programs where small lines or gaps are there but can't be easily seen. Double lines are much harder to see and require the student to be more vigilant to changes in colour of the line or thicker lines. A further error that students seem to struggle with is additional lines that prevent the model from being a whole shape and therefore being able to be extruded. Imagine that you are trying to fill a bottle with water but the bottle has a barrier halfway down. The bottle would obviously never fill beyond the barrier. The program would be confused by this 'barrier' seen as an unnecessary line in the example in Fig. 2. The program is unable to determine what you are trying to do so this stops it doing anything until you remove the line.

A final common error, which is relevant to several digital technologies, is that students often do not see that a product may need to be built up in layers. For example, if they are drawing a car in CAD, they will draw the details such as wheels and windows as part of the drawing as if they are drawing a picture on paper rather than as a separate component or extrusion. In image manipulation software, the layers may be there but be hidden by an additional layer or given the wrong priority. In Fig. 3, a pink ellipse is present but it can't be seen because an incorrect sequence of layers means it is covered with the yellow rectangle. Coding experiences a similar common problem as students use the correct commands but apply incorrect sequencing or syntax.

This can be a very difficult concept to teach as it goes against how the students have been taught to draw and paint up until this point. It also requires the ability to be able to visualise the product as layers to create it as well as being able to see and rotate a 3D object in their imagination.

Not all students will need help with this concept. For those that do, two possible options present themselves to improve the ability of separating the model into components and features:

1. Use plasticine or Playdoh to model the idea and note each stage as they make it. If you take the car model, for example, students will generally create the car body shape, then add wheels and then add windows. If they can see each stage as a new feature, they can begin to develop the awareness needed for the task (Fig. 4).
2. Draw the design on paper but create each new 'feature' as a new colour. This can be less successful at first as it still relies on the student's ability to be able to think in 3D and convert 2D to 3D images and back. This is actually a very difficult skill that requires practise.

If students can identify common errors and be able to visualise their ideas in 3D as different features or as layers, then they begin with an excellent grounding in how to use the common digital technologies.

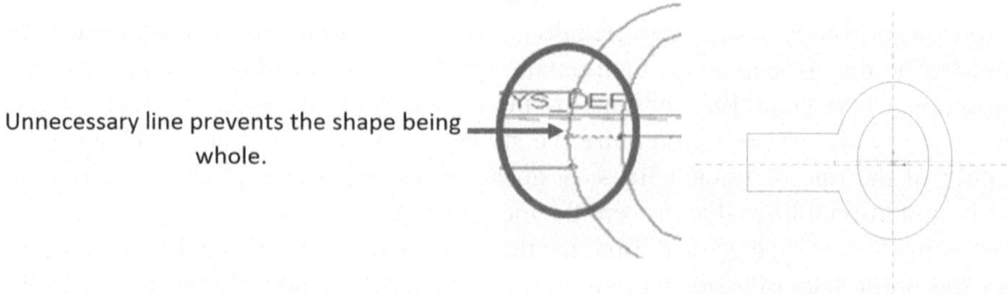

Fig. 2 Common errors such as additional lines cause the model to fail

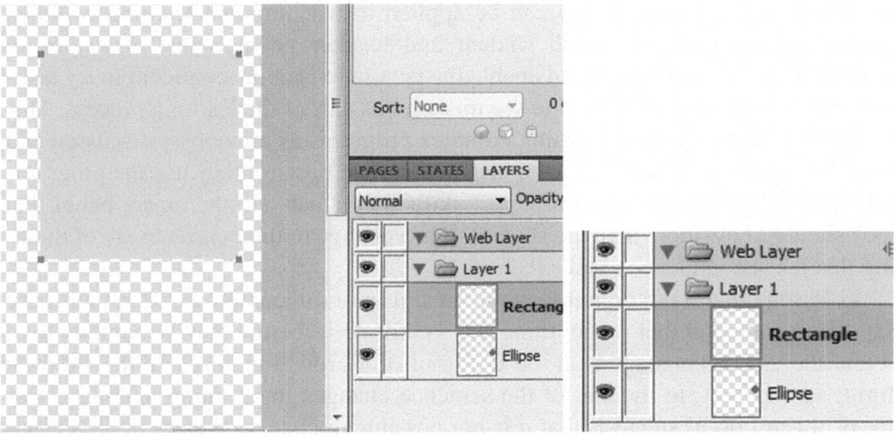

Fig. 3 Common layer or syntax errors cause many digital technologies to fail

Fig. 4 Three different features are represented by different colours. A common error is for students to draw all three together

2 How Do We Foster Resilience?

I briefly mentioned mindsets earlier in this chapter. It is a term used to describe someone's attitude to their intelligence and attributes. Dweck (2008) describes a *fixed* mindset as one where the person believes they are either talented at something or not and intelligence is set by their genetics and cannot be changed. In a *growth* mindset, the person realises that their qualities can be developed through hard work and is willing to learn.

It is not an entirely new concept. Bandura's theory of self-efficacy (1997) highlights that if you have the confidence and belief that you can do something you are more likely to succeed at it. Some believe this confidence is more important than talent and equally that talent is somewhat pointless without confidence. As Henry Ford allegedly put it, whether you believe you can or can't, you are right. Developing a growth mindset takes this theory one stage further by providing positive methods to look beyond talent and understand you can develop the skills needed to become confident to try as well as not become fixed by setbacks along the way.

This growth mindset theory can be applied to teaching and learning complex digital technologies from both student and teacher perspectives. Strategies to develop a growth mindset would enable the person to become confident to try tasks they perceive to be difficult and to see mistakes as part of the learning process. This is a very important part of learning complex programs as previously discussed and equally important for teachers who lack confidence because they use the programs infrequently due to the wide variety of skills they teach and the innate belief that they should know more than the student. It is a change to their current way of thinking that takes confidence to take the risk.

In lessons involving digital technologies, it is not uncommon to hear someone say 'I'm no good at that sort of thing' or 'I can't do it' before they've actually tried it and the use of language is at the forefront of a growth mindset. In the simplest form, adding 'yet' to the end of the sentence changes its meaning making it less fixed. 'I can't do it' suggests that it is not possible to change. 'I can't do it yet' has an entirely different meaning implying that with effort you will be able to do it. Gershon (2016) warns however that overuse of adding the 'yet' can dull its impact as the students may start to see it as meaningless and 'just something that Miss/Sir does'.

Writing about humanist teaching methods, Legge and Harari (2000) state that 'succeeding at a very easy task will not improve a child's self-esteem but asking students to undertake tasks that are too difficult for them will simply make them feel like failures'. Therefore, having a range of tasks and learning outcomes is a useful strategy to ensure students are working at the appropriate level for them to progress whilst not being faced rather than a single objective based on a successful model. Differentiation in a task is also not a new concept to a teacher but the tasks set considering failure of the model as part of the learning process is likely to be.

Setting a learning objective is a standard practice that every teacher understands for every lesson. It sets out for the student the intention of the lesson, what they need to achieve and gives a point of reference for the teacher. At times in teaching, there may also have been sub-levels to this objective in the form of all must achieve, most should achieve and a few could achieve. An improvement in this practice that has been encouraged in schools is the use of SOLO taxonomy. SOLO was developed by Biggs and Collis (1982) after researching samples of students' thinking in different subjects and different levels. Later publications by Hook and Mills (2011) focused on ways in which SOLO could be implemented in the classroom. SOLO stands for Structure of the Observed Learning Outcome and seeks to provide differentiated learning outcomes based on five levels: no idea, one idea, a few ideas, related ideas and extended ideas. This is particularly helpful when teaching digital technologies as the students are likely to progress at wildly different rates and will require different starting points and objectives to ensure all students are presented with an appropriate level of challenge for them and make appropriate progress in the lesson.

Vignette

Imagine students A and B in a class about to learn the extrusion feature in a complex CAD program. Student A is nervous of CAD, has no confidence in their ability, does not recognise the term extrude and knows they are going to fail before they start. Student B also doesn't recognise the term but they've always been good at this sort of thing so is looking forward to the lesson. Previously, a learning objective may have been something like 'create a dice using the extrusion feature'. Student A has had their fears confirmed. They are never going to be able to do that. To ensure all students achieve the objective, the teacher must make sure all students create the dice which is an additional pressure for the teacher. Many students have no idea why they are doing it or what they did to achieve it but they were able to follow the instructions. In contrast a sample of a SOLO objective for a digital technology, lesson could be as follows:

Pre-structural – I have not yet attempted the task or may need the vocabulary/features explained to me or help to start.

Uni-structural – Pre-declarative – I know what the words mean, e.g. mirror, extrude and transparency.

Multi-structural – Declarative – I know what the modelling techniques are.

Relational – Strategic – I know how to create models and can explain why the model worked or failed.

Extended abstract – Strategic expertise – I know the best ways to combine features to allow for flexibility in the design for editing, assembly or manufacture.

Both students start the lesson on pre-structural level. Student A feels relieved as they are confident that they can understand what the feature is by the end of the lesson. In reality, they leave on the uni-structural or multi-structural level. Student B races ahead, they are confident in their abilities but miss a step and the model fails but they know where they went wrong. With the first objective, this would mean they had failed the task. Would they be so confident next time? With a more tiered objective not based on a successful model, the student leaves and confidence remains intact on the relational level.

At the start of the next lesson, each student has a differentiated objective suited to their abilities and confidence level, which supports the Fujitsu report finding stating that 94% say a personalised digital learning approach is important.

3 What Approaches Can be Used to Teach CAD?

Realistically, you can't split yourself among the 20–30 students you may have in the classroom to provide individual support, and a 60-min lesson would only allow 2–3 min per student for individual instruction and is unlikely to be helpful. There are several ways to provide tasks. From experience, the one *not* to attempt with younger students is a step-by-step follow-my-leader style method! A few things tend to happen in this scenario. You show a step and ask the students to repeat the step themselves then:

1. Some students didn't understand or lack the confidence to try therefore you stand next to them and repeat the instruction. They complete the task but may have no understanding of why they did it this way.
2. Whilst you were doing this, some other students have got bored and no longer have any interest in the subject matter.
3. Others also got bored but have lots of confidence and tried to skip ahead. For a small amount, they would have gotten it right. For many others, they are now in a muddle and need you to problem solve for them. They may or may not now get frustrated and potentially give up but either way, you are now 15 min into the lesson and you have only just finished the first step.

A more successful method I have employed in the past is to make the task part of a game in which certain items need to be made in order to free a trapped wizard. The tasks increase in level of complexity, and both success and failure are rewarded with points according to the learning potential. To complete the tasks, the students have access to video clips that they can play and repeat as they need to until they are competent enough to break away from the example and complete a more creative product that earns extra points. Students are also provided with a book of 'magic' that shows the common errors and how to avoid them and recover from them. Some students never get past the basic features of extrude, revolve and shell but the concept is understood as are the common errors and seeds for progression have been sown. Others are able to really extend themselves and move into far more complex products as the final task is an open task to rebuild the castle that has been destroyed in an attack (Fig. 5). The additional advantage of this method is that the guide given

Fig. 5 Sample of outcomes following self-led game-based learning methods

to the students to aid problem-solving in each task is also useful for teachers who are less confident and lack the time to become experts in this as well as the many other aspects of Design Technology.

Another possibility that may aid the teaching of CAD work is to allow group or pair work. Research has shown that computer-supported collaborative learning may facilitate deep learning and provide motivational benefits (Lipponen, Rahikainen, Lallilmo, & Hakkarainen, 2003). Webb, Troper, and Fall (1995) state that 'students can benefit from both giving help to and receiving help from their peers. The process of giving explanations may encourage explainers to clarify or reorganise material in new ways, recognise and fill in gaps in understanding, recognise and resolve inconsistencies, develop new perspectives, and construct more elaborate conceptualisations than they would when learning material by themselves'. More recently, 92% of respondents from the teaching profession say that collaborative learning is an important digital learning strategy (Fujitsu 2017), so it is an option worth exploring even if it is only for selected students. Care would certainly need to be taken when monitoring the pair or group to ensure one student doesn't do all of the work whilst the other sits back.

At the very least, having student 'experts' is helpful to reinforce the learning for the 'expert' and answer simple questions for those with less confidence. This was shown to be an effective way of working in a study conducted by Barlex and Miles-Pearson (2009). All parties in the study reported benefits to some level.

4 Conclusion

Teaching digital technologies is seen by some teachers as an onerous task and some others simply follow a scheme of work that focuses on the outcome. In reality, when you really think about what would be most useful to the students, the important aspects are to encourage an interest and confidence in the subject, and develop strategic knowledge. Students then become more confident to try new technologies as they emerge and develop, they also become more informed consumers able to make reasoned choices and enjoy using digital technologies. The least important skill would be for the student to be able to have knowledge of how to complete an outcome, such as a 3D CAD model, from memory. It is the process, and sometimes failures, which have the most value.

The evidence laid out in this chapter suggests that the most important factors to consider when developing a pedagogical approach to digital technologies are as follows:

- To inspire the students to have an interest in using digital tools.
- To teach students to know what digital tools can be used for and therefore have a context as to why they are learning it.
- To teach students to have an understanding of the language used in the programs so as to instil confidence when they use the program.

- To teach students to recognise and solve the common errors that make an outcome fail.
- To have tasks that focus on individual progress rather than a successful end product.

References

Barlex, D., & Miles-Pearson, S. (2009). Assessing the impact of peer-peer mentoring on pupils' engagement with CADCAM. In C. Benson, P. Bailey, S. Lawson, J. Lunt, & W. Till (Eds.), *Seventh international primary design and technology conference 2009* (pp. 25–29). Birmingham, UK: CRIPT at Birmingham City University.

Biggs, J. B., & Collis, K. F. (1982). *Evaluating the quality of learning: The SOLO taxonomy*. New York: Academic Press.

Bransford, J. D., Brown, A. L., & Cocking, R. R. (Eds.). (2000). *How people learn: Brain, mind, experience and school*. Washington, DC: National Academy Press.

Cox, S. in Livingstone, H. (2012). http://www.cadalyst.com/cad/get-'em-while-they're-young

Design Council. (2017). *Designing a future economy, executive summary.*

Dweck, C. S. (2008). *Mindset: The new psychology of success*. New York: Ballantine Books.

Fujitsu Technology Solutions. (2017). *The road to digital learning, Research report*: Insight Avenue.

Gershon, M. (2016). *How to develop growth mindsets in the classroom*. Createspace Independent Publishing Platform, South Carolina.

Hook, P., & Mills, J. (2011). *SOLO Taxonomy: A guide for schools*. Essential Resources Educational Publishers Limited, New Zealand.

Jones, S. (2018). http://www.readingrockets.org/article/multisensory-vocabulary-instruction-guidelines-and-activities. Accessed 4/01/18.

Lane, H. B., & Allen, S. (2010). The vocabulary-rich classroom: Modelling sophisticated word to promote word consciousness and vocabulary growth. *The Reading Teacher, 63*(5), 362–370.

Legge, K., & Harari, P. (2000). *Psychology and education*. Oxford: Heinemann.

Lipponen, L., Rahikainen, M., Lallilmo, J., & Hakkarainen, K. (2003). Patterns of participation and discourse in elementary students' computer-supported collaborative learning. *Learning and Instruction, 13*, 487–509.

Musta'amal, A. H., Norman, E. W. L., & Hodgson, T. (2009). Gathering empirical evidence concerning links between computer aided design (CAD) and creativity. *Design and Technology Education: An International Journal, 14*, 53–66.

Rynne, A., Gaughran, W. F., & Seery, N. (2010). Defining the variables that contribute to developing 3D CAD modelling expertise. In E. Norman & N. Seery (Eds.), *Graphicacy and modelling* (pp. 161–233). Loughborough, UK: The International Conference on Design and Technology Educational Research and Curriculum Development.

Webb, N. M., Troper, J. D., & Fall, R. (1995). Constructive activity and learning in collaborative small groups. *Journal of Educational Psychology, 87*(3), 406–423.

Winn, D. (2014). *CAD and creativity at key stage 3: Towards a new pedagogy*. PhD Thesis. Milton Keynes: The Open University.

Developing a Pedagogy of Critiquing as a Key Dimension of Design and Technology Education

Steve Keirl

Abstract This chapter advances a case for the pedagogy of critiquing as a central player in quality Design and Technology (D&T) education. Critiquing is seen as a partner to designing – the two being mutually advantageous to D&T learning. However, as with designing, the case is also made for how critiquing serves not only the D&T field itself but also students, teachers and society alike. Thus, it benefits many players and interests, and contributes to D&T's advocacy both as subject and as component of general education of all students.

While the chapter explores what is entailed in nurturing a pedagogy of critiquing, it does so by taking a 'critical' approach to many of the orthodoxies of D&T education such as narrow, teacher-centred, instrumental pedagogies. It necessarily offers critiques of tokenistic interpretations of design pedagogy, of constrained understandings of the nature of technology, and of the overall purposes of what counts as 'education'. The chapter includes a theoretical base to critiquing; an examination of the kinds of contextual and curriculum issues which face D&T; an exploration of the all-important critiquing-designing interplay; and some concluding cautionary notes.

1 Introduction

Key to any rich Design and Technology (D&T) practice is a curriculum partnership of critiquing and designing. These two speak to each other for student, teacher and society alike. They also speak to each other pedagogically. This is not to conflate the two as they have their distinctive qualities and roles. They are educationally symbiotic. This chapter explores what is entailed in nurturing a pedagogy of critiquing.

Narrow, teacher-centred instrumental pedagogies of technology education have always concentrated on matters such as doing, making and skilling. In turn, the take-up of design by the field has been varied. A spectrum might be sketched from

S. Keirl (✉)
Goldsmiths College University of London, London, UK
e-mail: S.Keirl@gold.ac.uk

P. J. Williams, D. Barlex (eds.), *Pedagogy for Technology Education in Secondary Schools*, Contemporary Issues in Technology Education,
https://doi.org/10.1007/978-3-030-41548-8_8

total design avoidance; through tokenistic closed design briefs ('designing with the corners knocked off'); to prescribed but open design briefs; to student-determined research-based designing. I mention this design pedagogy issue because if the design has been actively eschewed, ignored, undermined or tokenistically taken up, then there is little hope for nurturing critiquing as a key D&T practice. To understand the importance of critiquing and designing to D&T education, we must look to the context in which Design and Technology educators are trying to educate and we should reflect on the kind of education that students and society need.

2 Context and Scope of this Chapter

From the outset, any call for a pedagogy of critiquing must accommodate a fundamental issue:

> As a species we are unable to define or describe ourselves without reference to technologies. Our very existence is dependant on, and inter-dependent with, technologies. The quality of our co-existence with other species and the planet cannot be determined without technological critique. Why is it then, when the phenomenon of technology constitutes such a pervasive and hegemonic part of life on the planet, that it is so ill-addressed in education? (Keirl, 2007: 77)

All technologies have been designed and with designs come consequences. No technology is neutral, values-free or universally good. In fact, all technologies are problematic and, like education, a matter of politics. Our local and global challenges such as climate alteration, pollution, resource depletion and inequitable economic systems are all technological in nature *by design*.

Thus, more comprehensive engagements with ethics, sustainability and democratic politics than currently happen are needed. The ways that designing occurs and that technologies are manifested warrant ethical, environmental and political interrogation at all phases – at the point of *initial conception* or *intention*, during *design*, during *realisation*, during *use* and for their *consequences* (whether anticipated or otherwise) (Keirl, 2009). Such scrutiny must come from more sectors of the public than technocratic experts alone. Who holds the innovation and design decision-making powers and voice is now a matter for citizens and public at large (Mitcham, 1994; Sclove, 1995). As Feenberg has consistently and cogently argued: '(t)he fate of democracy is…bound up with our understanding of technology' (Feenberg, 1999:vii).

None of the above can be accommodated without a call for a particular type of education and one that is for all people as the issues are of common global concern. Such an education will have to be democratic, that is, teaching about democracy, serving democracy and modelling democracy. But it will also be an education in, for and about technologies and their design. For both the democratic and the technological aspects of education, ethics and a culture of questioning are necessary. Clearly, Design and Technology education has a role to play here though just what the extent and nature of that role should be is a matter for professional determination. Key to all of the issues noted and to democratic, and technological, education is what might be called *The Critical*.

3 Critiquing and Its Co-concepts

Critiquing is the preferred term adopted from a family of related words, to articulate a particular dimension of Design and Technology. First, as an active word, it salutes the *doing* or praxical nature of our field. Here, it is intentionally distanced from the term 'critique' which, like 'design', is ambiguous – each being both a verb (denoting action) and a noun (a naming word, denoting a thing). Thus, when we *critique* this or that technology or design, the product of that critiquing is *a critique*. Second, drawing on this distinction, *critiquing* actively resists tendencies to generate critiques-as-products, that is, as entities potentially separated from practice (and becoming objects of assessment focus). This is pedagogically important. Third, *critiquing* not only resists positivistic separations of practice and theory but also, as a form of praxis, models '...purposive and purposeful human activity...(and can be)...used to...consciously interrupt...the hegemonic status quo' (Buchanan, 2010: 385).

Critiquing clearly bears a relationship to *criticism*, a term which, when used in its holistic rather than its negative sense, amounts to *acts of judgment*. Just as criticism is often related to judgments about works of art and the arts in general, so it has a key role in the problematic and contested fields of technology and design. Where simple 'right' or 'wrong' answers are inadequate judgment is called for. Equally, critiquing means far more than offering tokenistic platitudes which can amount to being demeaning or patronising. The judgments that are sought will be of the articulate, informed and reasoned kind. Critiquing can be viewed as being both an inward-directed reflective practice and an outward-directed intentional act addressing phenomena and the actions of others. The concept of *reflective practice* (Schön, 1983) is now well established in professional fields while *critical thinking* enjoys an abundance of attention ranging, in education, from sound-bite quick fixes peddled by fly-in/fly-out 'trainers' to properly researched and theorised applications that can be pedagogically embedded across curricula. In its robust philosophical form, critical thinking embraces logic, metacognition and creativity alike in how it serves education. It serves in other ways too. Peters (1967: 19) argued the incompatibility of critical thinking and *indoctrination* showing the former is a necessary counter to the latter. In the same period, Passmore (1967) drew the relationship between critical thinking and *imagination* (and, unusually, technology):

> Critical thinking as it is exhibited in the great traditions conjoins imagination and criticism in a single form of thinking; in literature, science, history, philosophy, or technology the free flow of the imagination is controlled by criticism and criticisms are transformed into a new way of looking at things...The educator is interested in encouraging critical discussion, as distinct from the mere raising of objections; and discussion is an exercise of the imagination. (Passmore, 1967: 201)

Summarising, the praxis of *critiquing* embraces a lexicon of related terms such a critique (as noun and verb), criticism, critical reflection, critical thinking, imagination and interpretation. To these can be added critical stance, critical distance, critical disposition, scepticism, empathy and ambiguity. However, we soon find broader critical perspectives too. Matters get richer when we explore critiquing as a blend of *hermeneutics*, as the art of *interpretation*, and *argumentation* (it is certainly greater than either of these alone). Addressing narrative theory, Kaplan discusses *critical hermeneutics* and how '(h)ermeneutics and argumentation are coextensive' (Kaplan, 2009:87). As he says: 'Determining which interpretations are more plausible than others requires that we argue by offering relevant reasons in order to communicate the virtues of one interpretation over another' (Kaplan, 2009: 86). He contrasts a story about technology that:

> …might be banally true, partially true, or true in a way that merely reinforces existing values and world views, as opposed to a story that has real insight, imagination, and helps us to understand things more clearly. The difference hinges on a distinction between an interpretation of things according to accepted norms and conventions and an interpretation of things that is critical, discerning and evaluative. (And he reflects)…is there a meaningful distinction to be made between a conventional and a critical reading of technology? Is there such a thing as an *uncritical* reading of technology? (Kaplan, 2009: 90–91)

He provides examples of conventional readings of technologies and of conventional understandings of technologies' broader social settings, and he takes us from looking at the 'technology-in-itself' to considering 'technology-in-the-world' commenting that 'Anyone who engages in the politics of technology has already stepped out of the conventional view, regardless of one's political convictions' (p. 91). In inviting critical interrogations of technologies, he draws on the work of Feenberg (e.g. 1991, 2002), Ihde (1990) and Haraway (1991), amongst others, discussing *critical theory of technology*.

4 Critical Theory, Critical Literacy and Critical Pedagogy as Context for Critiquing

Three further constructs can be discussed to support the theoretical framing of any pedagogy of critiquing. First, as a backdrop, critical theory has informed all of critical literacy, critical pedagogy, critical technological literacy and critical practice in general. For education, Freire (1972, 2001) and Habermas (1971) have been particularly influential. McLaren (1989/2009) notes that critical theorists see the world as '…rife with contradictions and asymmetries of power and privilege' and that people are 'essentially unfree' and argues that:

> Critical educators argue that any worthwhile theory of schooling *must be partisan.* That is, it must be fundamentally tied to a struggle for a qualitatively better life for all through the construction of a society based on non-exploitative relations and social justice. The critical educator doesn't believe that there are two sides to every question…(but that)…there are *many* sides to a problem…. (McLaren, 1989/2009: 62)

(Such is the multifaceted nature of design problems!) McLaren also says that: 'The critical educator endorses theories that are, first and foremost, *dialectical*; that is, theories that recognize the problems of society as more than simply isolated events of individuals or deficiencies in the social structure' (McLaren, 1989/2009: 61). Any consideration of a pedagogy of critiquing must engage the notion of the dialectical. For this chapter, we can draw on one of the established understandings of dialectics as relations enacted through *thesis>>antithesis>>synthesis*. A simple example of this is a dialogue of argument, counter-argument and subsequent resolution bringing new understandings for both protagonist and antagonist. Dialectical practices challenge what is and bring forth new possibilities.

Because our designed and technological worlds are non-neutral, complex and ever-dynamic, education systems struggle to prepare students adequately for technological citizenship. This is part of the reason (I contend) that the idea of Design and Technology in the compulsory years of schooling having a prescribed body of ('subject') knowledge is a mirage. First, the scope of knowledge/s that could be drawn upon is absolutely immense. Second, what knowledge should be privileged and by whom? Policy-makers and examination boards are neither teachers nor trustworthy arbiters of education for technological citizenship. Third, as a consequence of these two, what is appropriate knowledge must ultimately be a matter of the school context, the teacher's pedagogical aims and the students' design engagements. Fourth, as a consequence of these three, knowledge in relation to D&T (a) is never wholly capable of prescription; (b) is accessed as needed; (c) takes multiple forms – propositional, procedural, philosophical, psychological, sociological and more; and (d) is as much revealed through D&T as it can be identified as a prerequisite for D&T.

Habermas (1971) is helpful here when he invites us to understand knowledge through critical theory. He suggests that we hold different kinds of *interests* in knowledge. One interest is of an *instrumental-technical* nature. It is empirical-analytic and reflects factual and functional ends. A second is *practical-hermeneutic* and is concerned with interpreting and understanding the world in ways that bring about consensus on knowledge through meaning-making. The third is *critical-emancipatory* and serves critical interrogations of one's lifeworld in ways that bring about personal autonomy and fulfilment. Morrison (2001), applying Habermas to curriculum in general, sees: first, bureaucracy-driven, test-focussed, skills-constrained curriculum as *instrument*; second, 'curriculum as *practice*' – humanistic, interpretive and pragmatic, rich in experiential learning, and privileging understanding over outcomes; and, third, '*curriculum-as-praxis*' an 'existential, empowering and ideology-critical' approach that is emancipatory in nature. Here, the curriculum is problematised by all players (Morrison, 2001: 218). (For a readily accessible text on applying Habermas to curriculum, see Smith & Lovat, 2006).

The second construct, well articulated out of Freire (1972), is that of *critical literacy*. Famously resisting the idea of 'banking education' which '…treats students as objects of assistance', he advanced education through literacy that led students to challenge and interrogate the status quo as 'critical thinkers' and '…persons as beings only when engaged in inquiry and creative transformation' (Freire, 1972: 56). A critical literacy movement grew out of his work and blossomed showing how, as with Habermas, different political and social interests are served by different forms of literacy. In turn, in 2000, Petrina published on *critical technological literacy* while the South Australian curriculum (DETE, 2001a, b) emerged with its Habermasian framing of, respectively, the *operational*, the *cultural* and the *critical* as dimensions of a holistic technological literacy. This approach was further developed with an ethical-democratic focus in Keirl (2006a).

The third construct to note is that of *critical pedagogy* which also has roots in Freire, Habermas and critical theory. Darder et al. (2009) attribute the first appearance of the term 'critical pedagogy' to Giroux (1983), a leader in bringing critical theory to education, who argued the importance of critiquing the positivist (privileging 'facts', laws, rules and sense-based truth claims) and functionalist (positioning education as an instrument of the economy) rationality dominant in schools. (This is even more the case today and would have technology education built solely around scientific knowledge, using assessment criteria of 'does it work?', and serving the needs of industry and economy.) Giroux argued that dialectical thought should replace positivist forms of social inquiry and that students' personal experience was ever significant. He saw '...dialectical thinking as critical thinking...' (Giroux, 1983: 35) and urged that teachers '... place the notions of critique and conflict at the center of their pedagogical models...' (Giroux, 1983: 62). (For lucid feminist, knowledge, ecoliteracy and social justice critical pedagogy perspectives, see Gore (1993), Kincheloe (2008/2010), Kahn (2010) and Smyth (2011), respectively.)

From what has been said, we can see that any discussion of critiquing in the context of D&T is not without its companion travellers. How, then, might we apply the wealth of background theoretical research, argument and curriculum experience that can support critiquing as a valid and valued dimension of Design and Technology education?

5 Curriculum Considerations for Articulating a D&T Pedagogy of Critiquing

When powerful ideological forces (Apple, 1979, 2001; Pinar, 2003; Keirl, 2015a, b) are driving education to mould students to ideological ends so it is that a curriculum that celebrates critiquing becomes so necessary. When democracy is being re-cast as equating 'the market' (i.e. capitalism, see, e.g. Galbraith, 2004),citizens are being re-cast as consumers, and students are being re-cast as economic resources, then teachers find themselves as mere technicians maintaining the machinery of curriculum delivery (of knowledge packages or skill sets) and testing regimes. In these times, by political design, there is little room for teachers' professional judgment in determining curricula that support democracy, citizenship and the critiquing of technologies.

D&T teachers often find themselves seemingly powerless in such a situation. Indeed, many are wittingly or unwittingly supporting neoliberal ideology. However, when, as professionals, we consider the big ethical challenges facing our existences, the well-being of the planet and all species, resource (human and natural) exploitation and so on, we are drawn to reflect on the values we hold; to question what is the right thing to do professionally; and to consider how we might achieve worthwhile outcomes even when the odds suggest otherwise.

From a general perspective, D&T teachers can consider just what kind of technological and design literacy would best serve a rich and informed democracy. Applying Habermas, do we stay with the instrumental knowledge interests – technical in delivery and outcome alike? Or do we strive to meet all three knowledge interests by developing designerly behaviours and critical dispositions in every student? I say ‘every student’ because our core concern should surely be that of the *general education* of the whole populace, delivered in the compulsory years of schooling. So long as D&T as a ‘subject’ constrains itself to a particular knowledge set (e.g. data, food, concrete); or a particular vocation or profession (e.g. machinist, programmer, engineer, architect), it sells itself and the majority of school students short.

Given such seeming constraints, what course can be steered between the big social, political and global challenges and the professional pressures to reproduce specialists who are vocationally or career constrained to very particular aspects of the technological realm? The answer cannot come from a Design and Technology curriculum serving ideological ends or the pursuit of a prescriptive body of knowledge. However, the answer can emerge from curriculum approaches as follows: first, respect the ‘doing’ nature of our field – which is a defining trait of our educational existence; second, reach out across the curriculum (as any ‘literacy’ should) and contribute to general education in and for democratic participation in the designed technological world; and, third, are holistic in nature, ethical, critical and respectful of all three knowledge interests that support democracy, citizenship and the critiquing of technologies.

To achieve such a curriculum reorientation verbs (indicative of *process* rather than *content*) become key. (This has other significant advantages since no particular medium or technology is privileged – such process words hold true for *any* technology, and verbs help accommodate *technological change* and maintain curriculum as both dynamic and sustainable.) While *creating*, *producing*, *manufacturing* and *making* remain core to our field, it is their interdependence with *designing* and *critiquing* that works holistically to empower students, teachers, curricula and schools alike to meet the educational challenges that D&T continually faces. Thus:

> ...critiquing cannot be seen as an isolated piece of a curriculum jigsaw. It must be understood *as permeating all* designerly and technological behaviours and circumstances. Critiquing is necessarily of the complex and of the holistic. It was never some isolated concept plucked from the air, chosen to be fashionable, or as a quick fix for a particular curriculum problem. (Keirl, 2017: 112)

For Design and Technology, there are senses in which critiquing is *dialogical*, generating conversation amongst persons, processes and products. It brings forth a discussion between designer and their design process throughout the iterations that lead to a design outcome. It may be in the mind, or between student and teacher, or it may concern how products ‘speak’ to people – or even to each other.

More powerfully, critiquing is *dialectical*. From every interaction between observer and observed, between subject and object, between defence and advocacy, new discourses and opportunities emerge. In turn, design imagination and creativity are stimulated. By nurturing a critiquing disposition, values are exposed and opened up to challenge and revision – whether they be the values of self, the designer or the values that have been 'designed into' artefacts and systems. In fact, inasmuch as philosophy is meta-critique, critiquing technologies and designs can introduce students to philosophical concepts and issues. Ultimately, all engagements in critiquing are consciousness-raising engagements because of what is demanded by critical acts and, as consciousness is raised, so are potentials for imagination, reflection and action on our worlds. At the personal level, one's consciousness of one's own capabilities, of being human, of the kinds of issues raised by designs and technologies, and, of understanding humanity, become ever more refined. At the public level, society's heightened consciousness of technological and design issues leads to new ways of addressing those very issues and to stronger, more inclusive, democratic participation.

To better educate about our designed and technological lifeworlds, broad perspectives are necessary. We must resist the kinds of 'orthodoxies of technology' (Keirl, 1999) that still inhibit D&T curriculum; namely, seeing technologies as: tools; applied science; hi-tech; being neutral/value-free 'things'; or, as inevitable, as 'progress'. D&T pedagogies of critiquing can do much to counter such narrow interpretations of technology but it can also serve as a leader for cross-curricular critical technological literacy. A whole-school approach maintains dispositions for critiquing technologies in all subjects rather than allowing them to perpetuate the orthodoxies. If a critical citizenry is needed for our designed and technological global coexistence, then a different curriculum imperative emerges. As Kaplan says: 'The first step towards reading things critically is reframing questions concerning technologies within a broader narrative setting' (Kaplan, 2009: 92; see also Sclove, 1995; Keirl, 2006a, b, 2015a; Feenberg, 2002, 2010).

6 Exploring Pedagogy of Critiquing More Closely

Having considered critiquing's curriculum potential – for both D&T and for general education – a deeper exploration of the phenomenon opens up the pedagogical potential.

When critiquing takes place, all of the following can develop: values-weighing; judgment-making; imagination; managing discomfort; meaning-making; advocacy and defence of design and production decisions; scepticism towards the status quo; empathy towards other people and other species; futures-thinking; consciousness-raising; understanding holism; question-asking as routine; and critical thinking. This is not to say that all of these should be *taught* – at least in the prescriptive syllabus content sense – but they are certainly desirable educational attributes for participants in democratic globally oriented societies. All of these in their educationally valid ways can be engaged in holistic Design and Technology education, that is, as integrated traits of a critical technological literacy (Keirl, 2000) that is both design-rich and concerned with creating ethically defensible products, processes and systems. Such an approach greatly enriches D&T for its own sake and, in turn, enriches the general education of all students. This said, it should be noted that this cannot be the remit of the compulsory years of secondary education alone. Indeed, the South Australian curriculum anticipated legitimate engagements in critiquing beginning at the age of 3.

Importantly, critiquing, as with designing, needs what might be thought of as a particular kind of educational climate or culture – one that permeates the classroom, the hearts and minds of teachers, and (ideally) the school. The scene can be set by returning to Freirean thought when McLaren and Giroux (1994) note that Freire emphasised knowledge as an *act of knowing*. Here, knowledge ceases to be (positivistic) 'stuff' or content to be identified, packaged and delivered. The knower (or learner) is at once co-constituted with what counts as knowledge. McLaren and Giroux also note another of Freire's 'important conceptual themes' through which Design and Technology can be read:

> ...the process by which, in Freirean praxis, the realms of the ethical and the rational become dialectically re-initiating and mutually constitutive. Equally edifying is Freire's conceptual understanding of how the power of institutionalised schooling finds its correlative in particular regimes of knowledge that stress technocratic reasoning and an introduction of a model of citizenship based on an individualist and consumerist ethics.... Freire has also consistently attacked forms of pedagogy that attribute great hermeneutical power to the figure of the master, preferring instead to situate teaching in the performative mode of dialogue. (McLaren & Giroux, 1994:xv–xvi)

Thus, the teacher ceases to be the *transmitter* of knowledge. Rather, *transformative* and *transactional* acts situate teacher and learner as co-creators of knowledge. As Gadotti (1994:13) reports, Freire saw as virtues indispensable for the learner such qualities as '...action, critical reflection, curiosity, a demanding questioning, worry, uncertainty'. He also reports Freire's development of a *transitive critical conscience* which is '...both challenging and transforming (and calls for) critical dialogue, talk, and experience of living together...' (Gadotti, 1994:49). (Here, 'living together' can be read against all four of our *realms of co-existence*: other people, other species, the planet and technologies Keirl, 2010.)

Regarding the (inward and outward) interrogational nature of critiquing, Gadotti (1994:89) contends that Freire '...showed that the act of questioning is tied to the act of existing, of being, of studying, of building, of researching, of knowing. He talked about the validity of all kinds of questions and stated, "no reply is definitive". Because of this we should always continue to question, as asking is the essence of knowing'. Here, for D&T, remembering that good design pedagogy cannot entertain a 'right answer' approach to learning can illuminate how designing is actually both a form of knowledge acquisition and a form of knowledge creation. Recalling Giroux, such educational journeying is not only dialogic but it both necessitates and facilitates dialectical-critical thinking.

Developing both students' and teachers' critiquing dispositions occurs across the spectrum of Design and Technology practices – whether in phases of creation, manufacture, design or review. Also, several audiences are served well – the self, peers, teachers and public. Whatever the practice, critically reflective behaviours enhance capabilities for dialogue, and from such dialogues, the dialectics of critiquing contribute to heightened design and technological consciousness of self and towards others.

In developing pedagogies of critiquing, the following can be informative:

- Just as there is no one way (or process) to design, so critiquing must eschew formulaic or prescriptive approaches. Critiquing is an excellent servant to design. As a dialectical praxis critiquing is ever creating new beginnings.
- Critiquing is also something more in itself. In D&T, it helps clarify needs-wants issues, values issues, highlights the contestable nature of technologies and their multiple effects, and heightens student designerly and technological consciousness. Such clarifications have social value well beyond the classroom and readily serve democratic life.
- Critiquing always *responds to* something that exists or has happened – whether an idea in the mind, a process or an artefact – it is reactive, after the fact. Rather than answering 'problems' critiquing raises and clarifies questions through reformulation and reassessment.
- Critiquing is deconstructive but not destructive. While it has limited problem-solving capacity, it does have excellent problem-finding or fallacy-exposing capacities when used to check and recheck validity, integrity, worth, accuracy and fairness in designs or technologies.
- Critiquing calls for an understanding of the audience for the critique (self, team, assessor, public). It may lead to close interrogation of assessment criteria and rationales or of one's own methodology of practice – on time management, design procedure chosen or research options taken.
- Critiquing aids *selection* of thinking styles. Sophisticated critiquing is a form of metacognition and highly applicable to D&T's varying calls for imagining, analysing, researching, synthesising, judging, advocating or arguing a case. At deeper levels, many technological issues are ultimately philosophical; see, e.g. Hickman (2001) for background and the Philosophy for Children [p4c:n.d.] movement for pedagogical inspiration.
- In being reflective and deconstructive, critiquing may involve discomfort but that is an aspect of critical purpose. The 'discomfort' of self-critiquing is not a matter of positive or negative criticism. It is an attribute like the risky nature of design and recognition of the uncertain nature of things. (Ihde (1990) reminds us of the *multistability* of technologies). Critiquing as experience-building is the interplay of one's personal experience and knowledge with that of others and the greater the critiquing experience(s) the greater the critical disposition – for which, critical friendship is an asset. *Educated discomfort*, which embraces uncertainty and risk, contributes to a critically aware and engaged citizenry. As Saul says: '(t)he virtue of uncertainty is not a comfortable idea, but then a citizen-democracy is

built upon participation, which is the very expression of permanent discomfort. The corporatist system depends upon the citizen's desire for inner comfort' (Saul, 1995:195). Elsewhere, *discomfort* has been described as one of three 'curriculum characteristics' of a sustainable-democratic D&T curriculum (Keirl, 2015a).
- Critiquing contributes to ethical education. An exploratory paper (Keirl, 2006b) looked at how D&T might engage with the 'ultimate (ethical) question' of 'How should we live?' and tentatively concluded that critiquing can embrace intangibility in ways liberating for the student; serve the interests of both design and ethics; be guided by the ultimate question; offer personal empowerment for all students and contribute to their identities; contribute to *en-visioning*; help students understand and explore possibilities beyond current location, time and knowledge; and help understand personal, cultural and political technological values.
- Finally, a few key points on the critiquing-designing relationship. While both critiquing and designing can enhance student consciousness, thinking styles and confidence, such traits remain undervalued in many situations. While both invite students to be (positively) critical, both also reject rote-learning, fact-learning or linear/staged procedures. However, there are important differences between the two. While critiquing happens *after* an idea, event, argument or product, designing *brings into being* these circumstances. While designing is pro-active, critiquing is reactive and, while critiquing is focussed, designing is holistic and dynamic. So far as their working arrangement is concerned, critiquing is a tool that serves the design enterprise. Good designing demands good critiquing.

7 (The Dangers of) Sampling the South Australian Critiquing Innovation

Just as Gore (1993:155) concludes her text acknowledging readers potential disappointment 'because of lack of prescriptive guidance' so it is with this chapter. Necessary to any critical pedagogy is the invitation to the teacher, as worker and professional, to embrace dialogue and dialectics critically in ways that challenge the orthodoxies of the status quo. Teachers' personal professional judgment, political stance and values are as necessarily prone to constant critique as the systems of which they are part. In a pedagogy of critiquing, there is no place for passive acquiescence towards content prescription, insular 'subject' boundaries or intellectual laziness. Thus, the following sampling should be read with critical care.

The South Australian D&T curriculum advanced five critiquing *outcomes* articulating the progression of learning from students' early years to age 17, namely, that the student:

- Makes judgments about the significance of different characteristics of products, processes and systems made by themselves and others.
- Identifies a range of ways in which the design of everyday products, processes and systems is related to those who use them.

- Describes the significance to diverse groups of people of the various criteria used in the design of particular products, processes and systems.
- Explains the decisions and choices made in designed and manufactured products, processes and systems, and identifies alternative possibilities.
- Examines critically the competing values embodied in designed products, processes and systems; clarifies relationships amongst people, products and quality of life; and presents ethical analyses of various possible technological futures (DETE, 2001a, b).

Achieving such outcomes might have generated *examples of evidence* such as follows – noting that (i) these example are not necessarily aligned with the outcomes above; (ii) that the outcomes and the examples can be readily re-written (after Bruner, 1960) in 'intellectually honest' age/development aligned ways; and (iii) that there are thousands of other possibilities. So, students might:

- Identify commonalities and differences in a variety of chairs.
- Describe the significant factors and considerations when buying a pair of shoes.
- Identify the relationships between climate, culture, people and resources in the designs of different forms of shelter.
- Explain why a variety of milk-based products have been developed (e.g. profit, health, taste, choice) and suggest ways of discriminating amongst them.
- Explain why items of sports equipment are made to particular standards.
- Explain the design considerations behind a range of contraceptives.
- Provide a critical analysis of ethical dimensions of information technologies, nanotechnologies, genetic engineering, robotics or combinations of these in terms of human displacement and depersonalisation.

These samples are drawn from a curriculum *framework* intended to be interpreted by teachers using their professional judgment in ways that serve appropriately their students and communities. The very real danger is that teachers merely follow or replicate such lists. Not only is that a negation of professional judgement but it would also negate the development of the very culture and pedagogy of critiquing that has been espoused in this chapter. Here lies the key point of pedagogy as critical empowerment of student and teacher alike – in the service of true education (not training or indoctrination) and for better communities and societies across the planet.

8 Concluding Cautions

I conclude with the following cautions:

- Pedagogy of critiquing should be a continuum from the early years throughout schooling.
- Assessment regimes are themselves technologies and critiquing must always be maintained towards their political aims and uses – in both D&T and beyond.

- To ensure a rich D&T education, we should maintain respectful distance (a) from 'the dominant three' school subjects (first language, mathematics and science) and their associated testing straitjackets and (b) from pernicious, constraining curriculum constructs such as STEM, training-for-jobs, computing and so on.
- An overcritical education is neither a balanced nor a healthy education. Our futures are bound up in how we design and in what designs we embrace.
- The aim of Postman and Weingartner (1971:204) still holds true, namely, that we '…help all students develop built-in, shockproof crap detectors as basic equipment in their survival kits'.

References

Apple, M. W. (1979). *Ideology and curriculum.* London: Routledge and Kegan Paul.

Apple, M. W. (2001). Educating the "right" way: Markets, standards. In *God and inequality.* New York: Routledge Falmer.

Bruner, J. S. (1960). *The process of education.* Cambridge, MA: Harvard University Press.

Buchanan, I. (2010). *A dictionary of critical theory.* Oxford: Oxford University Press.

Darder, A., Baltodano, M. P., & Torres, R. D. (Eds.). (2009). *The critical pedagogy reader* (2nd ed.). London: Routledge.

Department of Education, Training and Employment (DETE). (2001a). *South Australian curriculum standards and accountability framework (SACSA).* URL: http://www.sacsa.sa.edu.au

Department of Education, Training and Employment (DETE). (2001b). *South Australian curriculum standards and accountability framework, (hardcopy version).* Adelaide, Australia: DETE Publishing.

Feenberg, A. (1991). *Critical theory of technology.* Oxford: Oxford University Press.

Feenberg, A. (1999). *Questioning technology.* London: Routledge.

Feenberg, A. (2002). *Transforming technology: A critical theory revisited.* Oxford: Oxford University Press.

Feenberg, A. (2010). *Between reason and experience and experience: Essays in modernity and technology.* Cambridge, MA: MIT Press.

Freire, P. (1972). *Pedagogy of the oppressed.* London: Penguin.

Freire, P. (2001). *Pedagogy of freedom: Ethics, democracy, and civic courage.* Lanham, Maryland: Rowman & Littlefield.

Gadotti, M. (1994). *Reading Paolo Freire, his life and work,.* Trans: Milton. Albany, NY: SUNY Press.

Galbraith, J. K. (2004). *The economics of innocent fraud.* London: Penguin.

Giroux, H. A. (1983). *Theory and resistance in education: A pedagogy for the opposition.* London: Heinemann Educational Books.

Gore, J. M. (1993). *The struggle for pedagogies: Critical and feminist discourses as regimes of truth.* New York: Routledge.

Habermas, J. (1971). *Knowledge and human interests.* Boston: Beacon.

Haraway, D. (1991). A cyborg manifesto: Science, technology, and socialist-feminism in the late twentieth century. In *Simians, cyborgs and women: The reinvention of nature,* (1991, pp. 149–181). New York: Routledge.

Hickman, L. A. (2001). *Philosophical tools for technological culture: Putting pragmatism to work.* Bloomington, Indiana: Indiana University Press.

Ihde, D. (1990). *Technology and the lifeworld: From garden to earth.* Bloomington, Indiana: Indiana University Press.

Kahn, R. (2010). *Critical pedagogy, ecoliteracy, & planetary crisis: The ecopedagogy movement*. New York: Peter Lang.

Kaplan, D. M. (2009). How to read technology critically. In J. K. B. Olsen, E. Selinger, & S. Riis (Eds.),. 2009 *New waves in philosophy of technology* (pp. 83–99). Basingstoke, UK: Palgrave Macmillan.

Keirl, S. (1999). Determining technology education: Knowing the orthodox, the interests, and the potential. In B. Johnson & A. Reid (Eds.),. 1999 *Contesting the curriculum* (pp. 74–89). Sydney, Australia: Social Science Press.

Keirl, S. (2000). Critiquing as a dimension of technological literacy. In *1st Biennial International conference on technology education research 2000 proceedings*. Queensland: Technology Education Research Unit, Griffith University.

Keirl, S. (2006a). Ethical technological literacy as democratic curriculum keystone. In J. R. Dakers (Ed.),. 2006 *Defining technological literacy: Towards an epistemological framework* (pp. 81–102). Basingstoke, UK: Palgrave Macmillan.

Keirl, S. (2006b). The ultimate design question: "How should we live?" as catalyst for 'visioning values' in design and technology education. In H. Middleton, M. Pavlova, & D. Roebuck (Eds.),. 2006 *Values in technology education: Proceedings of the 4th Biennial International conference on technology education research, surfers paradise, Australia, 7–9 December 2006*. Queensland, Australia: Centre for Learning Research, Griffith University.

Keirl, S. (2007). 'Within-it/without-it' and the search for ethical technological literacy. In *Curriculum perspectives*. Vol. 27, no. 3, pp. 77–80, Sept. 2007. Australian Curriculum Studies Association, Deakin West, ACT.

Keirl, S. (2009). Seeing technology through five phases: A theoretical framing to articulate holism, ethics and critique in, and for, technological literacy. In *Design and technology education: An international journal*. (2009), vol 14, no. 3, pp. 37–46. URL: http://jil.lboro.ac.uk/ojs/index.php/DATE/article/view/1274/1239

Keirl, S. (2010). Critiquing and designing as thinking tools for technology education for sustainable co-existence. In R. Hansen, & S. Petrina, (Eds.), *Proceedings of the technological learning and thinking: Culture, design, sustainability, human ingenuity conference* (pp. 531–540). Vancouver: BC, 17–19 June, 2010. URL: http://m1.cust.educ.ubc.ca/conference/index.php/TLT/2010/paper/view/57/5

Keirl, S. (2015a). Against neoliberalism; for sustainable-democratic curriculum; through design and technology education. In K. Stables & S. Keirl (Eds.),. 2015 *Environment, ethics and cultures: Design and technology education's contribution to sustainable global futures* (pp. 153–174). Rotterdam: Sense.

Keirl, S. (2015b). Global ethics, sustainability, and design and technology education. In K. Stables & S. Keirl (Eds.),. 2015 *Environment, ethics and cultures: Design and technology education's contribution to sustainable global futures* (pp. 33–52). Rotterdam: Sense.

Keirl, S. (2017). Critiquing as design and technology curriculum journey: History, theory, politics, and potential. In J. Williams & K. Stables (Eds.),. 2017 *Critique in technology education* (pp. 109–133). Dordrecht: Springer.

Kincheloe, J. L. (2008/2010). *Knowledge and critical pedagogy: An introduction*. Springer, Springer.com.

McLaren & Giroux. (1994). Foreword Gadotti, M.. (1994). Reading Paolo Freire, his life and work, Trans: Milton, pp. xiii–xvii, SUNY Press, Albany, NY.

McLaren, P. (1989/2009), Critical pedagogy: A look at the major concepts, Darder, A., Baltodano, M. P. & Torres, R. D.., (2009), The critical pedagogy reader, (2nd), pp. 61–83, Routledge, London.

Mitcham, C. (1994). *Thinking through technology: The path between engineering and philosophy*. Chicago: University of Chicago Press.

Morrison, K. (2001). Jürgen Habermas. In J. A. Palmer (Ed.),. 2001 *Fifty modern thinkers on education: From Piaget to the present* (pp. 215–224). London: Routledge.

p4c. (n.d.). Philosophy for children. URL: p4c.com

Passmore, J. (1967). On teaching to be critical. In R. S. Peters (Ed.),. 1967 *The concept of education* (pp. 192–211). London: Routledge & Kegan Paul.
Peters, R. S. (Ed.). (1967). *The concept of education*. London: Routledge & Kegan Paul.
Pinar, W. F. (Ed.). (2003). *International handbook of curriculum research*. Mahwah, NJ: Lawrence Erlbaum Associates.
Postman, N., & Weingartner, C. (1971). *Teaching as a subversive activity*. Harmondsworth, UK: Penguin.
Saul, J. R. (1995). *The unconscious civilisation*. Toronto, ON: Anansi.
Schõn, D. A. (1983). *The reflective practitioner: How professionals think in action*. San Francisco, CA: Jossey-Bass.
Sclove, R. E. (1995). *Democracy and technology*. New York: The Guilford Press.
Smith, D. L., & Lovat, T. J. (2006). *Curriculum: Action on reflection* (4th ed.). South Melbourne, Australia: Thomson/Social Science Press.
Smyth, J. (2011). *Critical pedagogy for social justice*. New York: Continuum.

Question-Think-Learn: A Pedagogy for Understanding the Material World

Belinda von Mengersen and Terry Wilkinson

Abstract Material selection in design is complex. In terms of sustainability principles, materials can't simply be labelled as "good" or "bad", but some students have trouble making sense of highly technical, contradictory, and confusing information. So how then do we engage and learn with students to question, think critically, and better understand the ecological, social, political, and economic implications of current production-consumption-throwaway practices? In this chapter, we present a framework of key elements and strategies drawn from the educational research literature that can support teaching and learning in Design and Technologies education. Specifically, we show how case studies could be used effectively to spark interest, inspire authentic open-ended questions, and provoke meaningful classroom talk about controversial technology-related issues concerning the material world. Content summaries for a number of engaging online resources are also provided.

1 Introduction

What can materials teach us about the designed world? How do we engage and learn with students to critically question and think about the ecological, social, and economic costs of our material choices? In this chapter, we identify a number of key elements of project-based learning (PBL) and cooperative learning structures, and describe how they can be put to work in an inquiry-oriented, design-based technology classroom. In particular, we focus on the pedagogical role of dialogue, peer collaboration, and media resources to foster critical awareness of current production-consumption-throwaway practices and their effects. We propose that critical life cycle thinking can enable students not only to choose appropriate materials for

B. von Mengersen (✉)
Faculty of Education and Arts, Australian Catholic University, Sydney, NSW, Australia
e-mail: Belinda.VonMengersen@acu.edu.au

T. Wilkinson
Faculty of Education, York University, Toronto, ON, Canada
e-mail: Terry_Wilkinson@edu.yorku.ca; twilkinson@edu.yorku.ca

P. J. Williams, D. Barlex (eds.), *Pedagogy for Technology Education in Secondary Schools*, Contemporary Issues in Technology Education,
https://doi.org/10.1007/978-3-030-41548-8_9

school projects but also encourage them to "think and act differently in terms of the ways they use, consume and design technologies" (Elshof, 2009).

As we were writing this chapter, the United Nations' Intergovernmental Panel on Climate Change (IPCC, 2018) released *Global Warming of 1.5 °C*; a landmark special report that claimed unequivocally, "Climate change represents an urgent and potentially irreversible threat to human societies and the planet". We have been put on notice: Based on present-day emission levels, we have 12 years left until the Earth's average temperature reaches a threshold of 1.5 °C above pre-industrial levels. Once this "tipping point" is reached, scientists predict greater risks associated with extreme weather events, drought, rising sea levels, floods, massive wildfires, and permanent loss of natural ecosystems (IPCC, 2018). Dire consequences for human societies are also projected – including an intensification of climate-related poverty and increased migration due to food and water insecurities (IPCC, 2018; Landler & Davenport, 2018; Watts, 2018). If we acknowledge that many of the knock-on effects of global warming will "fall disproportionately on the poor and vulnerable" (IPCC, 2018; Worland, 2016), we must also recognize that these are problems that cannot be easily or quickly solved by technological innovation.

Now the good news is that this forecast is not inevitable. Our so-called Hothouse Earth trajectory could still be reversed if humans collectively embrace a stewardship role of the planet (Steffen et al., 2018). As committed D&T educators, we take the UN's urgent call to action very seriously. However, unless we educate ourselves and raise the consciousness of young people who will be most affected by the consequences of climate change in the future, we will be implicated in creating "the next generation of uncritical consumers and producers of things" (Petrina, 2000). This has particular resonance in light of IPCC reports that the increase of "fossil-fuel-based material consumption and changing lifestyles is a major driver of global resource use, and the main contributor to rising […] emissions" (Fleurbaey et al., 2014; IPCC, 2018).

Two guiding principles – environmental stewardship and social justice – will inform our efforts to raise awareness regarding technoscientific issues that curtail universally agreed-upon human rights and freedoms (Levinson, 2009) – or indeed, pose an existential risk to intelligent life on Earth (Bostrom, 2002). It is our contention that any education that stops short of these considerations will be an inadequate response. Fundamentally, we are advocating a "planetary ethic" of care and respect that takes into account the effects of our actions on the lives and well-being of "all people in the world community" (Laszlo, 2001).

2 Pedagogy for Understanding the Material World

For the purposes of this chapter, we take *pedagogy* to mean instructional practice. Given the multifaceted and creative process of classroom teaching, we certainly appreciate and acknowledge that there is no absolute "best way" to teach; what works for one teacher and a particular group of learners may not be effective or appropriate for others (Bennett, 2010). Having said this, however, we offer here a modest number of approaches and resources that do work for us, and could support teaching and learning in your school as well.

When choosing materials for projects, students need to have relevant scientific and technical knowledge about, for example, the physical and chemical properties, aesthetic, and expressive qualities of candidate materials (de Vries, 2005; Karana, Pedgley, & Rognoli, 2014; Pedgley, Rognoli, & Karana, 2016). Fact-based knowledge is certainly very important for developing design and technological capability. Often in D&T classrooms, students learn about the way materials and products perform through teacher-guided, hands-on experimentation. Investigations may be structured around questions like: Does this material/product design represent best use of materials and techniques? Is it fit for purpose? Is it cost-effective? Technocratic questions about efficiency, performance, fitness, availability, price, and profitability implicitly reflect technical and economic values (Madge, 1997; Pavlova, 2005; Trimingham, 2008). But from an ecological sustainability standpoint, a functionalist model of teaching and learning fails to account for *all* the costs of design decisions – not simply economic, but environmental, cultural, political, and social consequences (Dakers, 2005; Petrina, 2000).

In light of the current climate crisis, we must consider what students really *need to know* about materials. Accordingly, the pedagogy we wish to advance here requires different questions to seek out a much wider range of material-related factors and their (sometimes disturbing) consequences (de Vries, 2005; Werberger, 2016). Student-led inquiries into processes within technological systems can deepen their understanding of the complex interactions among technology, science, society, and the environment, as well as sensitize them to the personal, social, and ethical implications of specific technologies (Hodson, 2003).

3 Questioning as an Essential Democratic Skill

In a democratic society, responsible citizenship requires having the ability and courage to ask questions (Morgan & Saxton, 2006). Effective teachers engage their students as productive questioners and as answerers with the aim to nurture "habits of thought that go beyond thinking *about* to thinking *why* and thinking *what if*" (Morgan & Saxton, 2006). As Greene (1995) pointed out, asking questions lies at the core of a participatory system that regards democracy as "an open possibility, always in the future, not as an achievement in the past". These arguments underscore the crucial role teachers play as mediators of learning when posing questions, encouraging young people to do the same (Hamilton, 2007), and creating a collaborative culture of mutual respect in order for learners to take intellectual and emotional risks (Dakers, 2005).

4 Design Pedagogies: Past, Present, and Future

A pedagogy for understanding the material world must also account for speculative material futures. Examples of sustainable materials currently in research and development include bio-engineered products like naturally biodegradable plastics, new regenerated materials, and composites created from organic waste materials including proteins (Thompson, 2013). Also, a range of "maverick materials" (Quinn, 2012) including smart, multifunctional, shape-changing products, materials that integrate electronics or generate power, and materials of the "synthetic age" (Preston, 2018) including nanotechnologies and molecular manufacturing techniques are emerging. These rapid advances in material science technologies make it impossible to know what will be available in the future. But what we can do is provide learners with a questioning framework that will enable them to launch their own inquiries – by generating new questions, modelling how to search for resources, evaluating multiple sources, testing ideas, and drawing their own conclusions.

In the next section, we take an "unproblematic" everyday situation to illustrate how the choice between a plastic or paper bag is not as straightforward as one might think. We then key elements of a question-think-learn approach, followed by two classroom lessons on the life cycle of materials to show how different kinds of questions and tactics could help focus discussion and provoke further inquiry. Our focus is on how to facilitate collaborative talk. The activities we describe should not be interpreted as a prescribed lock-step sequence but as a repertoire of dialogic thinking tools for testing ideas, and co-constructing knowledge and values (Dakers, 2005). Lengthy discussions and debates can be extremely productive thinking tools, although we recognize that these particular forms of discourse might seem unfeasible for time-constrained programmes. Creating a culture of collaboration and interdependence is a process that requires continuous teacher support and time for the consistent practice of explicitly taught social and communication skills, and reflection (Hennessy & Murphy, 1999; Kolodner et al., 2003). Nevertheless, we present the following for your consideration, fully recognizing that collaborative learning can seem at odds with current definitions of student success which equate ability with independent thinking and speedy solutions (Murphy & Hennessy, 2001).

5 Paper or Plastic?

According to Greene (1995), devising situations in which young people will "move from the habitual and ordinary to consciously undertake a search" (p. 24) is a difficult but necessary task, for once "the given or the taken-for-granted is subject to questioning," we may then have "an opportunity to posit alternative ways of living and valuing and to make choices" (p. 23).

Material selection in design is complex. Global supply chains make tracing the flows of energy and materials a real challenge. So how will students come to know the ecological and social consequences associated with resource streams and production cycles when they have been made invisible (Petrina, 2000)? Conducting research on a product's life story is one way to better understand materials and the processes of production. In order to assess potential benefits and harmful effects, teachers and students may find themselves co-investigating material extraction, energy types, manufacturing and assembly processes, distribution methods, externalized costs, patterns of consumer use, disposal, and potential for reuse.

Materials cannot be simply labelled "good" or "bad" in terms of ecological sustainability or human justice because there are usually benefits *and* drawbacks. Furthermore, information available to assist students in making sound judgements can be misleading, contradictory, and confusing. Consider, for instance, these apparently simple questions: Should I select paper or plastic bags for my shopping? Which is more environmentally friendly: paper or plastic? (McGrath, 2018). While many people assume that paper is a more sustainable choice, once additional factors are known (including comparing water and energy consumption, inefficient recycling, biodegradability challenges, waste generation, and dangers to wildlife), the decision becomes far less clear-cut (McGrath, 2018).

A question could focus their search: Should plastic shopping bags be banned? A comparison table can help students collect and organize their findings (Table 1).

Table 1 Paper or plastic?

Paper	Plastic
Renewable resource (plantation timber)	Non-renewable resource (crude oil)
Can be recycled	Can be recycled with little loss of quality
Naturally biodegradable	Recycling process takes 91% less energy than paper bags
Weak, non-durable, non-abrasion-resistant, and non-water-resistant	Non-biodegradable; it takes 20–10,000 years for plastic to decompose; toxic elements remain in the environment
Higher rate of recycling (10–15%)	Durable, strong, and abrasion- and water-resistant
Large amount of water required for production of paper pulp	Very low rate of recycling (1–3%) because it is much harder to recycle at material recovery facilities (MRFs)
Use of bleaches	High rates consumption, between half a billion and one trillion globally
Only biodegradable under the right conditions; i.e. not in landfill	Ten per cent of plastic bags end up in the ocean; a square mile has approximately 46,000 plastic bags floating
Can't be recycled if composite or plastic coating used in manufacture	Plastic bags are hazardous to animals on land and water
Ten billion bags per annum used in USA = 14 million trees required	Process of cleaning up plastic bag waste is expensive; the US budget is more than 1.7 trillion dollars
Expensive manufacturing process, far more energy required to make paper bags than plastic	
Seventy per cent more air pollutant produced	
Fifty per cent more soil and underground water pollution than plastic	

This table was developed from consumer, journalistic, and non-academic sources to show consumer perceptions prevalent in the media. Environmental science researchers surprisingly conclude: "plastic bags are found to be a little better in terms of environmental impacts compared to paper bags" (Muthu, Li, Hu, & Mok, 2009). In the next section, we ponder the bewildering virtues of polyester.

6 Polyester Paradox: A Material Case Study

Are polyester textiles "good" or "bad" or "less bad"? Polyester textiles are "produced from synthesized polymers" (Hallett & Johnston, 2010); derived from non-renewable fossil fuel sources (including coal, petroleum, crude oil, and natural gas (Hallett & Johnston, 2014)), they are non-biodegradable (Fletcher, 2013), and yet, under the right conditions, polyester (one of the synthetic fibres which dominate [represents 70%] global fibre production textiles can be recycled with very little or no* loss of quality (McDonough, 2005), "almost to its virgin state" (Hallett & Johnston, 2014), potentially in an infinite recycling loop (Hallett & Johnston, 2014). Polyester requires "virtually no land use" in production in comparison to natural fibres; "less than 1% of petroleum is used for the global production of all man-made synthetic fibers" and, "polyester requires only a few cubic tons of water for its production" [whereas] "in excess of 20,000 cubic tons is required to produce the same amount of cotton"(Hallett & Johnston, 2014). The "most common microfiber is made from polyester" (Hallett & Johnston, 2014). A polyester fleece jacket which meets the international, independently tested Oeko-Tex® 100 (Grummt, 2019) standard whereby "guarantee of the safety of textiles and dyestuffs to human health and to the environment" (Hallett & Johnston, 2014) and be produced by a company with an increasingly transparent supply chain like Patagonia (2018); yet, later in the garment's life cycle, through consumer wear and care (washing) as it becomes more worn it releases millions of polyester (plastic) microfibers into the ocean which could ultimately be ingested through the food chain (Patagonia, 2018; Sach, Deluna, & Wilson, 2017). Somewhat ironically, through experimental approaches to textile life cycle thinking, polyester has become the poster child of designers aiming to shift from linear to circular economy (Ellen MacArthur Foundation, 2017a) where polyester textile waste becomes one nutrient which can be repurposed in a cradle-to-cradle (Institute, 2018) material ecology. However, this innovative product Climatex® has required a complete re-think of sustainable materials, processes, and production methods, a leading example in what has been called a "material revolution" (Peters, 2014).

The "paper vs. plastic" debate (Acaroglu, 2013) and the "polyester paradox" material case study demonstrate the complexity of weighing up positive and negative aspects of materials. A conclusion might be drawn about which is "less bad" – but this conclusion is relative and must take into account local contexts and specific conditions in which the product will be made and used. Wrestling with messy problems like these is central to responsible design(ing). Moreover in technology education, we contend it's vitally important that all learners—as future citizen-consumers—have opportunities to critically appraise conflicting information skewed by diverse perspectives, values, and worldviews. If education is to be more than a means to maintain existing problematic practices of Western industrialized countries, we need a broader range of tools to help young people engage in research-informed, ethical decision-making practices. One way to help students learn about materials and their effects on the world will be detailed next.

7 Question-Think-Learn: Collaborative Discussions

Grounded in constructivist pedagogy, Q-T-L draws on instructional principles of problem-based learning to guide group and individual learning. Structured activities are devised to foster dialogue, discussion, critical thinking, and reflection about complex, real-life issues. Key features of this approach will be explained along with specific suggestions on how they can support and deepen students' understanding of the consequences of material selection in design.

7.1 Authentic Context

Unless students perceive the problem as real – with potentially real consequences for their lives – they are unlikely to "own" it (Murphy & Hennessy, 2001). User-centred design briefs in local contexts encourage students to evaluate designed products which they know and use, enabling them to define their own design problem/s. To ask: How can this product be improved?

7.2 Entry Event

One of the hallmarks of a good project is an "entry event" that presents a perplexing, problematic, or controversial situation that is personally relevant and meaningful for students (Dillon, 1988; Larmer & Mergendoller, 2010; Werberger, 2016). It can be just about anything that sparks interest, initiates questioning, and promotes discussion: an intriguing observation, a video, news story, field trip, or guest speaker (see Resources Table? in Appendix). Historical and contemporary case studies can also play instructive roles by illustrating how market, political, juridical, and/or ethical factors can influence product design in the real world (de Vries, 2005; Pedretti & Little, 2008).

7.3 *Driving Question*

Open-ended or "authentic" questions are a prominent feature of PBL that invite ongoing thinking and inquiry, and stimulate discussion and debate about unsettled issues (McTighe & Wiggins, 2013; Wiggins & Wilbur, 2015). For example: Does it matter where our shoes are made? Should environmental policy be based on the "polluter pays" principle? In technology education, student projects are often organized around a problematic situation or driving question (Larmer & Mergendoller, 2010, 2015). Crafted in student-friendly language, an engaging design challenge or question *brings the problem home* (Savery & Duffy, 1995), makes learning more meaningful, and aligns with curriculum content, knowledge, and skills (Larmer, 2018). Examples include "How can we improve…, or create a…?" or "How might we reduce landfill waste (or our water footprint) when it comes to …?" Students are motivated to learn when they have a genuine "need to know" so they can answer a question, solve a problem, or perform a task that matters to them (Larmer, Mergendoller, & Boss, 2015). Before choosing project materials, students might ask: What are the benefits and hazards for ecology and community?

7.4 *Researching Life Cycles*

Storytelling is a creative approach to problem-finding. Starting with an everyday item, a life cycle thinking scaffold prompts students to identify what they know, need to know, and would like to know:

- What is this soda can/cell phone/soccer ball/chocolate bar made of?
- Where did it come from?
- How was it made?
- How did it get to the retailer?
- Where does it go when we throw it away?

After researching their artefact's "story", students could be introduced to a zero-waste vision which "denormalizes" our assumption that waste is inevitable (Sheehan & Spiegelman, 2010). Asking: *How could we reduce or prevent products and packaging from becoming waste?* poses a new challenge to create interventions that could transform our linear mindset towards a more circular way of thinking.

A profusion of graphic images is available online, which explain the concepts of linear and circular economies (see Fig. 1 (Ellen MacArthur Foundation, 2017a)). The simplicity of a diagram is both a strength and a weakness. Reduced text and detail make it easy to "read" but people and communities along the chain have been removed. This calls for another round of questions to bring them back into view! Hodson (2003) following (Laszlo, 2001) suggests:

- Does my way of living interfere with the rights and freedoms of others?
- Does it take away basic resources from them?
- Does it have harmful effects on living beings in nature?

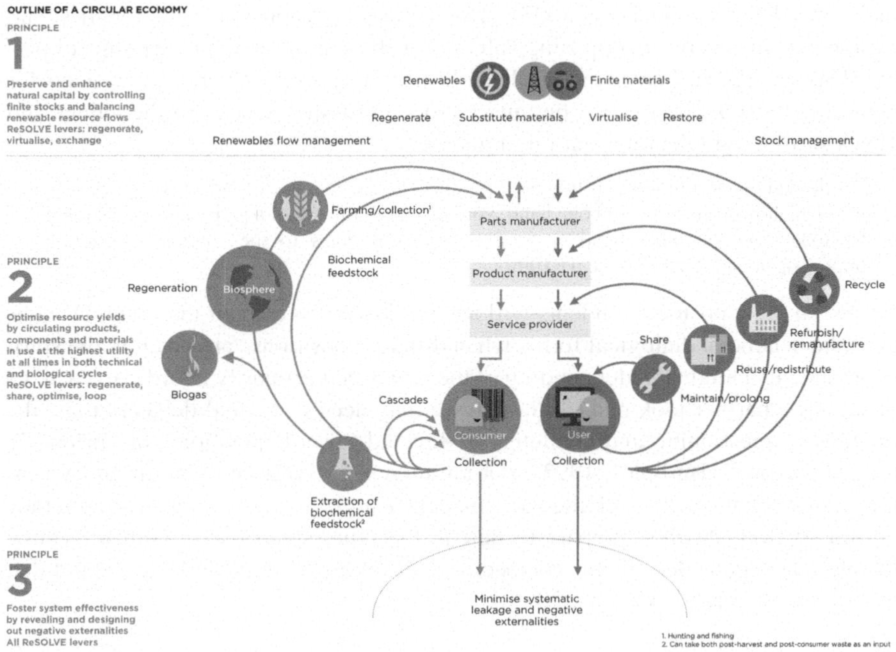

Fig. 1 Circular economy overview (Ellen MacArthur Foundation, 2017a). (Source: Ellen MacArthur Foundation, SUN, and McKinsey Center for Business and Environment: Drawing from Braungart &McDonough, Cradle to Cradle (C2C))

As they search for answers, students may be surprised to learn there are both beneficial and harmful effects. Levinson (2009), for example, illustrates through a series of "interlocking narratives" how manufacturing processes have global effects. Market fluctuations, government policy, foreign investment in recycling, improvements in housing and healthcare for rubbish-pickers in Brazil, and reduced pay and lay-offs for smelter workers in Jamaica are all interconnected. Life cycle thinking helps students understand how the everyday decisions they make (as young consumers, designers, and technology users) can affect the quality and longevity of life in terms of health and security of people and the environment – both now and for the future.

7.5 *Dialogue and Debate*

Authentic questions elicit input from students through meaningful classroom talk and discussion groups which serve as a means of "active inquiry" (Nystrand & Gamoran, 1991). Classroom debates also encourage students to actively take part in their own learning (Dakers, 2005). They provide opportunities to identify and explore tensions between opposing values as well as question the viewpoints of others (Dakers, 2005). For example, the following question prepares the ground for exploring tensions between economic factors and issues related to the values and ethics surrounding the importation of goods.

> Would you support the sale of imported products made from such tropical woods as rosewood, mahogany, or teak? On one hand, you help the economy of a poor, developing country today; on the other hand, your purchase contributes to the country's long-term impoverishment. (Galbraith et al., 1999)

Before taking a position, students will need to know more about the issue. Debates can take beneficial dialogical forms when different positions "are taken in the spirit of advancing a mutual understanding of the issues, not primarily in order to 'win' or make one's partner look bad", although we may need to remind debaters that "the process of questioning and challenge is directed towards positions, not primarily toward persons" (Burbules, 1993). For an interesting variation, you might try running a mock debate which challenges students to argue in favour of a viewpoint they do not support. As they prepare to defend the other side of an argument, young people must "consider others' perspectives, develop empathy, and understand that few ideas are binary" (Collins, 2018).

7.6 Articulation (Discussion, Clarifying/Co-constructing Values, Critique)

As Solomon (1992) points out, stories-as-dialogues can support the processes of argumentation and reflection and thereby serve as key mechanisms for learning. Students share personal anecdotes to either confirm or counter previous assertions in an effort to persuade other listeners (Solomon, 2002). Through repeated small-group discussions, different perspectives on social issues can be explored, as concepts and values are clarified (Solomon, 1992).

7.7 Application (Product Redesign)

Acaroglu (2019) has developed a set of thought-provoking, easy-to-use design for sustainability play cards which challenge students to apply design for sustainability (DfS) strategies including design for environment (DfE); reduce, reuse, recycle (3Rs); environmentally sustainable design (ESD); cradle-to-cradle (C2C); design for disassembly (DFD); and extended producer responsibility (EPR) in response to design problems (VCAA, 2017). This discussion-based "game" fosters questions that enable players to put their knowledge into practice through product redesign.

Asking: Which sustainability strategy would be the most effective solution to this design problem?

7.8 Evaluation

Students can write their own unique design evaluation criteria for a redesigned product as part of a PBL approach, including specific criteria related to sustainable material selection. The evaluation can be separated into two parts; for instance, some PBL authors differentiate evaluation (as critique and revision) from reflection. Reflection involves students thinking back on their process (i.e. what and how they learned, and what they have accomplished) (Larmer et al., 2015; Savery & Duffy, 1995).

8 Driving Questions: Two Examples

This chapter is in part a response to Hendley and Lyle's (1995) call for D&T teachers to clearly articulate underlying procedures and principles of collaborative group work. In what follows, we will illustrate how the interplay of two instructional approaches – cooperative learning (CL) and project-based learning (PBL) – can support learning in a D&T classroom. Founded on theoretical principles of social interdependence, CL is one of the most widely accepted and used instructional practices in schools and universities around the world (Johnson & Johnson, 2009). We draw from Bennett and Rolheiser's (2001) work to show how Place Mat and Round Robin, if effectively applied, can focus students' attention, promote active listening, and support small-group discussion. We then describe an entry event (watching a video) to illustrate how these instructional tactics afforded students the opportunity to use meaningful talk to develop their own understandings of problems associated with the linear flow of materials. The approach dovetails nicely with several essential elements of a PBL framework including, for example, a driving question, authenticity, collaboration, reflection, and critique (Larmer et al., 2015). However, for the sake of brevity, we will limit our focus to how small-group discussion – informed by a quasi-case study – seemed to raise D&T students' critical awareness and level of concern for other people and the planet.

8.1 Driving Question 1: Why Does This T-Shirt Cost Only $3?

Entry Event Two provocative short videos provided background information to generate interest about a real-world problem:

- TEDEd: The life cycle of a t-shirt: https://ed.ted.com/lessons/the-life-cycle-of-a-t-shirt-angel-chang.
- Fast Fashion: The Business of Fast Fashion: https://www.youtube.com/watch?v=ZhkBfbwCzxc.

Authentic Context Basic clothing items like jeans and cheap cotton t-shirts offer relatable examples of material complexity. This approach also works best where the value and ecological and/or social footprint are not obvious: so that the product, upon discussion and interrogation, is revealed to the student as only *deceptively* simple. The point is that simplicity is only a perception, like a cotton t-shirt – such items are ubiquitous, yet the more closely you look, the more impacts on distant geographies and communities you find. Granted, there are many uncertainties, contradictory facts, and values influencing debates about sustainable development. However, it seems clear to us that unless consumers of fast fashion are made aware that their purchases actually link them to impacts such as water depletion and pollution in remote locations, they have little incentive to take responsibility for their hidden water footprint (Chapagain, Hoekstra, Savenije, & Gautam, 2006).

Aim of Discussion Ideally, the students' investigation is rewarded by a gradual realization of the complexity of making simple judgements about material selection. Students were encouraged to speculate about economics and investigate the externalized costs of production for profit. For example:

- How is it that the cotton t-shirt the student is wearing only costs $3?
- If they are hidden, who or what is paying the balance of the real cost? And what are those costs?

T-shirts are wardrobe "staples"; yet, the supply chain is likely environmentally and socially unsustainable and may include more than one instance of social injustice. For example:

- How did these students think and feel if they know that both their t-shirts and jeans were made from cotton picked by child slave labour in Uzbekistan (Goldsmith, 2017)?
- How did these students feel when they saw the photographs of synthetic indigo dye pouring out of a dye house, first into a river and then directly into the ocean in China (Greenpeace, 2010)?

Researching Life Cycles Questions like these provoke initial discussions. Time is needed to develop a comprehensive understanding of the social nature of technology, and increasingly critical questions to form opinions, and make decisions. Time and opportunities for discussion are equally important to help students develop empathy, compassion, and moral sensitivity as they engage with different scenarios (Fowler, Zeidler, & Sadler, 2009).

Life cycle thinking uncovers some of the issues associated with cotton t-shirts and jeans including the following:

- Environment (ecological aspects).

  Water use (approx. 2700 litres for one t-shirt) (Bain, 2018; Chapagain et al., 2006).
  Agrichemical use damages the environment (cotton uses more herbicides and pesticides than any other crop).
  Genetically modified seeds (like Bollgard®3) reduce water and agrichemical use but are expensive.
  Better cotton initiatives (educating farmers): https://bettercotton.org/.

- Society (social justice aspects).

  Child slave labour used to pick cotton in Uzbekistan and India: http://www.cottoncampaign.org/uzbekistans-forced-labor-problem.html.

- White gold: the true cost of cotton https://vimeo.com/1708935.
- Agrichemical use causes serious illness and death: http://www.pan-uk.org/cotton/.

So, what might a more sustainable t-shirt supply chain look like? See (Fig. 2) Teemill: follow the product journey, seed to shop https://teemill.com/the-journey/.

And organic, locally made-to-measure, ethical t-shirts at Citizen Wolf: https://www.citizenwolf.com/.

We suggest taking an entry event (like this example) as a provocation and using the Q-T-L sequence (Authentic Context, Entry Event, Driving Question, Researching Life Cycles, Dialogue and Debate, Articulation, Application and Evaluation) enables students to investigate issues through discussion and action.

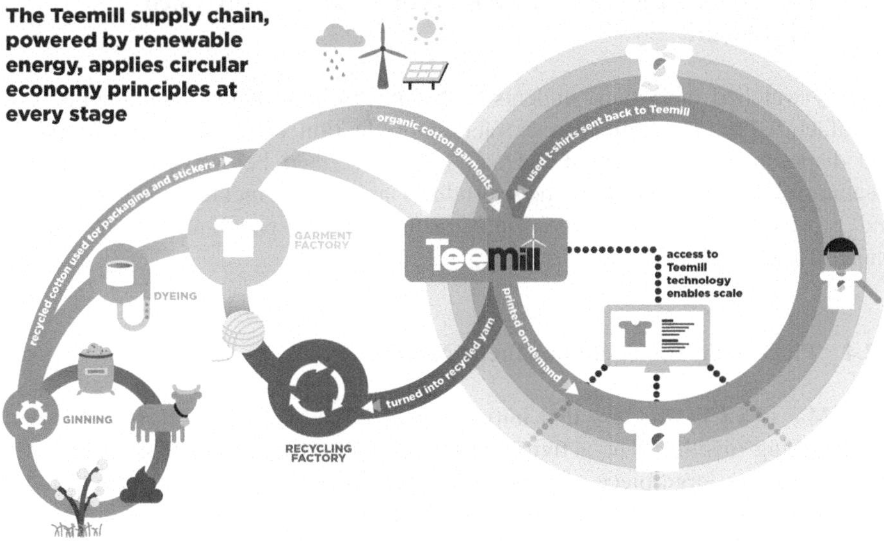

Fig. 2 The Teemill supply chain (Ellen MacArthur Foundation, 2017b)

8.2 *Driving Question 2: Why Are All These Locker Shelves Being Thrown Away?*

8.2.1 Authentic Context: What Is the Problem?

Picture a stack of broken collapsible free-standing locker storage shelves which had been discarded in the hallway in the early days of the school term. The teacher held one up and asked her students why they thought this might be happening, how did these items get here, who made them, where might the different materials have come from, and what would likely happen to them now (i.e. where does our garbage go after it leaves the curb)? The students responded enthusiastically by wanting to share their own personal stories, disappointments, and criticisms about these "cheap products". The problem of poorly designed locker shelves set the stage for an authentic design challenge that students would find personally relevant and emotionally engaging (Larmer, 2018; Lave, 1988).

8.2.2 Entry Event

Before introducing the project, the teacher showed a video called *The Story of Stuff* (Leonard, Fox, & Priggen, 2007). Using humour and animation, this quasi-case study exposes problems associated with global supply chains and the hyper-consumption of everyday electronic gadgets and fast fashion (Bain, 2018; Barenblat, 2017; Whitehead, 2014).

8.2.3 Researching Life Cycles: Need to Know–Why Is This Such a Problem?

As they watched the video, students took notes on one of five components of a manufactured product's lifespan: extraction, production, distribution, consumption, and disposal. The purpose of the activities (described below) was to generate meaningful discussions about how products are made, including the resources and processes involved in their manufacture and disposal after they are no longer useful.

8.2.4 Place Mat: Instructional Tactic Number 1

Place Mat is a CL structure that initially involves individuals working alone and then working together. The activity works best with groups of two to four people sitting around a large piece of paper. Each person is given a pen or pencil and a designated space on the paper to write down key points (see Fig. 3). It's imperative that they work individually first. Following the video, participants are given

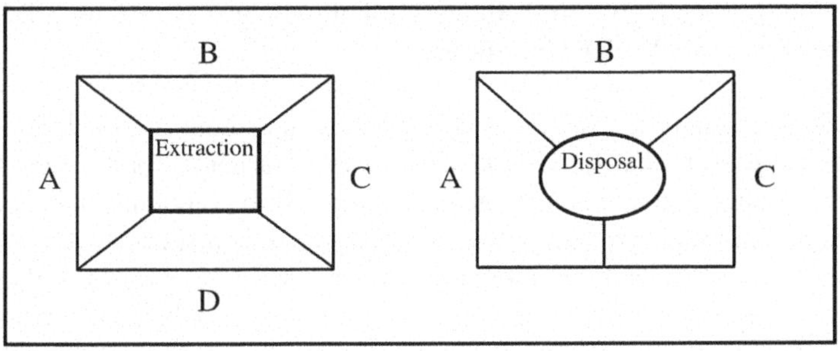

Fig. 3 Place Mat diagrams. (Adapted from Bennett & Rolheiser, 2001)

sufficient time to review their personal notes and reflect on their thinking. After reviewing their personal notes on one of the stages of the linear materials economy, they then draw circles around what they deem are three key points and indicate the most significant with an asterisk. The next step is to verbally share with the rest of their group and then present it to the class.

Before we proceed, it's important to point out the potential benefit that comes with providing private time to think, imagine, or visualize before making the shift to more overt forms of participation. Bennett and Rolheiser (2001) explain that opportunity for covert participation "increases success and a sense of safety because students can rehearse" before they speak publicly. Indeed, respecting silence and privacy is one of several important social skills that contribute to a positive and non-threatening learning environment.

Murphy and Hennessy (2001) also note that young learners need to feel secure during D&T collaboration. To that end, they stress that the purpose of the joint task, the desired outcomes, and the particular roles that students are to play must be clearly understood at the outset of the activity. Working together to achieve a mutual learning goal helps set an expectation that individuals take responsibility for their own learning as well as encourage others to do the same (Johnson & Johnson, 2009). This concept, sometimes referred to as individual accountability, is an essential element of effective group work.

8.2.5 Dialogue: Round Robin: Instructional Tactic Number 2

To encourage risk-taking without fear of failure or being ridiculed, teachers can (or better yet, with student input) set up explicit guidelines for constructive communication. This involves, for example, attentive listening, speaking in a courteous manner, and respecting different points of view. In a round-robin session, each member shares one of her pre-selected ideas to be entered in the middle space of the Place Mat and gives a rationale. The group space provides evidence of mutual agreement arrived at through a process of consensus-building (Medway, 1994).

8.2.6 Articulation: How Can I Share My Knowledge? How Can I Tell the Story to a Wider Community?

Large group presentation: The function of classroom discourse now shifts from figuring out, or *exploratory talk,* to *presentational talk* (Barnes & Todd, 1995). Students are told in advance that someone from each team will be randomly selected to represent and explain their group's collective thinking. When speakers are not chosen ahead of time, most students are more likely to stay actively engaged so they will be prepared if called upon. This is one more way to increase individual accountability.

8.2.7 Application: A PBL Approach: Redesign a Product for Increased Sustainability

An authentic school *problem* situation was presented to students: when poorly made commercial products cannot be repaired, they are thrown in the garbage. After a lively discussion, they watched a video to learn more about issues associated with the linear materials economy. Key concepts identified in the video were discussed in small groups, then presented to the whole class. As novice designers, they were tasked with addressing a personal need for a stable and durable locker space organizer that would not end up in the landfill.

8.2.8 Evaluation: Throwing Things Away—A Real Context for a Design Challenge

The students responded enthusiastically by wanting to share their own personal experiences and criticisms about these "cheap" and "badly designed" products. Judging from the level of excitement, it was evident that the expression "designed for the dump" (Leonard et al., 2007) had resonated with them. The notion that products were deliberately "made to break" (Slade, 2006) opened up the possibility for designing alternatives to an unsustainable cradle-to-grave approach to waste management. Through a storytelling structure that reveals how things come to be from distinct points of view (Solomon, 2002), employing authentic, open-ended questions (Nystrand, Wu, Gamoran, Zeiser, & Long, 2003), analysing commercial products, and discussion, students were able to develop critical understandings of present-day material practices in the fields of design and technology.

9 Concluding Remarks

We opened this chapter with the IPCC's startling report on the imminent threats of human-caused environmental changes – which young people in schools today will inherit. To grapple with these escalating problems, this generation will need leaders and citizens who can "think ecologically, understand the interconnectedness of human and natural systems, and have the will, ability, and courage to act" (Stone & Barlow, 2010). Contemporary design solutions need to include a critical awareness of closed-loop systems and renewable, restorative, sustainable supply chains. Understanding the material world can enrich students' and educators' knowledge of and ability to critically evaluate materials in design practice.

In our view, an important aim of D&T in its general educational role is to develop a critical awareness and understanding that materials are intricately linked with how we make and use things. When young people understand that our technological world is not a given but has been *designed* this way, they can begin to imagine how things could be done differently. To this end, we propose a question-think-learn approach whereby educators and their students engage in collaborative inquiry: to question and unpick assumed knowledge, to break things down into their basic parts or raw materials, to recognize and critique certain values that come into play (Rowell, 2004), and then decide for themselves whether these systems are desirable, ethically defensible, humane, and just. There is much work to do!

10 Resources for Educators

1. *Paper or Plastic?*
   - https://www.ted.com/talks/leyla_acaroglu_paper_beats_plastic_how_to_rethink_environmental_fol
   - http://www.wrap.org.uk/category/subject/carrier-bags
   - https://climatekids.nasa.gov/paper-or-plastic/
   - https://www.reusethisbag.com/articles/plastic-bag-bans-worldwide/
2. *Cotton T-Shirt*
   - Teemill. The journey: Follow the product journey, seed to shop: https://teemill.com/the-journey/
   - TedEd + sustainability: https://ed.ted.com/search?utf8=%E2%9C%93&qs=sustainability
   - The Business of Fast Fashion: https://www.youtube.com/watch?v=ZhkBfbwCzxc
   - Cotton Australia: https://cottonaustralia.com.au/
   - Ethical fashion guide: https://baptistworldaid.org.au/resources/2018-ethical-fashion-guide/
   - The true cost: https://truecostmovie.com/
   - White gold: the true cost of cotton: https://vimeo.com/1708935
   - Better cotton initiative: https://bettercotton.org/
3. *Entry Event – Table of Resources*

Circular economy explained	Format	Description
Closed loop systems		
Ellen MacArthur Foundation. (2013, Jan 7). The circular economy: From consumer to user. https://www.youtube.com/watch?v=Cd_isKtGaf8	Video animation (3:11 minutes)	Concept of performance-based circular economy is explained. Manufacturers assume responsibility for maintenance, repairs, and running costs. Premature obsolescence is eliminated and energy needs are reduced
Ellen MacArthur Foundation. (2011). Re-thinking progress: The circular economy. https://www.youtube.com/watch?time_continue=49&v=zCRKvDyyHmI	Video animation (3:48 minutes)	How can our waste build capital? Separate the circular economy into two types of materials: biological (avoids all toxic chemicals) and technical (polymers, metals, etc.); need to rethink the way we view ownership
Cradle to cradle		
C2C World – by EPEA. (2014, Apr 30). Introduction to Cradle to Cradle. https://www.youtube.com/watch?v=QMsF1P-_vWc	Video animation (5:49 minutes)	A short introduction to the cradle to cradle concept
Ellen MacArthur Foundation. (2012, Mar 21). Katja Hansen - The Cradle to Cradle concept in detail. https://www.youtube.com/watch?time_continue=5&v=HM20zk8WvoM	Lecture style presentation (23:20 minutes)	C2C concept applied to waste water management; profitable bio-nutrient recycling into crops; utilization of biomass energy
McDonough, W. (2005, Feb). Cradle to cradle design. https://www.ted.com/talks/william_mcdonough_on_cradle_to_cradle_design	Lecture style presentation (19:53 minutes)	Explains C2C template: put materials in closed cycles, analyze them for potential harmful effects, know how they are made and used in production. Eco-effective design criteria includes cost, performance, aesthetics, ecological intelligence, justice, and fun
Food		
EAT. (2015). Dame Ellen MacArthur: Food, health and the circular economy. https://www.youtube.com/watch?v=M6MLFJDddM4	Lecture style presentation (10 minutes)	Designs food waste out of the system; describe cases where biological materials are used to restore and regenerate land
Plastics		
National Geographic. (2018, May 18). Plastics 101. https://www.youtube.com/watch?v=ggh0Ptk3VGE	(6 minutes)	Explains how plastics are made, and problems associated with single-use synthetic plastics. Identifies alternative solutions (behavioural, scientific, industrial) to reduce future impacts

Circular economy explained	Format	Description
Textiles/fashion		
Ellen MacArthur Foundation. (2018, Nov 21). Make fashion circular. https://www.youtube.com/watch?time_continue=38&v=3iKHr-JnWYA	Video animation (2:02 minutes)	Three elements needed for the future: business models that extend use (sharing, swapping, rental, repair, resale); eco-safe and renewable materials; redesign and reuse materials
Teemill. [2019]. The journey: Follow the product journey, seed to shop. https://teemill.com/the-journey/	Webpage includes photos, video clips	Describes "high-tech" practices using renewable energy and natural materials for products and packaging
van Son, B. (2015, May 5). Revolution of circular economy in clothing industry. https://www.youtube.com/watch?v=Y_qmdC9cJr4	Stage presentation (9:18 minutes)	TEDxYouth@Maastricht. Describes how one company's fair trade organic cotton jeans are "leased," recycled, repaired, upcycled
Linear materials system		
Linear materials economy explained		
Leonard, A. (2007). The Story of stuff. https://storyofstuff.org/movies/story-of-stuff/	Video animation (21:24 minutes)	Explains how planned and perceived obsolescence, externalized costs of production, and consumption have been deliberately designed into a system. Other videos focus on microfibers, ewaste, bottled water, cosmetics
Acaroglu, L., & Kallincos, N. (2012, Jun 17). This is Your Life Cycle. https://www.youtube.com/watch?time_continue=4&v=01tF21O2iso	Video animation (4:58 minutes)	Humorous TV game show imitation format to teach why designers should assess product life cycles. Identifies three ecodesign strategies: disassembly, longevity, and de-materialization
Thwaites, T. (2010). How I built a toaster--from scratch. TED Salon. https://www.ted.com/talks/thomas_thwaites_how_i_built_a_toaster_from_scratch#t-632266	Stage presentation (10:45 min.)	Speculative designer tells amusing story about his project to reverse engineer a toaster. Determined to track down raw materials at their source, he shows how little we know about everyday mass-produced technologies
Food		
GRACE Communications Foundation. (2003). The Meatrix. Access: http://www.themeatrix.com/	Animated video (3:47 minutes)	Anti-corporate satire critiques problems of factory farming and promotes sustainable food and agriculture. Website provides a series of video and educator resources

Circular economy explained	Format	Description
Plastics		
PBS News. (2018, Sep 26). Plastic lasts more than a lifetime, and that's the problem. https://www.pbs.org/newshour/series/the-plastic-problem	Investigative report (6 minutes)	First segment of a series. Plastic pollution is a global problem that is threatening the lives of humans and animals. Different perspectives shown including economics and design innovations to reduce waste
The Story of Stuff Project. [2019]. The story of plastic. https://www.storyofplastic.org/watch	Mini-documentaries (4 minutes)	How plastic production pollutes small American towns; the Indonesian plastic bag diet; Manilas's zero waste neighbourhoods
Electronics		
Acaroglu, L. (2010, July 12). The secret life of things– Life pscycle-ology. https://www.youtube.com/watch?v=OKyrB2Jn2Zs	Video animation (5:59 minutes)	Cleverly written story anthropomorphizes a discarded phone undergoing regression psychotherapy to discover his origins. Alternative solutions are identified, including design for disassembly and resource recovery
Preshoff, K. [2018]. What's a smartphone made of? https://ed.ted.com/lessons/what-s-a-smartphone-made-of-kim-preshoff	Animated video (4:55 minutes)	Explains how material components pose risks to environments and communities. Petroleum use links phones to climate change. Solutions proposed: reclaim metals, reuse, recycle, refurbish
Pulitzer Centre. (2011). Congo's Bloody Coltan. https://www.youtube.com/watch?time_continue=35&v=3OWj1ZGn4uM	Documentary video (4:36 minutes)	A critical focus on the role a vital mineral for cell phones plays in Congo's civil war. Considers who benefits and who pays the price
Weigensamer, F., & Krönes, C. (2018). Welcome to Sodom. http://www.welcome-to-sodom.com/?page_id=289	Documentary film trailer (2 minutes)	E-waste from developed countries is illegally shipped to the world's largest electronic waste dump, Accra. Powerful images of a toxic landscape of trash, people, and animals
Textiles/fashion		
Chang, A. (2017). The life cycle of a t-shirt. https://ed.ted.com/lessons/the-life-cycle-of-a-t-shirt-angel-chang	Video animation (6:03 minutes)	Traces the environmental impacts of t-shirt production. Offers suggestions for reducing negative effects
National Public Radio. (2013). Planet money makes a t-shirt: The world behind a simple shirt. http://apps.npr.org/tshirt/#/title	5 vVideo shorts with voiceover narration	Uncritically celebrates high-tech advancements in American-grown cotton, genetically modified seeds, yarn production, and shipping. Supplemented with scroll-down information, including production costs of a t-shirt

Circular economy explained	Format	Description
Environmental Justice Foundation. (2008, Sep 11). White gold--The true cost of cotton. https://vimeo.com/1708935	(7:47 minutes)	Uzbekistan's cotton production has created devastating environmental, social, and economic problems: human rights abuses, forced child labour, corrupt politicians, complicity of Western businesses are some of the issues
CBC News. (2018, Jan 19). How fast fashion adds to the world's clothing waste problem. Marketplace. https://www.youtube.com/watch?v=elU32XNj8PM	Investigative report (22:23 minutes)	An exposé on textile waste. Identifies retailer greenwashing, "clothing deficit myth" (Cline, 2013), and undercutting of local textile industries in developing countries
Minute MBA. (2013, Mar 11). The Business of Fast Fashion. https://www.youtube.com/watch?v=ZhkBfbwCzxc	Illustration with voiceover (1:47 minutes)	Rapid production of clothing negatively impacts the market and the environment. Sustainable Clothing Action Plan tries to reduce waste and improve fair labour relations

References

Acaroglu, L. (2013). *Paper beats plastic? How to rethink environmental folklore*. TED Talks. https://www.ted.com/talks/leyla_acaroglu_paper_beats_plastic_how_to_rethink_environmental_folklore#t-13717

Acaroglu, L. (2019). *Design play cards*. https://www.leylaacaroglu.com/toolkits/design-play-cards. Accessed 3 June 2019.

Bain, M. (2018). The dirty industry of fast fashion is causing an environmental "emergency". *Quartz*. https://qz.com/1222569/fast-fashion-is-causing-an-environmental-emergency/

Barenblat, A. (2017). Fast fashion is a disaster for women and the environment. *Forbes. Quora*. https://www.forbes.com/sites/quora/2017/07/26/fast-fashion-is-a-disaster-for-women-and-the-environment/#3dbc974c1fa4

Barnes, D., & Todd, F. (1995). *Communication and learning revisited: Making meaning through talk*. Portsmouth, NH: Boynton/Cook.

Bennett, B. (2010). The artful science of instructional integration. In R. Marzano (Ed.), *On excellence in teaching* (pp. 64–91). Bloomington, IN: Solution Tree Press.

Bennett, B., & Rolheiser, C. (2001). *Beyond Monet: The artful science of instructional integration*. Toronto, ON: Bookation.

Bostrom, N. (2002). Existential risks: Analyzing human extinction scenarios and related hazards. *Journal of Evolution and Technology, 9*, 542.

Burbules, N. C. (1993). *Dialogue in teaching: Theory and practice*. New York: Teachers College Press.

Chapagain, A. K., Hoekstra, A. Y., Savenije, H. H. G., & Gautam, R. (2006). The water footprint of cotton consumption: An assessment of the impact of worldwide consumption of cotton products on the water resources in the cotton producing countries. *Ecological Economics, 60*, 186–203.

Cline, E.L. (2012). *Overdressed: The shockingly high cost of cheap fashion*. New York, NY: Penguin.

Collins, B. R. (2018). *Teaching students to disagree productively*. https://www.edutopia.org/article/teaching-students-disagree-productively. Accessed 2019.

Cradle to Cradle Products Innovation Institute. (2018). *Cradle to cradle certified*. https://www.c2ccertified.org/. Accessed 15 June 2019.

Dakers, J. (2005). The hegemonic behaviorist cycle. *International Journal of Technology and Design Education, 15*(2), 111–126. https://doi.org/10.1007/s10798-005-8275-3

de Vries, M. J. (2005). *Teaching about technology: An introduction to the philosophy of technology for non-philosophers*. Dordrecht, Netherlands: Springer.

Dillon, J. T. (1988). *Questioning and teaching: A manual of practice*. New York: Teachers College Press.

Ellen MacArthur Foundation. (2017a). Circular economy overview. https://www.ellenmacarthurfoundation.org/circular-economy/overview/concept. Accessed 12 June 2019.

Ellen MacArthur Foundation. (2017b). *Teemill circular supply chain Cowes, Isle of Wight*. Cowes, UK: Ellen MacArthur Foundation. https://www.ellenmacarthurfoundation.org/case-studies/an-open-access-circular-supply-chain-for-fashion

Elshof, L. (2009). Toward sustainable practices in technology education. *International Journal of Technology and Design Education, 19*, 133–147. https://doi.org/10.1007/s10798-008-9074-4

Fletcher, K. (2013). *Sustainable fashion and textiles: Design journeys* (2nd ed.). Florence, KY: Taylor and Francis.

Fleurbaey, M., Kartha, S., Bolwig, S., Chee, Y. L., Chen, Y., Corbera, E., et al. (2014). Sustainable development and equity. In O. Edenhofer, Y. Pichs-Madruga, E. Sokona, S. Farahani, P. E. K. Kadner, A. Seyboth, I. Adler, et al. (Eds.), *Climate change 2014: Mitigation of climate change (pp. 283–350). Contribution of working group III to the fifth assessment report of the intergovernmental panel on climate change*. Cambridge, UK: Cambridge University Press.

Fowler, S. R., Zeidler, D. L., & Sadler, T. D. (2009). Moral sensitivity in the context of socioscientific issues in high school science students. *International Journal of Science Education, 31*(2), 279–296. https://doi.org/10.1080/09500690701787909

Galbraith, D., Gue, D., Bullard, D., Bullard, J., Clancy, C., & Kiddell, B. A. (1999). *Sciencepower 7*. Toronto, ON: McGraw-Hill Ryerson.

Goldsmith, B. (2017). *Can going green help pick the slavery out of cotton?*https://www.reuters.com/article/us-india-cotton-slavery/can-going-green-help-pick-the-slavery-out-of-cotton-idUSKBN15P011. Accessed 5 June 2019.

Greene, M. (1995). *Releasing the imagination: Essays on education, the arts, and social change*. San Francisco, CA: Jossey-Bass Publishers.

Greenpeace. (2010). *The dirty secret behind jeans and bras*. http://www.greenpeace.org/eastasia/news/stories/toxics/2010/textile-pollution-xintang-gurao/. Accessed 2019.

Grummt, C. (2019). *International Association for Research and Testing in the field of textile and leather ecology Oeko-Tex® Oeko-Tex® Standard 100*. https://www.oeko-tex.com/en/business/certifications_and_services/ots_100/ots_100_start.xhtml. Accessed 15 June 2019

Hallett, C., & Johnston, A. (2010). *Fabric for fashion: A comprehensive guide to natural fibres*. London: UK Laurence King.

Hallett, C., & Johnston, A. (2014). *Fabric for fashion: The complete guide: Natural and man-made fibres*. London: Laurence King.

Hamilton, J. W. (2007). Enhancing learning through collaborative inquiry and action. *Design and Technology Education: An International Journal, 12*(3), 33–46.

Hendley, D., & Lyle, S. (1995). The potential of collaborative group work to increase pupil learning in the implementation of design and technology in the National Curriculum. *Curriculum Journal, 6*(3), 363–376. https://doi.org/10.1080/0958517950060306

Hennessy, S., & Murphy, P. (1999). The potential for collaborative problem solving in design and technology. *International Journal of Technology and Design Education, 9*(1), 1–36. https://doi.org/10.1023/A:1008855526312

Hodson, D. (2003). Time for action: Science education for an alternative future. *International Journal of Science Education, 25*(6), 645–670. https://doi.org/10.1080/09500690305021

IPCC, Intergovernmental Panel on Climate Change. (2018). *Global warming of 1.5°C. An IPCC special report on the impacts of global warming of 1.5°C above pre-industrial levels and related global greenhouse gas emission pathways, in the context of strengthening the global response to the threat of climate change, sustainable development, and efforts to eradicate poverty.* [Masson-Delmotte, V., P. Zhai, H.-O. Pörtner, D. Roberts, J. Skea, P.R. Shukla, A. Pirani, W. Moufouma-Okia, C. Péan, R. Pidcock, S. Connors, J.B.R. Matthews, Y. Chen, X. Zhou, M.I. Gomis, E. Lonnoy, T. Maycock, M. Tignor, and T. Waterfield (eds.)]. Geneva, Switzerland, Switzerland World Meteorological Organization. Retrieved from: https://www.ipcc.ch/site/assets/uploads/sites/2/2019/06/SR15_Full_Report_Low_Res.pdf

Johnson, D. W., & Johnson, R. T. (2009). An educational psychology success story: Social interdependence theory and cooperative learning. *Educational Researcher, 38*(5), 365–379.

Karana, E., Pedgley, O., & Rognoli, V. (2014). *Materials experience: Fundamentals of materials and design.* Oxford: Elsevier Science.

Kolodner, J. L., Camp, P. J., Crismond, D., Fasse, B., Gray, J., Holbrook, J., et al. (2003). Problem-based learning meets case-based reasoning in the middle-school science classroom: Putting learning by design(tm) into practice. *Journal of the Learning Sciences, 12*(4), 495–547. https://doi.org/10.1207/S15327809JLS1204_2

Landler, M., & Davenport, C. (2018, October 8). Climate warning hits silent wall on Trump's desk. *The New York Times.* Accessed at: https://www.nytimes.com/2018/10/08/us/politics/climate-change-united-nations-trump.html

Larmer, J. (2018, July 13). A tricky part of PBL: Writing a driving question. [Blog] Accessed 17 Aug 2018 at: http://www.bie.org/blog/a_tricky_part_of_pbl_writing_a_driving_question

Larmer, J., & Mergendoller, J. (2010). Seven essentials for project-based learning. *Educational Leadership, 68*(1), 34–37.

Larmer, J., & Mergendoller, J. (2015). Why we changed our model of the "8 Essential Elements of PBL". *Buck Institute for Education.* https://my.pblworks.org/system/files/documents/Why_We_Changed_8EEs_article.pdf

Larmer, J., Mergendoller, J., & Boss, S. (2015). *Setting the standard for project based learning.* Alexandria, VA: ASCD.

Laszlo, E. (2001). *Macroshift: Navigating the transformation to a sustainable world* (1st ed.). San Francisco: Berrett-Koehler Publishers.

Lave, J. (1988). *Cognition in practice mind, mathematics, and culture in everyday life.* Cambridge, MA: Cambridge University Press.

Leonard, A., Fox, L., (Director), & Priggen, E. (Producer). (2007). *The story of stuff.* [Video]. Accessed at: https://www.storyofstuff.org/

Levinson, R. (2009). The manufacture of aluminium and the rubbish-pickers of Rio: Building interlocking narratives. *School Science Review, 90*(333), 119–124.

Madge, P. (1997). Ecological design: A new critique. *Design Issues, 13*(2), 44–54.

McDonough, W. (2005). Cradle to cradle design. In *TED Talks.* https://www.ted.com/talks/william_mcdonough_cradle_to_cradle_design?language=en

McDonough, W., & Braungart, M. (2010). Cradle to cradle: Remaking the way we make things. New York: North point press.

McGrath, J. (2018). *Which is more environmentally friendly: Paper or plastic? HowStuffWorks.* https://science.howstuffworks.com/environmental/green-science/paper-plastic1.htm. Accessed 2019.

McTighe, J., & Wiggins, G. (2013). *Essential questions: Opening doors to student understanding.* Alexandria, VA: ASCD.

Medway, P. (1994). Language, learning and 'Communication' in an architects' office. *English in Education, 28*(2), 3.

Morgan, N., & Saxton, J. (2006). *Asking better questions* (2nd ed.). Markham, ON: Pembroke Publishers.

Murphy, P., & Hennessy, S. (2001). Realising the potential – and lost opportunities – for peer collaboration in D&T setting. *International Journal of Technology and Design Education, 11*(3), 203–237. https://doi.org/10.1023/A:1011286331859

Muthu, S. S., Li, Y., Hu, J.-Y., & Mok, P.-Y. (2009). An exploratory comparative study on eco-impact of paper and plastic bags. *Journal of Fiber Bioengineering and Informatics, 1*(4), 307–320.

Nystrand, M., & Gamoran, A. (1991). Instructional discourse, student engagement, and literature achievement. *Research in the Teaching of English, 25*(3), 261–290.

Nystrand, M., Wu, L. L., Gamoran, A., Zeiser, S., & Long, D. A. (2003). Questions in time: Investigating the structure and dynamics of unfolding classroom discourse. *Discourse Processes, 35*(2), 135–198. https://doi.org/10.1207/S15326950DP3502_3

Patagonia. (2018). *Materials and technologies*. https://www.patagonia.com/materials-tech.html. Accessed 15 June 2019.

Pavlova, M. (2005). Knowledge and values in technology education. *International Journal of Technology and Design Education, 15*(2), 127–147. https://doi.org/10.1007/s10798-005-8280-6

Pedgley, O., Rognoli, V., & Karana, E. (2016). Materials experience as a foundation for materials and design education. *International Journal of Technology and Design Education, 26*(4), 613–630. https://doi.org/10.1007/s10798-015-9327-y

Pedretti, E., & Little, C. (2008). *From engagement to empowerment: Reflections on science education for Ontario*. Toronto, ON: Pearson Education.

Peters, S. (2014). *Material revolution 2: New sustainable and multi-purpose materials for design and architecture*. Basel, Switzerland: Walter de Gruyter GmbH.

Petrina, S. (2000). The political ecology of design and technology education: An inquiry into methods. *International Journal of Technology and Design Education, 10*(3), 207–237. https://doi.org/10.1023/A:1008955016067

Preston, C. J. (2018). *The synthetic age: Outdesigning evolution, resurrecting species, and reengineering our world*. Cambridge, MA: The MIT Press.

Quinn, B. (2012). *Fashion futures*. London: Merrell.

Rowell, P. M. (2004). Developing technological stance: Children's learning in technology education. *International Journal of Technology Design Education, 14*(1), 45–59.

Sach, J., Deluna, R., & Wilson, S. (2017). *The story of microfibers*. [Video]. Accessed at: https://www.storyofstuff.org/movies/story-of-microfibers/

Savery, J. R., & Duffy, T. M. (1995). Problem based learning: An instructional model and its constructivist framework. *Educational Technology, 35*(5), 31–38.

Sheehan, B., & Spiegelman, H. (2010). Climate change, peak oil, and the end of waste. In R. Heinberg & D. Lerch (Eds.), *The post carbon reader* (pp. 363–381). Healdsburg, CA: Watershed Media.

Slade, G. (2006). *Made to break: Technology and obsolescence in America*. Cambridge, MA: Harvard University Press.

Solomon, J. (1992). The classroom discussion of science-based social issues presented on television: Knowledge, attitudes and values. *International Journal of Science Education, 14*(4), 431–444. https://doi.org/10.1080/0950069920140406

Solomon, J. (2002). Science stories and science texts: What can they do for our students? *Studies in Science Education, 37*(1), 85–105. https://doi.org/10.1080/03057260208560178

Steffen, W., Rockström, J., Richardson, K., Lenton, T. M., Folke, C., Liverman, D., et al. (2018). Trajectories of the earth system in the Anthropocene. *Proceedings of the National Academy of Sciences, 115*(33), 8252–8259.

Stone, M. K., & Barlow, A. (2010). Smart by nature: Schooling for sustainability. In R. Heinberg & D. Lerch (Eds.), *The post carbon reader: Managing the 21st century's sustainability crises* (pp. 409–418). Healdsburg, CA: Watershed Media.

Thompson, R. (2013). *Sustainable materials, processes and production*. London: Thames & Hudson.

Trimingham, R. L. (2008). The role of values in design decision-making in design and technology education. *Design and Technology Education: An International Journal, 13*(2), 37–52.
VCAA, The Victorian Curriculum and Assessment Authority. (2017). *VCE product design and technology 2018–2022.* https://www.vcaa.vic.edu.au/Documents/vce/productdesign-and-technology/ProductDesignTechnology_SD_2018.pdf
Watts, J. (2018). We have 12 years to limit climate change catastrophe, warns UN. *The Guardian* 8.
Werberger, R. (2016). *From project-based learning to artistic thinking: Lessons learned from creating an unhappy meal.* Lanham, MD: Rowman & Littlefield.
Whitehead Lohr, S. (2014, Aug 19). 5 Truths the fast fashion industry doesn't want you to know: *The Huffington Post.* https://www.huffpost.com/entry/5-truths-the-fast-fashion_b_5690575
Wiggins, G., & Wilbur, D. (2015). How to make your questions essential. *Educational Leadership, 73*(1), 10–15.
Wilkinson, T. & Bencze, J.L. (2015). With head, hand, and heart: Children address ethical issues of design. In K. Stables & S. Keirl (Eds.), *Environment, ethics and cultures: Design and technology education's contribution to sustainable global futures* (pp. 231–243). Rotterdam, NL: Sense Publishers.
Worland, J. (2016). How climate change unfairly burdens poorer countries. *Time Magazine.* https://time.com/4209510/climate-change-poor-countries/

Pedagogy for Technical Understanding

Torben Steeg and David Hills-Taylor

Abstract This chapter starts with a summary of what we understand by technical knowledge, encompassing materials, manufacturing and functionality. Various theories of learning are summarised included those based on constructivism, constructionism, situated cognition and cognitive psychology. Based on these understandings of how learning happens, the chapter explores approaches to teaching technical understanding that include a mixture of just-in-time and just-in-case approaches based on four broad kinds of activity: making without designing, designing without making, designing and making, and considering consequences. The use of chooser charts to help reconceptualise scientific knowledge is explored in some detail along with product analysis and the use of systems ideas to support children's designing and making in technical contexts.

1 Technical Understanding

We want to start by trying to be clear about how we use the words 'technical' and 'technological' in this chapter. Though there is often overlap between them in spoken English, we take 'technological' to refer to broad systems of technology (e.g. mobile phone technology, auto technology) and 'technical' to refer to the elements of a technology (e.g. aspects of electronics or structures) that together make it work. The chapter title refers to 'technical' understanding and that is its focus, but the nature of D&T education means that some of what we say will refer to supporting the development of technological understanding.

We should also mention the ideas of technical and technological 'literacy'; often used interchangeably, these terms relate to the ability of an individual to effectively use, respectively, technological understanding and technical skills, to achieve an

T. Steeg (✉)
Manchester Metropolitan University, Manchester, UK
e-mail: torbensteeg@me.com

D. Hills-Taylor
Attainment in Education Ltd, London, UK
e-mail: dit38@hotmail.com

P. J. Williams, D. Barlex (eds.), *Pedagogy for Technology Education in Secondary Schools*, Contemporary Issues in Technology Education,
https://doi.org/10.1007/978-3-030-41548-8_10

end. This chapter does not discuss such literacies specifically, although the desired result of an individual's increased technical and technological understanding is an allied increase in literacy.

A way to summarise the above is to describe the purposes of design and technology education as being to develop *technological/technical perspectives* and *technological/technical capabilities* (Barlex & Steeg, 2017a, p. 13) where perspective provides insight into how technologies work (understanding) and capability is designer-maker proficiency (literacy) in using that understanding to intervene in the made and natural worlds.

Technical understanding is one of a linked group of domains of knowledge that together supports the development of perspective and capability – other domains (including the understanding of making, design and critique) are explored in other chapters in this book. Following Owen-Jackson and Steeg (2007, p. 170–185), we take technical understanding to relate to pupils' understanding of how things work and how to make things work, including understanding about the materials they use, scientific and mathematical concepts related to materials and methods, and understanding of mechanical, structural and control systems.

Although the description above of technical understanding includes scientific and mathematical concepts, technical understanding is not the same as either mathematical or scientific understanding. More specifically, technical understanding relates to

> …an autonomous body of knowledge, identifiably different from the scientific knowledge with which it interacts. …technology, though it may *apply* science, is not the same as or entirely *applied science*.
>
> (Vincenti, 1990, p. 3–4, emphasis in original)

In particular, scientific understanding is focussed on describing the physical, chemical and biological (broadly defined) domains, and the aim of scientific endeavour is to understand and describe these domains in ever more precise ways. The purpose of technical understanding is very different, being to inform designer-maker capability and to underpin technological perspective. This means that very often scientific concepts have to be reworked or transformed to be directly useful as technical ideas – that is ideas that can be used in the service of practical capability (Layton, 1988, 1993).

It is common to categorise knowledge into factual, conceptual or procedural (Barak, 2013). Factual knowledge ('knowing that') "includes knowledge of terminology, names or symbols of components, technical vocabulary or names of processes" (ibid). Conceptual knowledge encompasses the abstractions that tell us how facts are related (classifications, theories, definitions, equations, models, etc.). In the context of technical understanding, factual and conceptual knowledge support thinking about technological activity (either in the context of designing and making or of critique), McCormick (1997).

Procedural knowledge is 'knowing how' to do things (i.e. how to carry out procedures) and includes, at a strategic level, knowing which procedures are applicable to a situation and when they should be employed (McCormick, 2002). Three catagories have been previously introduced; we are bringing in a fourth category of knowledge is tacit knowledge (Polanyi, 2009), which is closely linked to 'knowing how'

Table 1 Big ideas of design & technology

Knowledge of materials	Sources, properties, footprint and longevity
Knowledge of manufacturing	By subtraction, by addition, by forming, by assembly and with finishing
Knowledge of functionality	Powering, controlling, structuring and systems
Knowledge of design	Identifying peoples' needs and wants; identifying market opportunities; generating, developing and communicating design ideas; and evaluating design ideas
Knowledge of critique regarding the impact	For justice, for stewardship

(Ryle, 1949), being the knowledge that the holder is able to use effectively but may find difficult to explain.

To take a simple context as an example; it is one thing to know *that* wood can be cut with a saw and to even know which saws should be used in which situations, but it is a rather different thing to know *how* to cut effectively and efficiently with a saw. An experienced sawyer will know tacitly, from the feel and the sound of the sawing, how the sawing procedure is going and, for example, if a knot has been encountered, but may find it hard to articulate how she knows this.

To expand slightly the definition from above, technical understanding in design & technology should provide a foundation on which pupils can draw:

- When making design decisions
- To develop novel (for the pupil) and interesting ways for things to work
- To inform making
- To help them understand what might be going wrong when something isn't working
- To help them repair something that isn't working
- To improve the design of something so that it works better
- To develop a technological perspective by helping them understand the technologies they engage with in everyday life

Barlex and Steeg (2017b) have outlined the 'big ideas' of design & technology, shown in Table 1, that provide the breadth of technical understanding required for the above.

The technical understanding that is the focus of this chapter encompasses *materials*, *manufacturing* and *functionality* from Table 1. Other chapters in this book focus on designing and on critique.

2 Learning and Technical Understanding

A pedagogy for technical understanding clearly needs to respond to the nature of technical understanding as outlined above. It also needs to be attuned to what we understand about how learning happens, while being aware that much of this is contested territory and that very little research in cognitive psychology examines learning in D&T or related disciplines.

Perspectives on learning and teaching that are especially relevant to a practically oriented subject like design & technology are constructivism, constructionism, situated cognition and some of the recent findings of cognitive psychology and neuroscience.

2.1 Constructivism

Constructivism, including social constructivism, is a collection of ideas about how learning happens that makes a number of claims (Wood, 1998), including that:

- Learning is an active process of constructing knowledge in context rather than simply acquiring knowledge from others.
- Learners construct knowledge based on their personal experiences and hypotheses based on the physical and social environments that surround them.
- Learners continuously test these hypotheses through social negotiation.
- Because everyone's life experiences are different, each learner interprets their environment in their own way and builds their own individual set of knowledge structures. (Though there are a lot of commonalities, for example, everyone lives in a gravity well and that inevitably influences the way that they interpret their world.)
- The learner is not a 'blank slate' but instead brings past experiences and cultural factors to bear on each new learning situation.
- The models that learners construct should be seen as 'correct' in the sense that they account rationally for the experiences of an individual, but they may be based on restricted data or false assumption, so may not agree with scientific models.

An example of such an 'alternative conception' is that it is quite common for young children to believe that it is the motion of trees that causes the wind to blow (Trundle & Saçkes, 2015).

- These mental models are remarkably robust once they have been established because they are based on direct experience.
- Change to a well-established cognitive model will only happen if there is persuasive evidence that it is incorrect.

We want to be clear that constructivism is a theory of *how* learning happens that appears to be pretty robustly supported by evidence, not, in itself, a theory of teaching. Theories of teaching built on constructivism may or may not be effective and require their own evidence to support them (for critiques of certain supposedly constructivist teaching approaches, see, for example, Mayer, 2004; Kirschner, Sweller, & Clark, 2006). However, at the very least, constructivism suggests that teachers need to be aware that learners come to them with existing and robust cognitive schema in place that may, or may not, be aligned with the knowledge that the teacher is planning to teach. So, the job of teaching has to include seeking and using ways of teaching that allow learners to actively reconstruct and/or consolidate their schema based on new evidence that the teacher introduces.

2.2 Constructionism

Constructionism is a theory of learning built on constructivism by Seymour Papert and others at the MIT Media Lab (Kafai & Resnick, 1996; Papert, 1994). The core argument of constructionism is that people learn best (that is, develop robust cognitive schema) when they are making (constructing) something, be it a sandcastle on the beach or a theory in physics, because of the powerful interaction between thinking and action during making. In this view, learning is most powerful when two conditions apply; the construction environment is rich and there is ample opportunity to view the success of one's construction efforts (feedback).

Papert initially developed constructionism as an argument for putting children in control of computers through the use of LOGO, a programming language with a 'low floor and a high ceiling' (that is, easy to get started with but almost limitless in its applications (Papert, 1980)). His argument was that the child should control the computer, rather than the computer controlling the child (this was at a time when computers were increasingly being touted as teaching machines that would 'deliver' learning to children through tightly controlled on-screen curricula). This work soon grew to encompass robotics, especially with the LEGO 'Mindstorms' programmable brick, where the programming of the computer controlled not simply what happened on screen but also events in the real world. More recently, the Lifelong Kindergarten group in the Media Lab has continued to develop constructionist inspired approaches to supporting learning resulting in such technologies as the Scratch programming language[1] and the MakeyMakey controller[2] (Resnick, 2017).

Constructionism does not see itself tied to any particular discipline, but rather as an approach that can support deep learning in all subjects, in school and outside the formal curriculum. Brightworks School in San Francisco[3] organises its whole curriculum around constructionist approaches, and the maker education 'movement' takes constructionism as one of its foundational pillars (Steeg & Barlex, 2018).

However, in the context of formal (traditional) education systems, the link to work in design & technology should be clear; the nature of learning activity in design & technology means that it is the subject whose teaching approaches are most naturally aligned to constructionism. What is needed is for this link between constructionist learning theory and D&T practice to be made explicit to teachers, so

[1] https://scratch.mit.edu

[2] https://makeymakey.com

[3] http://www.sfbrightworks.org

that, when they are focussed on teaching technical understanding, they have a clear mental model of the ways in which the practical experiences that they are providing support the ways that their pupils are constructing new mental knowledge schemas.

2.3 Situated Cognition

Lave and Wenger (1991) built on social constructivist thinking by observing that, since learning takes place in a particular context (situation), it therefore becomes bound in learners' mental schema to that situation. This means that it is difficult for learners to transfer their knowledge from one domain to another. An implication for teaching technical understanding is that the contexts in which pupils meet technical knowledge need to vary so that as their understanding develops it also develops in a range of contexts, increasing the likelihood that, in time, they will be able to apply their understanding to novel contexts.

Evidence from neuroscience confirms that "the development of our brains and the storage of information in them is hugely context-dependent" (Hobbiss, 2018), and explains why transfer of understanding between domains is so hard for humans.

2.4 Cognitive Science

In recent years, findings from cognitive science have started to make a serious impact on the way that we think about how learning happens. In some cases, long-cherished beliefs have been challenged, in others existing understandings of cognition have been put on a firmer footing. Willingham (2010) outlines a number of cognitive principles that he says are

> …as true in the classroom as they are in the laboratory and therefore can reliably be applied to classroom situations. (p. 2)

There is only space here to discuss a few of these principles, one of which is that people are naturally curious, but we are not naturally good thinkers; unless the cognitive conditions are right, we will avoid thinking. To many teachers, this may seem like a statement of the obvious; however, it seems useful to have this everyday intuition backed by scientific understanding of the brain's working; understanding that this is the natural condition of humans rather than unnecessary obtuseness in children. The key phrase in this principle is '*unless the cognitive conditions are right*'. A crude summary of what 'right' means here is that tasks have to be difficult enough not to bore, but not so hard that they are unsurmountable. A task that is too easy doesn't satisfy curiosity or provide pleasure in the solution. A task that is too hard may stimulate curiosity, but if it can't be achieved, there is no reward. A just right (goldilocks) task stimulates the curiosity that helps drive motivation and is achievable, so it also provides the pleasure that comes with success. Although this again sounds obvious, it takes considerable thought and practice to devise tasks at the goldilocks level of difficulty that also support the particular learning that is desired.

This includes being clear about what cognitive work you want your students to do; what you want them to be thinking about and ensuring that this work is achievable within the cognitive limits of working memory and long-term memory.

Another principle is that factual knowledge must precede skill. Many find this counterintuitive; it is common to suggest that the most important thing we can do in education is to develop twenty-first-century skills such as creativity, critical thinking or problem-solving. However, it is easy to show that none of these skills can be practised in a vacuum; for example, both authors can be creative in the design of electronic systems because we have a large reservoir of knowledge to draw on, but only one of us (Hills-Taylor) can be musically creative, because the other lacks any serious musical knowledge, despite being an avid consumer of music. This underscores why it is important for technology educators to teach technical understanding (and, more broadly, the subject's big ideas); without it, students will not have the capacity for creativity in this arena. McCormick (2002) writing in the context of D&T learning also argues that skill in problem-solving is dependent on domain knowledge.

A further principle is that memory is the residue of thought. This means paying careful attention in the planning of a lesson to what it is that you want students to actually think about in a task you set them, as that is what they will remember.

The final principle noted here is that it is virtually impossible to become proficient at a mental task without extended practice. In this context, it is worth noting that Nuthall (2007) found that students need to be exposed to an idea between three and five times, in different ways, for them to truly commit it to memory. This appears to be a pretty robust finding, distilled from 40 years of highly focused, classroom-based studies of what actually happens in lessons. To be more precise, what students need is the experience of moving information from short-term to long-term memory and then being required to reinforce that long-term memory through retrieval. Spacing these retrieval experiences is beneficial (Karpicke, 2017).

In what follows, we will try to tease out some ideas about how these might apply to the teaching of technical understanding.

3 Teaching Technical Understanding

A tactical question in the design of learning experiences for design & technology has long been how best to balance a focus on practical designing and making activities with the need for adequate knowledge to support such activities.

Should you front-load the curriculum with knowledge that you think (hope) will be relevant to the task? Doing so ensures content 'coverage' (important when one has, for example, an examination specification to complete), but runs the risk that pupils will not transfer the understandings they have gained into the practical context. Additionally, if the designing and making context is reasonably open, there is a risk that pupils will pursue solutions that don't actually make use of the knowledge they have been taught and they may come to see the front-loaded teaching as irrelevant.

Gershenfeld (2005) describes this kind of front-loaded curriculum as 'just-in-case' teaching and contrasts it with 'just-in-time' teaching where students access the technical knowledge they need for carrying forward a design and make activity at the point they need it. This is a compelling distinction that is clearly intended to favour just-in-time learning. However, it is worth noting that Gershenfeld is describing work with graduate students and thus has the dual advantage over secondary school teachers that, first, his students are (as graduates) experienced in taking control of their learning and, second, he is not bound to ensuring his students learn the content of an externally imposed curriculum. The principle that factual knowledge must precede skill suggests that children shouldn't be sent off into open designing and making tasks unless there is either an adequate knowledge base to support the task, or a clear plan for providing that base within the structure of the task.

The possibility of adopting a mixture of just-in-time and just-in-case approaches is supported by the increasingly accepted idea that four broad kinds of activity are required to make up an appropriate pedagogy for design & technology: *making without designing*, *designing without making*, *designing and making*, and *considering consequences* (Barlex & Steeg, 2017a).

The first two of these, making without designing and designing without making, provide the opportunity to develop a wide range of factual, procedural and conceptual technical understanding in a practical context. These activities can be quite short, for example, contained within a single lesson or even as just one element of a lesson, aiming to efficiently introduce pupils to new concepts or skills; these are what the English National Curriculum for some years described as 'focused practical tasks' and the Nuffield D&T Project (1995) called Resource Tasks. Or they might be longer, encompassing a series of lessons in which pupils have a more immersive experience of making or designing; an example of this is the Young Foresight materials (Barlex, 2004) which provide an extended experience of designing without making in the context of novel technologies.

Importantly, these two sets of experiences are, clearly, not full-blown designing and making activities. A clear view of the distinctions between these activities enables teachers to plan for open designing and making experiences (that is, experiences where the context for designing and making is not narrowly defined, and pupils are able to make genuine decisions about the path of their response), supported by shorter activities that teach the understanding required. Some of these activities (in particular longer ones and those focussed on core concepts or skills) may be best met in preparation for the open task (just-in-case) while others can be accessed as the need for the knowledge is revealed in the course of the task (just-in-time).

From the big picture of how best to plan a curriculum to enable the acquisition of technical understanding, we turn to some practical approaches that can support this learning and that take seriously the nature of technical understanding. We describe three approaches to teaching technical understanding that we have found useful in teaching design & technology. First, the use of methods to encapsulate (reconceptualise) explicit scientific knowledge to make it useful technical knowledge. Second, product analysis where examining closely the designing and making of others reveals the application of technical knowledge in the development of products and systems. Third, the use of systems thinking to focus attention on the relevant level of detail when designing and making. These are not the only ways to approach the teaching of technical understanding, but, in our experience, they have particular utility.

These teaching approaches provide opportunities, between them, for the learning of procedural, factual and conceptual technical understanding. Tacit understanding grows from direct experience, which suggests that the teaching of design & technology needs to include providing pupils with lots of practical experience in which they develop mastery over an increasing range of both materials and processes. The nurturing of a comprehensive range of such understanding needs planning for rather than leaving to chance. But, since a pedagogy designed to develop procedural and conceptual understanding in design & technology will, of necessity, be highly practical, opportunities to also develop tacit understanding can be woven into the planning with low additional costs on time and resources.

3.1 Reconceptualising Scientific Knowledge

We cannot expect school pupils, novices in the practice of design & technology, to do the work of reconceptualisation by themselves. We need to both do this work ourselves and present the result in a way that is of practical use to a young person engaged in designing and making.

One approach, popularised through the Nuffield Design and Technology project (1995), is to create ‘chooser charts’ that encapsulate some aspect of scientific knowledge and present it in such a way that it can help with making design decisions. For example, the Key Stage 3 mechanisms chooser chart (see Fig. 1) summarises a lot of detailed knowledge about mechanical systems under the simple structure of implicitly asking ‘what kind of change in motion do you want to achieve?’ and then showing possible mechanisms that achieve this kind of change. This doesn’t get children off the hook of needing to understand technical ideas for two reasons. First, to make use of the chooser chart the pupil needs to understand that there are different kinds of movement and that one role of mechanisms is to change one kind into another. Second, they are presented with a number of alternatives from which to choose, which may require further investigation, supported by a suitable focussed task.

The net result of this work, coupled with the actual implementation of the mechanism within what they are making, is that pupils will have deepened their understanding of this particular class of mechanism and will also have learned that they are capable of using technical information to inform their decision-making, at least in this technical area. In terms of Willingham’s (2010) cognitive principles, used right, chooser charts provide factual knowledge to support the development of skills and also set up the cognitive conditions to support thinking – the thinking that underpins the development of understanding. In some cases, they support the development of understanding by providing a concrete context to build on.

Chooser charts are also intended to support design decision conversations between pupils and teachers. For example, pupils might have their own copies of a chooser chart, which they can annotate as they discuss possibilities with their teachers. This could be very helpful in evidencing the technical design decisions the

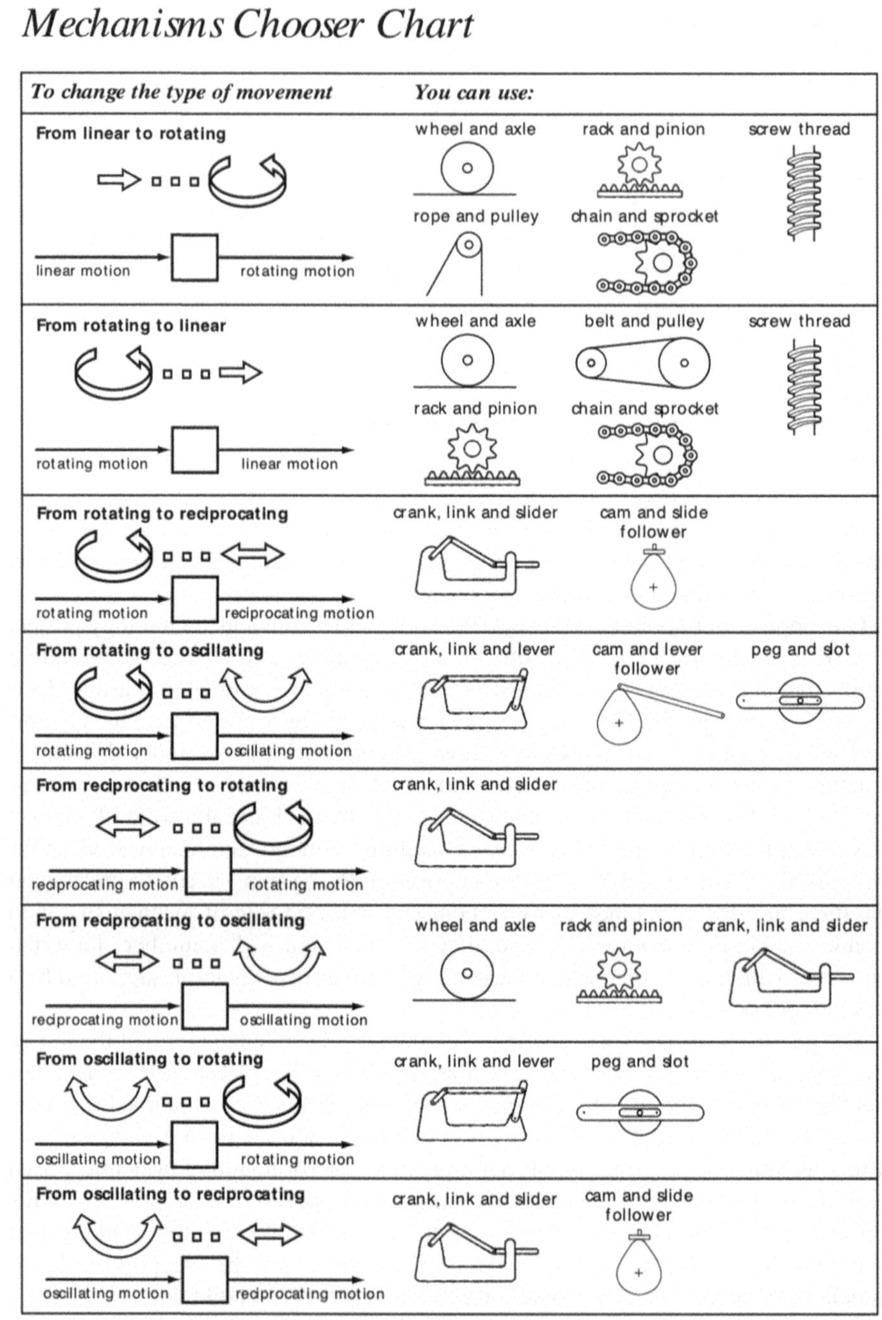

Fig. 1 Nuffield D&T Key Stage 3 mechanisms chooser chart (page 1 of 2) (The full suite of Nuffield KS3 chooser charts is available at https://dandtfordandt.wordpress.com/resources/nuffield-ks3-dt-resources/chooser-charts/)

LEARN-IT: Fasteners

Fasteners for metal

	Sheet metal screw	Pop rivet	Machine screw	Hex bolt
When to use this fastener	When not using a nut for joining thin-to-thin Joining thin to thick materials	Joining relatively thin metal	General assembly of **medium** pieces of **metal or wood**	General assembly of **larger** pieces of **metal or wood**
Composition	Steel	Aluminum or steel (Use with similar metal to reduce corrosion)	Steel	Steel
Required preparation	Clearance hole through the first layer, pilot hole in the second layer	Clearance hole	Tapped hole	Clearance hole
Key characteristics	Self-tapping tip **No countersink** required **Repeated assembly wears the hole**	Semi-permanent *the rivet must be **destroyed to remove** it* Very **low head profile** **Access to other side is not required**	Typically **fully threaded** Uses a **threaded hole** *or a nut – **can be used as a bolt***	Some not fully threaded Uses a **nut** *or a threaded hole – **can be used as a machine screw*** **Sizes vary immensely**

Fig. 2 A section of the Learn-IT: fasteners resource from MIT's D-Lab. (From https://d-lab.mit.edu/news-blog/blog/d-lab-pilots-learn-it-self-teaching-tools)

pupils make, for example, in the English Design & Technology GCSE's[4] new contextual challenge.

Another example of this kind of encapsulation of scientific knowledge for technical purposes comes from MIT's D-Lab which has made available, as high-resolution PDFs, three 'Learn-IT' kits that are, in essence, chooser charts of a high graphic quality; one each on Fasteners, Adhesives and Material Selection[5] (see Fig. 2). These are designed to be hung on a workshop wall and are more detailed than the Nuffield chooser charts. They are:

[4]The General Certificate of Secondary Education (GCSE) is a qualification generally achieved by young people aged 16 years

[5]https://d-lab.mit.edu/d-lab-learn-it-material-selection
https://d-lab.mit.edu/resources/publications/d-lab-learn-it-adhesives
https://d-lab.mit.edu/resources/publications/d-lab-learn-it-fasteners

> …self-guiding resources that provide an integrated introduction to basic mechanical design elements; they bridge the gap between superficial how-tos and super-detailed technical guides. They give people the right vocabulary to ask targeted questions in the workshop and online, while outlining detailed tips and explanations of physical phenomena driving how different mechanisms, tools, materials, and fasteners work. People are provided with enough information to critically select the right material, adhesive, or tool for their project.

One clear advantage of these devices is that they support both just-in-case and just-in-time learning, allowing the teacher to judiciously deploy them to best meet the demands of learners and the curriculum.[6] The second advantage is that they can be used to support resource management. That is, a teacher can give an open brief but only include the resources that they have available on the chooser chart, thus allowing choice but preventing pupils from continually asking for resources that cannot be provided.

However, they do introduce a need to build in the development of broader strategic understanding that allows pupils to successfully use these kinds of encapsulated

Chooser Charts Case Study

One of us, Hills-Taylor, created an open brief for his year 9 pupils based around the use of programmable microcontrollers. Chooser charts were used at each stage of the design process – investigation, design, manufacture and evaluation.

The first use was in the exploration of the context for the product to be designed. Pupils were given chooser charts for available input devices, output devices, driver circuits (such as transistors, Darlington pairs, etc.) and sections of code that could be used to produce a program for the microcontroller (e.g. creating sequences, pulsing, time delays and logic-based functions) (see Fig. 3).

Pupils used these initially to help them to create a brief based on the context of home lighting. That is, they were shown what technology they had available to them and were asked 'what could we create with this?' For some classes where classroom management could have become an issue with all pupils working on individual briefs, the decision on the design problem that they would tackle was taken as a group. For others, the pupils were allowed to make their own individual choices.

The second use of chooser charts was for pupils to decide which input, output, driver and output sub-systems were necessary to realise their solution. They were used to support both initial ideas generation and development of the chosen design. Some pupils were able to independently write code for the microcontroller, but for others the relevant chooser chart allowed them to 'piece together' an appropriate program. The teacher discussed these options at each stage with the pupils, ensuring that their choices were feasible and, if not, encouraging them to explore other potential ways forward.

[6] From https://d-lab.mit.edu/news-blog/blog/d-lab-pilots-learn-it-self-teaching-tools

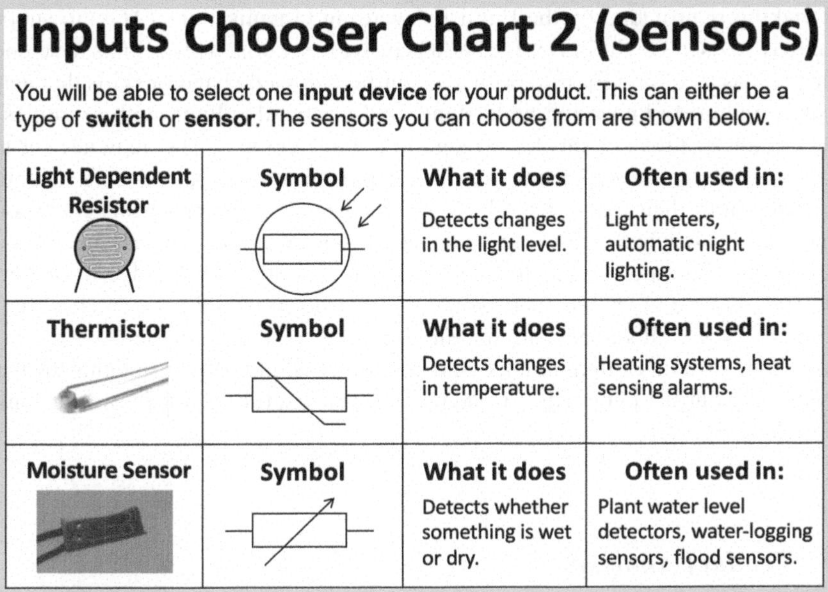

Inputs Chooser Chart 2 (Sensors)

You will be able to select one **input device** for your product. This can either be a type of **switch** or **sensor**. The sensors you can choose from are shown below.

Light Dependent Resistor	**Symbol**	**What it does** Detects changes in the light level.	**Often used in:** Light meters, automatic night lighting.
Thermistor	**Symbol**	**What it does** Detects changes in temperature.	**Often used in:** Heating systems, heat sensing alarms.
Moisture Sensor	**Symbol**	**What it does** Detects whether something is wet or dry.	**Often used in:** Plant water level detectors, water-logging sensors, flood sensors.

Fig. 3 The sensors chooser chart

After manufacturing, chooser charts were used to aid in evaluation of the prototype. Pupils tested how their chosen sub-systems worked and compared the results of this with their intended function. They were also able to use the chooser charts to assist in suggesting improvements to their product, for example, using LEDs instead of incandescent lamps to reduce energy consumption. This also ensured that improvements focussed on the specific technical aspects of the system design, rather than the vague, general responses that can often be found in product evaluations.

information, allowing technical understanding to be built through the purposeful use of technologies in practical design and make tasks, supported by focussed tasks that use the encapsulated knowledge.

3.2 *Reflection*

In the context of an actual design and make scenario that you use with students, think about the technical knowledge that they need to have to inform their designing and making. Think about how you could reframe this knowledge as a one or more chooser charts.

A second route to contextualising the use of scientific knowledge in designing and making is provided by the English Awarding Organisations (AOs) to support the newly revised GCSE in D&T. The AOs provide schools with a 'links to science' table that has a column for the specific scientific knowledge and/or skills that learners are required to know and understand, and a second column with examples of potential applications of this knowledge in a D&T context. The audience for this chart is teachers, to help them map the opportunities for introducing scientific knowledge into their units of work. A student-friendly version of this chart could be created by teachers, allowing students to look at their own learning and see where there are opportunities for them to apply scientific knowledge in that context. The purpose of such a chart would be twofold; it would act as a checklist of where and how well students are making the links to science in their work, and, if constructed with this aim in mind, could be a way to highlight for students the ways in which scientific knowledge is adapted to be useful in informing designing and making.

3.3 Product Analysis

Carefully scrutinising the products that fill the made world is a long-established teaching tool in design & technology education; indeed, for a while, it was a required feature of the English National Curriculum for D&T. Studying the results of others' designing and making can be used to focus students' attention on a wide range of the features of a product from the aesthetic to the technical. The key idea here is that of focus; there is a limit to what pupils, especially those new to product analysis, will learn simply by looking at or even dismantling a product. A fruitful exploration of a product needs to be guided so that pupils are actively engaged and have their attention directed to the features of the product that the teacher wishes them to learn about. This suggests the need for a commentary to guide the investigation, accompanied by questions that prompt reflection, discussion, practical activity (such as modelling) or further research.

One example of this approach is the James Dyson Foundation's Engineering Box. This delivers to schools

> …a Dyson DC39 vacuum cleaner, Carbon Fibre Turbine Heads and Tangle-free Turbine Heads. Students take these apart, using the screwdrivers provided, to better understand how the technology works. The box also contains a comprehensive teacher's pack, lesson plans, videos and posters. Schools loan the box for four weeks, free of charge – with delivery and collection included. (See Fig. 4)

However, not all products can be brought into the classroom or workshop for actual handling, for example, products and systems from other times and cultures as

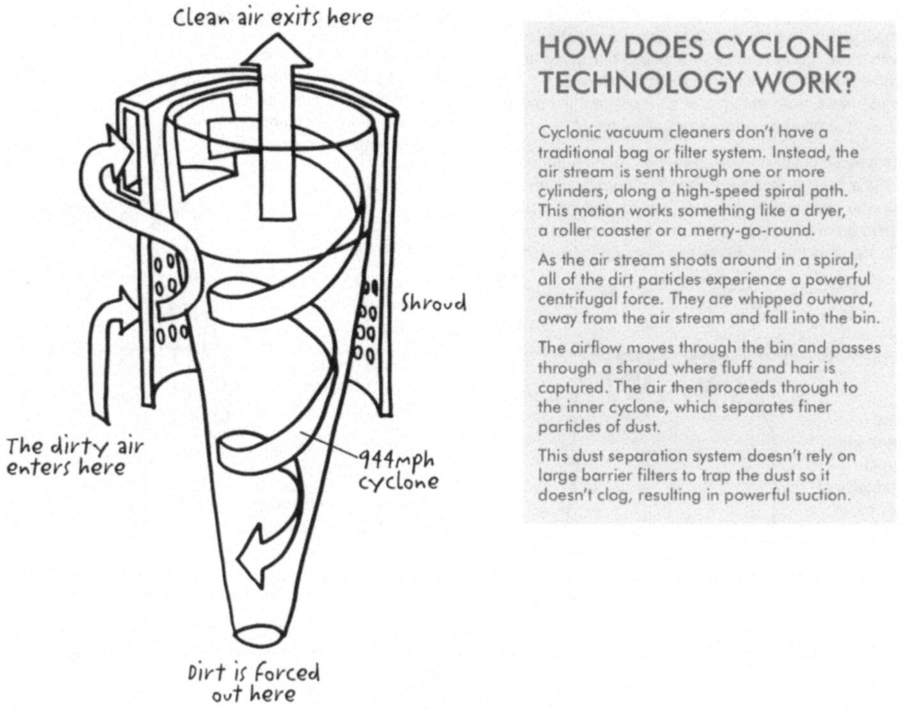

Fig. 4 Extract on product analysis from the Engineering Box's Teacher's Pack. Image copyright and credit: James Dyson Foundation

well as those too large, small, dangerous or expensive to put in pupils' hands.[7] For these, case studies that provide a clear description of the product or system, also accompanied by prompts for thinking and activity, are needed. The Furby case study from The Nuffield Key Stage 3 materials illustrates some of these features[8] (see Fig. 5).

Clearly, selection of an appropriate product and associated support materials by a teacher is important, to ensure that pupils are focussed on the features of the product that are pertinent to the learning aims.[9] Again, using Willingham's (2010) principles, well-focussed guidance for product analysis promotes deep thinking to support the development of understanding in a concrete context.

As with any other activity in practical D&T, product disassembly needs proper prior risk assessment and especial care needs to be taken with mains-powered devices.

---

[7] From https://www.jamesdysonfoundation.co.uk/resources/engineering-box.html

[8] From https://www.jamesdysonfoundation.co.uk/content/dam/pdf/FOR%20WEB%20Engineering%20Box%20Teachers%20Pack%20inside_Single%20Pages_Updated%20New.pdf

[9] The full suite of Nuffield KS3 case studies is available at https://dandtfordandt.wordpress.com/resources/nuffield-ks3-dt-resources/case-studies/

In the mid-1990s virtual pets were popular with children but one toy designer was not satisfied with them. Dave Hampton, a skilled hardware and software engineer who lives in the Californian wilderness, liked the idea of an intelligent toy pet but says, 'I wanted a cuddly, loving little thing'.

As a result of Dave's dissatisfaction, in the Autumn of 1998 a new toy was launched, 'The Furby', 13 cm of fur with gremlin ears, big eyes, a beak and the ability to move them as well as to talk and bounce up and down. On top of all this, the Furby can respond to touch, light and, perhaps most radically of all, other Furbys.

2

A Furby; designed like a cuddly toy

So how does a Furby work?

Under the fur it is a simple robot and, like any other computer-based system, it has a set of inputs (senses), a set of outputs (muscles) and, between them, some processing power (the nervous system).

The main inputs are:

- a switch on the back to detect stroking and petting;
- a switch on the tummy to detect tickling;
- a switch on its tongue to detect feeding;
- a tilt switch so it knows when it is upside down;
- a light sensor between its eyes so it can tell when it is dark;
- a microphone through which it detects sounds.

Naked Furby – what Furby looks like under the fur

It also has some internal switches that allow the micro-controller to monitor the position of its various outputs.

The main outputs are:

- a reversible motor that controls: the eyelids, the mouth opening and closing, the ears, the jumping motion;
- a loudspeaker through which it speaks;
- an infra-red sender so that it can 'speak' to other Furbys.

The control of the outputs is a cunning mixture of computer and mechanical control with at least two custom chips (one being a micro-controller) but just one motor to control all of the movements. This keeps the cost and size down, since motors are nearly as expensive as micro-controllers but are bulkier and use more battery power.

The motor controls each movement through a cam. By careful design of the shape of these cams and precise control of the motor, each motion appears to be independent. There are about 300 unique combinations of ear, eye, mouth and jump movements.

3

> **Q** ***Question***
>
> **3** Here are some of the combinations of motion of just two of a Furby's outputs:
> - the mouth can open and dose while the eyes are shut;
> - the mouth can open and close while the eyes are open;
> - the eyes can open and close while the mouth is shut;
> - the eyes can open and close while the mouth is open;
> - the mouth and eyes can open and close together.
>
> Can you think of other possibilities?
>
> **4** Try to design a pair of cams that could control the eyes and the mouth of a toy in the way described. Remember that the motor can be reversed and controlled quite accurately.

Marketing Furby

The makers of Furby want to make sure that they remain popular toys for some years. As well as designing them to be cute they have included:

Fig. 5 Extract from the Nuffield D&T Key Stage 3 Furby case study (pp. 2–3 of 4)

Product Analysis Case Study

An example of product analysis used in the classroom by one of us, Hills-Taylor, is an activity based around product disassembly. The aim of this year 10 product design lesson was to understand how mobile phone designs have evolved over time. This was part of a broader unit of work considering the wider issues that affect the design of products.

Pupils were split into groups of three and each group was given a mobile phone of a different age. Care was taken to ensure that each phone could be assembled and dissembled safely. For example, many modern smartphones cannot be taken apart due to the way that their casings are designed. Issues such as potentially dangerous components and the power supplies were also considered. In the case of very old phones, the battery was removed to prevent potential safety issues due to leakage.

Each group disassembled the main sections of the phone that they were given. At each stage of this disassembly, they drew a picture of the parts, labelled them and described what they thought their function was. This formed a basis for classroom discussion later in the lesson. ACCESS FM (aesthetics, cost, customer, environment, size, safety, function, materials/components) was used as a tool for analysing the product as a whole. This ensured that pupils considered a broad range of design issues when analysing the product. As an extension activity, some learners also produced a formal assembly drawing of the phone.

Each group reported back to the class on their findings, again using ACCESS FM as a way of focussing in on the specific design aspects. Groups were asked to place their phones in the order that they thought they were produced, thus creating a product timeline. This then formed the basis of further discussion surrounding the changes made over time and why they occurred.

4 Reflection

Think about how a focus on one or more existing products could help reinforce the technical understanding aspects of one of your units of work. This could be via physical engagement with actual products or through the use of a written case study. Think about the practical aspects of how you might focus the student's engagement with the product analysis to maximise learning.

4.1 Systems Thinking

Systems thinking is an approach to reconceptualising scientific knowledge that has particular power. Steeg (2000) provides a detailed background to the use of systems thinking for D&T, and a full exploration of systems thinking is provided elsewhere in this book. The focus here is on the use of systems thinking to help students

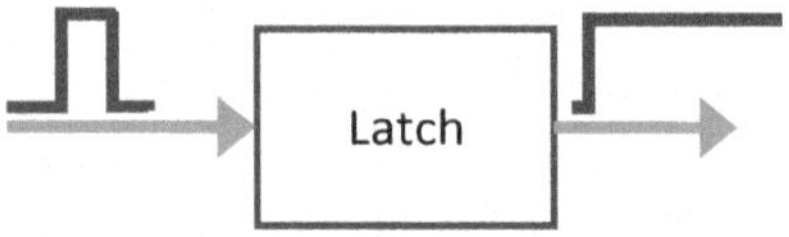

Fig. 6 A system-level diagram of a latch

develop technical understanding. Chapter 4 in this book, "Making the Invisible Visible: Pedagogies Related to Teaching and Learning About Technological Systems" has much more to say about the usefulness of systems thinking within Technology Education.

The core idea in a systems approach is that detailed concepts are abstracted to a 'higher' level, that is, for some purposes, more useful because it hides the detail. For example, take the idea of a latch; a device that when 'triggered' changes its state and remains in that changed state until reset.

Figure 6 shows a system-level diagram of a latch (these diagrams are often referred to as block diagrams). There are two key elements of the diagram: the arrows and the rectangular block. The arrows represent a signal going into the block (on the left) and a signal leaving the block (on the right). The block itself represents a 'function' that in some ways changes the input signal to produce the output signal. Above the signal arrows are icons representing the signal; in this case, the input signal momentarily changes state causing the output signal to change state and remain in its new state.

A key point to note is that this diagram says nothing at all about how this change from the input to the output signal is achieved or even what kinds of signal are involved. It could be achieved mechanically (e.g. the latch on a garden gate or on a door), or in various ways electronically including with discrete components (transistors, capacitors, thyristors, etc.), with integrated circuits (for example, a 555 timer or one of a number of types of flip-flop) or using a programmable microcontroller.

This level of abstraction may seem a bit impractical, but what it allows is a transformation in the way that electronics, for example, can be taught. In the absence of a systems-based approach, pupils are required to learn a great deal of rather theoretical technical information before being able to engage in any kind of design of electronic circuits. This might be characterised as having to learn quite a lot to be able to do just a little (or, in Papert's (1980) terms, having a high floor and a low ceiling). Using a systems approach allows pupils to approach the design of a circuit at the level of what it should *do* (e.g. detect a changing light level) without having to engage, at the design stage, in relatively complex scientific and mathematical ideas of *how* it should be done. Using this approach, a relatively small amount of technical knowledge can allow pupils to tackle a wide range of problems in a range of ways (so it has a low floor and a high ceiling).

Once again, we have a tool that satisfies some of Willingham's (2010) cognitive principles; they provide factual knowledge to support the development of skill and support the kinds of thinking that are likely to underpin the development of understanding.

One can conceptualise systems thinking used in this way as a kind of optical instrument that allows the pupil to zoom out from the (often too complicated) detail to approach the work of designing (especially) at a more generalised level. And then zoom back in again when necessary, for example, for implementation purposes. How this might look is explored in the following case study.

Systems Thinking Case Study

A common Key Stage 3 electronics project is a security alarm based around a simple transistor switching circuit. This might typically have an input sub-system consisting of an LDR or other sensor as part of a potential divider and a buzzer as output device.

One of us, Hills-Taylor, had identified several issues with the way that this unit was taught to the year 8 pupils in his school. The teaching was based on a component-by-component approach. Pupils would spend the first three or so lessons learning in depth about how potential dividers, transistors and buzzers operated, before being given the circuit PCB to construct. This teacher-led, content-heavy approach was demotivating for many pupils who struggled with the sheer volume of theoretical knowledge and lack of immediate practical application. In addition, because pupils had no design input into how the circuit was designed, they had little ownership over the completed product. To make matters worse, some pupils had ideas as to how the circuit could be improved but had no outlet for implementing their thoughts.

As a result, the teacher made several changes. First, he replaced the previous brief with a range of broader starting points, as shown below.

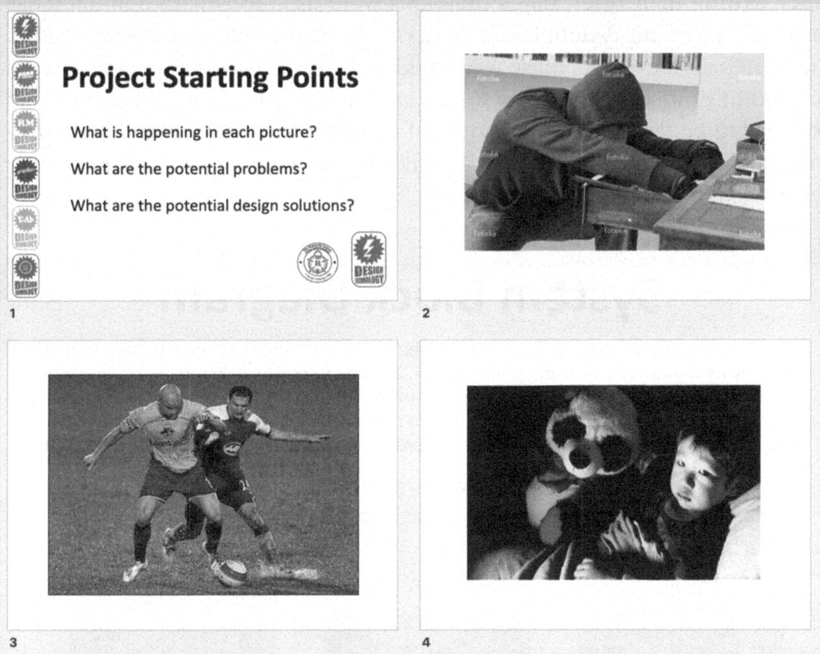

This immediately gave the pupils the chance to take ownership as they could create their own design brief.

Design and Make it!

Electronic Engineers have to **identify**, as well as **solve** problems that occur in everyday life. During this project you will learn how to **identify** possible ideas for **projects** and then **design and make** an appropriate **solution**.

PROJECT STARTING POINTS:

1) Look at the **'Project Starting Points'** PowerPoint and the images on each slide. Each picture shows a real life situation taking place.
2) Think about what is happening in each picture. What are the potential **problems** in each picture? What might be the potential **solutions?**
3) Using the **Project Starting Point Analysis** slide complete an analysis of two starting points that grab your interest. Which project starting point do you want to **develop** into a project for you to complete?
4) A **design brief** is a **short statement** about what your project is going to be about. Use the **Design Brief** slide to write an appropriate brief for the **project idea** that you have chosen.

After writing their brief, block diagrams were used to generate initial ideas for solutions at the system level – in terms of their input, process and output sub-systems. In essence, these were used in the same way that sketches might be used to draft initial designs for a materials-based product. As pupils did not need to be taught a large amount of technical knowledge to do this, they were able to design a variety of different ideas incorporating features such as timing, latching and even counting.

System Block Diagram

Electronic Engineers usually represent system ideas as a **block diagram. Inputs** turn a **real world signal** into an **electronic signal**. They are usually switches and/or sensors. The **process** in this case is the **PIC microcontroller** and is like the **'brain'** of the system. **Outputs** turn an **electronic signal** back into a **real world signal**, such as light, sound etc. The **arrows** represent the **signals**, whereas the blocks represent the **sub-systems.**

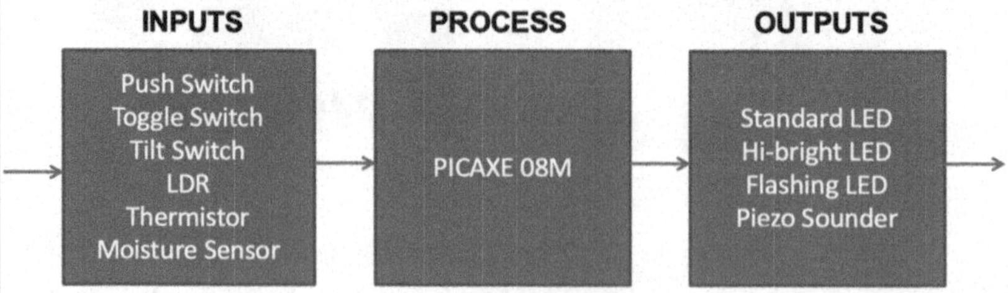

TASK:
Draw a **block diagram** for your system design, showing the **input, process and outputs** that you are going to use. Use the **chooser charts** to help.

They then used CAD software to model these ideas at a systems level and evaluate their effectiveness with a focus on them being potential solutions to the initial design problem.

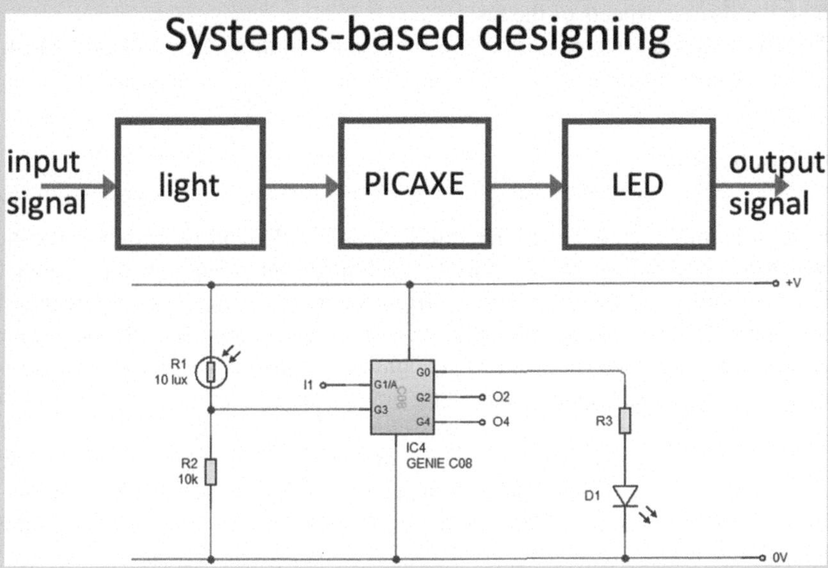

The same software was also used to automatically create a circuit diagram and PCB layout from their final design, which they then used to construct their circuit. At the point of construction, the focus returns to the components required to implement each sub-system and, at the very least, requires pupils to identify components, understand whether their orientation in the circuit matters and be able to identify values from component codes.

For some lower ability pupils, this was the limit of the technical knowledge taught; thus, they were still able to make real design decisions surrounding the electronics, without being overloaded with information that they would not have understood at this stage in their learning journey. Higher ability pupils were taught the additional technical details surrounding each sub-system through just-in-time techniques, such as the previously mentioned chooser charts. It was also observed that pupils generally now became far more interested in learning more about how their system functioned, as they understood the relevance to their own work. A version of this unit of learning was made available as a D&T Association STEM careers awareness resource.

4.2 *Reflection*

Take one of your units of work in which the focus is designing and making with mechanical, electrical, electronic or programmable elements.[10]

Use elements of systems thinking to redesign the task to support students as you open up the design aspects of the task.

Can you also use systems thinking to support the making elements of the task?

5 Conclusion

Building students' technical understanding is a core element of design & technology teaching. The risk is that teaching for technical understanding can overwhelm students with information, often derived from science, that cannot easily be put into practice in their design and making. Developing a pedagogy for technical understanding needs to take into account what cognitive psychology and neuroscience tell us about how learning happens. Knowing how learning happens helps us to build tools that better supporting learning. This chapter has described three sets of tools that we believe are particularly effective at supporting the development of technical understanding: chooser charts, product analysis and the use of systems thinking to inform technical design decisions and to support the making that arises from these decisions.

References

Barak, M. (2013). Teaching engineering and technology: Cognitive, knowledge and problem-solving taxonomies. *Journal of Engineering, Design and Technology, 11*(3), 316–333.

Barlex, D. (2004). *Young foresight*. London: Young Foresight, Available from https://dandtfordandt.wordpress.com/resources/

Barlex, D. & Steeg, T. (2017a). Big ideas for design & technology; a working paper, D&T for D&T. Available from dandtfordandt.wordpress.com/working-papers/big-ideas-for-dt/

Barlex, D. & Steeg, T. (2017b). Re-building design & technology in the secondary school curriculum; a working paper, D&T for D&T. Available from dandtfordandt.wordpress.com/resources/re-building-dt/

Gershenfeld, N. (2005). *FAB the coming revolution on your desktop – from personal computers to personal fabrication*. New York: Basic Books.

Hobbiss, M. (2018). Constructivism is a theory of learning, not a theory of pedagogy. Neuroscience explains why this is important. *Nature Partner Journals Science of Learning*. Available from https://go.nature.com/2G54t2V

Kafai, Y., & Resnick, M. (Eds.). (1996). *Constructionism in practice*. Mahwah, N.J.: Lawrence Erlbaum Associates.

[10] https://www.data.org.uk/resource-shop/stem-careers-resources-design-and-make-it/

Karpicke, J. D. (2017). Retrieval-based learning: A decade of progress. In J. T. Wixted (Ed.), *Cognitive psychology of memory, vol. 2 of learning and memory: A comprehensive reference* (J. H. Byrne, Series ed., pp. 487–514). Oxford: Academic.

Kirschner, P. A., Sweller, J., & Clark, R. E. (2006). Why minimal guidance during instruction does not work: An analysis of the failure of constructivist, discovery, problem-based, experiential, and inquiry-based teaching. *Educational Psychologist, 41*(2), 75–86.

Lave, J., & Wenger, E. (1991). *Situated learning: Legitimate peripheral participation*. Cambridge, UK: Cambridge University Press.

Layton, D. (1988). Revaluing the T in STS. *International Journal of Science Education, 10*(4), 367–378.

Layton, D. (1993). *Technology's challenge to science education – cathedral, quarry or company store?* Milton Keynes, UK: Open University Press.

Mayer, R. E. (2004). Should there be a three-strikes rule against pure discovery learning? The case for guided methods of instruction. *American Psychologist, 59*(1), 14–19.

McCormick, R. (1997). Conceptual and procedural knowledge. *International Journal of Technology and Design Education, 7*, 141–159.

McCormick, R. (2002). Capability lost and found? In G. Owen-Jackson (Ed.), *Teaching design and technology in secondary schools: A reader*. London: Routledge Falmer.

Nuffield Design and Technology Project. (1995). Harlow: Longman. Available from https://dandtfordandt.wordpress.com/resources/

Nuthall, G. (2007). *The hidden lives of learners*. Wellington, NZ: NZCER Press.

Owen-Jackson, G., & Steeg, T. (2007). The role of technical knowledge in design & technology. In D. Barlex (Ed.),. 2007 *Design & technology for the next generation*. Whitchurch, UK: Cliffco.

Papert, S. (1980). *Mindstorms: Children, computers and powerful ideas*. New York: Basic Books.

Papert, S. (1994). *The children's machine: Rethinking school in the age of the computer*. Hemel Hempstead, UK: Harvester Wheatsheaf.

Polanyi, M. (2009). *The tacit dimension* (revised ed.). Chicago: University of Chicago Press.

Resnick, M. (2017). *Lifelong kindergarten: Cultivating creativity through projects, passion, peers, and play*. Cambridge, MA: MIT Press.

Ryle, G. (1949). *The concept of mind*. Chicago: University of Chicago Press.

Steeg, T. (2000). Systems thinking and practice; a review and analysis of key ideas and their implications for practice. In R. Kimbell (Ed.), *Design and technology international millennium conference* (pp. 203–214). Wellesbourne, UK: DATA.

Steeg, T., & Barlex, D. (2018). Maker education in the English context. In N. Seery, J. Buckley, D. Canty, & J. Phelan (Eds.), *PATT 36 research and practice in technology education; perspectives on human capability & development*. Athlone, Ireland: Technology Education Research Group.

Trundle, K. C., & Saçkes, M. (Eds.). (2015). *Research in early childhood science education*. New York: Springer.

Vincenti, W. G. (1990). *What engineers know and how they know it; analytical studies from aeronautical history*. Baltimore: John Hopkins UP.

Willingham, D. T. (2010) *Why don't students like school? A cognitive scientist answers questions about how the mind works and what it means for the classroom*. San Fransisco, CA: Jossey Bass.

Wood, D. (1998). *How children think and learn* (2nd ed.). Oxford: Blackwell.

Capability, Quality and Judgement: Learners' Experiences of Assessment

Richard Kimbell

Abstract One of the most profound challenges for teachers of design & technology is the point at which they are required to engage with assessment. If design & technology involves learners in creative performance in pursuit of capability, then it involves the kinds of risk-taking that only happens when learners have complete trust in their teachers. And when teachers exercise honest and disinterested judgement, which is essential for good assessment, that trust is easily compromised. This conundrum lies at the heart of any effective pedagogy for design and technology.

In this chapter, I have constructed an argument in defence of assessment as a core element of teachers' pedagogy. The argument begins with learners' own in-the-moment qualitative judgements as a natural part of any designing activity and moves to the question of where learners' yardsticks of quality come from, and how they might be constructed. This discourse about *quality* is seen to depend on comparison, through which words like stronger, lighter, easier and quicker are embodied in real objects and can be better understood and interpreted. I argue that *constructs of quality*, informed by comparative judgement, are central both for learners developing their capability and for teachers making good judgements about learners' work. It is what Polanyi (1958) described as connoisseurship. Empirical studies are discussed in which learners are seen to be building their own constructs of quality.

Turning to the matter of summative assessments for awarding purposes, I argue, using international examples, that current practice is deeply flawed. When teachers are required to be coaches maximising the hit rate on externally constructed and atomised examination criteria, it is very difficult to see how they can simultaneously be developing learners' autonomous capability. But I argue that there is a way in which this could work. Whilst comparative judgement (for learners) can be seen as assessment *as* learning, so too in the realm of summative assessment (for awarding), it can operate to build communities of teachers that share and disseminate constructs of quality. This enables high reliability in the assessments but additionally, it develops cohesion within the community of teachers.

R. Kimbell (✉)
Goldsmiths, University of London, London, UK
e-mail: R.Kimbell@gold.ac.uk

P. J. Williams, D. Barlex (eds.), *Pedagogy for Technology Education in Secondary Schools*, Contemporary Issues in Technology Education,
https://doi.org/10.1007/978-3-030-41548-8_11

1 Capability as the Goal of Design and Technology

The pedagogical issues and the assessment approaches that I will explore in this chapter are all conceived as devices to aid learners towards the goal of autonomous technological capability. The concept of capability was well articulated in the Interim Report of the National Curriculum Working Group and at its heart lies the essence; to identify shortcomings and take creative action to improve the made world (DES/WO, 1988). There is a world of difference between *understanding*, that might properly be the domain of science, and *capability,* that is the province of design and technology:

> By capability, we mean that combination of ability and motivation that transcends understanding and enables creative development. It provides the bridge between what is and what might be. (Kimbell, Stables, & Green, 1996 p. 25)

Capability is a difficult but important phenomenon that explains how accomplished producers engage with their tasks so as to create complex works of high quality. The phenomenon is dominantly *procedural*, enabling learners to manage themselves through tasks, and this view of capability finds an interesting echo within the literature on self-regulated learning:

> …a style of engaging with tasks in which students exercise a suite of powerful skills: setting goals … deliberating about strategies to select … and as steps are taken and the task evolves, monitoring the accumulating effects of their engagement… Obstacles may be encountered … it may become necessary to adjust or even abandon initial goals and to adapt and occasionally invent tactics for making progress. In short, self-regulated learning (SRL) is a deliberate, judgemental, adaptive process. (Butler & Winne, 1995 p. 245)

In the context of learning in higher education, Stephenson addresses some of its constituents in the following terms:

> Capability is not just about skills and knowledge. Taking effective and appropriate action within unfamiliar and changing circumstances involves judgements, values, the self-confidence to take risks and a commitment to learn from experience. (Stephenson, 1992, p. 2)

Tuning this generic view of capability to our immediate concern with design & technology, Hicks – in the very early days of Design & Technology in England – alluded to the subtlety of it:

> Teaching facts is one thing; teaching pupils in such a way that they can apply facts is another, but providing learning opportunities which encourage pupils to use information naturally when handling uncertainty, in a manner which results in capability, is a challenge of a different kind. (Hicks, 1983, p. 1)

And a couple of years later, Black and Harrison added to Hicks' challenge:

> It is a continuous engagement and negotiation between ideas and facts, guesswork and logic, judgments and concepts, determination and skills. (Black & Harrison, 1985, p. 6)

When Bronowski described the qualities of humankind that gave rise to the civilisations that over millennia we have built (and subsequently destroyed), he did not produce lists of stuff that humans had to know in order to do it. Rather he points us to those nuanced elements of procedural capability:

> … His imagination, his reason, his emotional subtlety and toughness (that) make it possible for him not to accept the environment but to change it. (Bronowski, 1973, p. 19)

So there is the challenge for a capability curriculum. The real business lies in imagination, reason, emotional subtlety and toughness … and in learning how to make progress with the task (what to do next) when the answer is not immediately clear. The particular value of technological capability is that it places these elements within the sphere of practical, task-based, action so that they can be seen, tried out, shared and measured. In design & technology, our capability is *made explicit* through all the modes of expression and communication that we know so well. And this explicitness makes it so much more accessible to learners.

It is important to recognise, however, that autonomous technological capability does not exist as a state – but rather describes a journey upon which learners are embarked. So the goal is better described as 'towards autonomous technological capability'. In this chapter, the assessment issues with which we shall deal are to be viewed through – and only make sense in relation to – that lens.

2 Capability and Assessment: A Problem of Compatibility?

If the curriculum seeks to develop imagination, emotional subtlety and toughness, and if task-based learning is the chosen vehicle, and if these tasks contain degrees of uncertainty so that we quite deliberately require learners to struggle to find appropriate direction and gain a hand-hold on what they should do next, then the very least that is required by them of us is that they ***trust*** us to deal fairly with them. A capability curriculum is all about the *affective* – and specifically about trust. This is the touchy/feely domain where measurement struggles to get a bearing. It is not just that these qualities are difficult to assess, it is that, since capability acts are essentially uncertain, learners will not go out on a limb and take chances if they believe that – should they fail – they will suffer serious penalties. As Hoy and Tschannen-Moran (1999) suggest, 'willingness to risk is the degree of confidence one has in a situation of vulnerability'. Awareness of penalties (withdrawal of affection, loss of marks or whatever) acts as a serious and frequently terminal disincentive to imagination and emotional subtlety, particularly when learners know that the stakes are high. Jeffrey and Woods (1997) explored children's attitudes towards creative work in the classroom and their study:

> …..draws attention to the need for trust in a creative classroom. The emotional climate of the classroom needs to offer each child personal confidence and security. (Jeffrey & Woods, 1997 p. 15)

But if learners' willingness to be vulnerable is based on their confidence that the teacher is '…benevolent, reliable, competent, honest, and open…' (Hoy & Tschannen-Moran, 1999 p. 37), that trust is easily undermined when teachers are seen to behave as agents of external assessment, not making their own judgements but using others' yardsticks for others' purposes. And yet external judgement (in the sense of *disinterested* judgement) lies at the heart of reliable assessment. Could it be that reliable assessment depends upon a set of conditions that contradict the role of teachers as facilitators of learners' capability?

3 The Janus Virtuoso: Pedagogy as Duality

Janus is one of the more complex Roman gods, and there is always a duality in his representation. We might see him as looking to the future (as the guide helping learners to see their ideas into reality) and into the past (as the assessor looking back over what has been achieved). 'Two-faced' is often currently thought of as an insult that suggests deceitfulness or inconstancy, but in the context of teacher expertise, we should rather see it as a basic requirement. This Janus factor is well identified in Gardner's *Creating minds*, for which he studied the lives of a number of creative individuals (including Freud, Einstein, Picasso and Stravinsky) focussing on the times in their lives at which they made their most important breakthrough:

> … the creator required both *affective* support from someone with whom he or she felt comfortable and *cognitive* support from someone who could understand the nature of the breakthrough. (Gardner, 1983, p. 43)

Craft (1997) focuses on the critical role played by teachers in fostering self-esteem and self-confidence (p. 83), but Jones, Nettleton and Smith (2005) go further and identify the Janus factor. Analysing perceptions of mentoring in the context of educating nurses, doctors and teachers, they identify the two key mentoring factors as being the *adviser* role and the *supporter* role:

> While in teaching, mentors and mentees identified the role of 'adviser' as the most important mentoring role, they seem to differ in relation to 'supporter'. While in mentors' perception this role aspect appears to be of lesser priority, mentees seem to consider it highly important. (p. 7)

Subsequently, in the specific context of nurse education, Bray and Nettleton (2006) found that mentors participating in their study struggled with their dual role as assessor and mentor and found conflict within this responsibility, and moreover that the role of assessor was poorly recognised.

To be effective as a design teacher, we need to be, on the one hand, a sufficiently affective and trustworthy 'supporter' of the individual learner so as to draw out a rich and risky performance. But also to be, on the other hand, a sufficiently critical disinterested presence such that any challenges offered, or assessment made, can carry intellectual authority. This is teacher as virtuoso … the Janus virtuoso; looking one way towards learners and their emotional needs and responses, whilst simultaneously looking the other way towards the rigours of the discipline and the demands of expert judgement. Whilst this is a difficult duality for teachers to manage, it is also difficult for learners to understand and appreciate, so it is helpful for teachers to make these two roles explicit to learners. Teachers need to be self-aware as they are playing these roles and learners need to understand why the teachers might appear to be sending 'mixed-messages'.

Reflect on your recent teaching and see if you can identify instances where you have adopted a Janus role.

4 Assessment in the Moment as a Natural Part of Designing Practice

The reality in classrooms shows us that there is an interplay between the faces of Janus. This arises because of the nature of design, which has widely been described as an iterative or recursive process (Baynes, 1992; Gorman & Carlson, 1990; Kimbell, Stables, Wheeler, Wozniak, & Kelly, 1991). Design education is rich in accounts of the interdependence of emergent design ideas with associated modelling/representation of them:

> The conduct of design activity is made possible by the existence in man of a distinctive capacity of mind ….the capacity for cognitive modelling ….(the designer) forms images 'in the minds eye' of things and systems as they are, or as they might be. Its strength is that light can be shed on intractable problems by transforming them into terms of all sorts of schemata … such as drawings, diagrams, mock-ups, prototypes and of course, where appropriate, language and notation. These externalisations capture and make communicable the concepts modelled. (Archer, Baynes & Roberts, 1992 p. 15)

This iterative imaging/modelling phenomenon has within it a central requirement for assessment that learners readily acknowledge. It is not typically called assessment, or indeed anything at all. It is simply what learners do as they develop their practice as designers. As an example, the learner whose work is shown here in Fig. 1 lives on a farm and is developing a rough-ground skateboard with big wheels. The steering system is the immediate concern, and as the various proposals emerge, a set of associated issues also emerge in the form of reflective jottings beside the drawings. The rack & pinion option shown here has the comment '….. *useless when dirt builds up'*.

This assessment is an immediate, personal reflection by the designer at the moment of creation. It illustrates not just the interplay of idea generation with critical reflection, but also that the designer/leaner *sees the need for such reflection* within the generative process. It is assessment in the moment and is a very personal response.

The learner in this instance might see this jotting about the build-up of dirt as a reason to reject the rack & pinion notion, or alternatively, it might act as a starting point for other ideas concerned with some sort of protective cover to prevent dirt getting into the workings. The purpose therefore for this 'in-the-moment' assessment is simply to *steer* the work. It might provoke no further action, or it might add a side-branch to the development process. But either way, it's about steering clear of

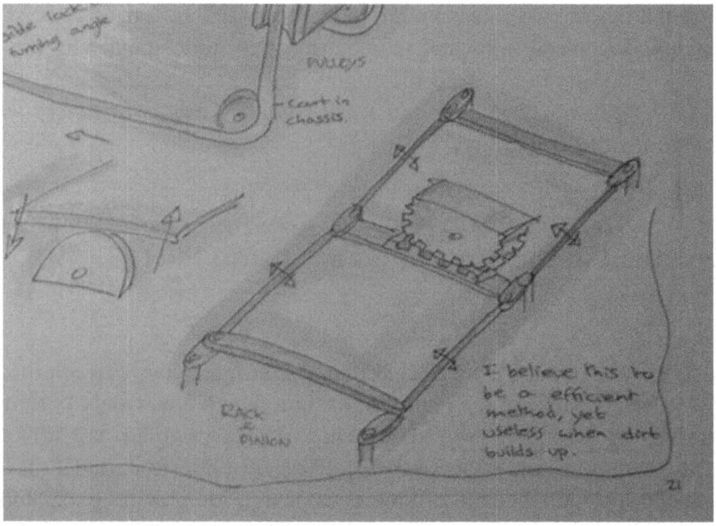

Fig. 1 Assessment as reflection in the moment

perceived pitfalls and towards manageable outcomes. It is, in the truest sense, 'formative' evaluation of the learners' ideas, by the learner and in the moment.

Interestingly, there is no evidence to suggest that this kind of reflection is seen by learners as destructive or confidence-damaging. Indeed, learners (as novice designers) might be expected to thrive on this kind of informal appraisal and equally when it comes from other sources ... especially their peers. The Assessment of Performance Unit team (1985–1991 at Goldsmiths University of London) focussed particular attention on *peer discussion* as a feature of design development. An activity administrator commented as follows:

> The pupils' response to each other's criticism was a major force in shaping the success or failure of the artefact in their own eyes. Pupils saw this (*discussion*) as a very rewarding activity and would frequently change the direction of their own thinking as a result. (Kimbell et al., 1991 p. 124, emphasis added)

One of the valuable facts of life about such early reviews is that they are low cost. It creates very little upheaval to modify ideas at this early stage, so good advice at this point is priceless. Later in the development process, when manipulating materials, the costs associated with any change of direction start to escalate. Coming to recognise this is part of the complex fabric of learners' developing capability.

Whilst I have labelled this in-the-moment appraisal as 'formative', I do not want readers to associate it with the whole panoply of formative assessment literature that is typically focused on teachers judgements:

> We use the general term assessment to refer to all those activities ***undertaken by teachers*** ... that provide information to be used as feedback to modify teaching and learning activities. Such assessment becomes formative assessment when the evidence is actually used to adapt the teaching to meet student needs. (Black & Wiliam, 1998 p. 140, emphasis added)

The key issue with appraisal in the moment within a design task as part of a capability curriculum is that the agency does not lie with *teachers* but with *learners*. They have, or are in the process of developing, a view about what they (for themselves) are trying to achieve, and their appraisal/review arises as an autonomous response to their pursuit of that purpose.

It is important for teachers to encourage this self-appraisal by learners so that it becomes second nature and part of the natural process of designing. One approach is for teachers to require a quick 'thumbs up … thumbs down' review at any moment. Learners (whatever they are doing) are asked to stop and jot down three good things (thumbs up) about their work and three areas of doubt or things that need more work (thumbs down). It can all be done in a couple of minutes and normal work then resumes. The importance of this is that (i) it provides some self-reflective steer to the designing, but critically (ii) it established the idea that self-critical appraisal is always an important part of the designing process.

Reflect on your recent teaching and see if you can identify instances where this thumbs up thumbs down approach to self-appraisal might have been useful.

5 Where Does a Sense of 'Quality' Come From? How Do We Build It?

Sadler (2013) is an assessment scholar at the University of Brisbane and his interest is in learning for complex outcomes, which is a broad category of learning types that certainly accommodates designing. He comments as follows:

> Three basic requirements for learners to become proficient in a given domain are that: (i) they acquire a concept of high quality and can recognize it when they see it; (ii) they can with considerable accuracy judge the quality of their works-in-progress and connect this overall appraisal with particular weaknesses and strengths; and (iii) they can choose from their own inventories of potential moves those that merit further exploration for improving quality. (p. 54)

These three elements might be seen as the cornerstones of a capability view of learning and of assessment and the key to it all is the first; that learners develop a sense of quality in the work that they see and do. But from where is this quality to emerge? How are they supposed to acquire it?

The sense of quality that informs the community of practice of design & technology is a tricky beast. The problem typically lies in the mistaken belief that if we write down enough qualities … and add them all together … we can say in the end that THIS is what it means to be excellent in design & technology. Such a process is mistaken, not least because, as Polanyi pointed out, 'connoisseurship, like skill, can be communicated only by example, not by precept' (Twietmeyer, 2012). Abstract prose will just not elicit the qualities that make Leonardo's work outstanding. The qualities are embedded in the work itself, and uncovering them is an inductive process rich in uncertainties and interpretations. It is not a deductive one susceptible to logic and lists.

So rather what we might do is to ask learners to look at some examples of work and decide what they think about them. What do they like/dislike about them? What would they change if they could? And from these inquisitions arise sets of qualities *that emerge from the work itself* and that (for learners) are seen to be indicators of quality. As Sadler comments:

> … they need to see as extensive a range of quality as possible, and also to see and appreciate how quite different works can be legitimately judged to be of about the same quality. Second, the identification of criteria followed (rather than led) the making of a judgement so that the role of criteria would remain important but subordinated to the main task of holistic appraisal. (Sadler, 2009, p. 62)

It is not just learners who need to develop a clear sense of quality. Teachers too need it, and particularly when they come across a completely new manifestation of performance. This was exactly the problem that we faced within the APU project (1985–1991) in which we had arranged for 10,000 15 year olds to undertake two design tasks. The particular problem was that for reasons beyond our control, the tasks were only 90-minute long, and, since real whole tasks are difficult to fit into such a short timescale, we arranged for individuals to tackle *elements* of those tasks and to be prepared then to pass on their 'so-far' solution to others. The result was that despite all our trialling, we were not confident about whether the kinds of performance we might derive from our 10,000 randomly drawn students would be at all familiar to our teacher/markers. How should we go about judging 'quality' in a 90-minute response? Well, interestingly we followed, largely intuitively, exactly the process described above (20 years later) by Sadler:

> Our experience of assessment in design & technology led us to the conviction that it is often easier to identify a high quality piece of design work than it is to say in detail why it is high quality. It is interesting that in the final analysis our markers were able to make these holistic judgements of excellence at a level of reliability that was significantly higher than that achieved for the assessment of individual aspects of capability. (Kimbell et al., 1991, p. 31)

These holistic judgements were on a 6-point scale, and using them, we created piles of work with reasonably high scores (4ers) and piles with reasonably low scores (2ers). We then took pairs of scripts (one 4 and one 2) and compared them to see what 'better' pieces were doing that 'poorer' pieces were not (and vice versa). And we did that many, many times:

> While the holistic mark enabled us to ***value*** a piece of work, the yes/no *(secondary analysis)* provided us with a composite ***description*** of it….We coined the expression 'fingerprinting' the scripts because, like a fingerprint each script was unique, but by building up a list of discriminators it became possible to describe that uniqueness in any particular script. (p. 32)

This process of drawing out discriminators from the work presented to them is one that learners need to experience within a capability curriculum. Whilst our APU challenge was very large scale (800+ schools and 10,000 learners), the principles that we used are applicable to every classroom, every teacher, every learner and every project. It was in essence 'look at this pair of portfolios … which do you think is better … and why?' And importantly, these strengths and weaknesses are drawn from the work itself and are a natural extension of learners' self-critical reflection in the moment.

Reflect on your recent teaching and see if you can identify instances where you would have been able to use this comparative pairs approach.

There is an important element here that is worth noting. The APU diagostic pursuit of an assessment framework lay in *comparison* … 'compare this 4 er with that 2 er' what's the difference? What is A doing that B is not … or B doing that A is not? Comparative judgements are inevitably simpler than absolute ones. It is easy to say (for example) that *this* carrier bag is heavier than *that* one … but far more difficult to say how many kilograms are held in each. Absolute judgements (in kilograms, or on any number-based mark scheme) are difficult, but direct comparison is easy. The same holds true for all our senses: brighter/darker colours; louder/softer sounds; and rougher/smoother textures. All can be discriminated very accurately simply by comparison. The result is that comparative judgement generates far better reliability statistics than conventional number-based assessments. And the bottom line is that it is far easier for learners to diagnose quality as, following Polanyi, the qualities are embodied in real examples (not abstract prose), and the *comparative* methodology makes them visible.

The challenge is to encourage a learner's discourse about quality. To kick-start the process, we present learners with two pieces of work that represent different levels of quality. Initially, we just invite them to identify which is better – and then to explain why they think that. Disagreement about this is not a difficulty; indeed, it is to be expected and even welcomed. For the point is not to seek consensus (as if we were assessing the work), it is rather that we are attempting to provoke a dialogue about quality. But on the other hand, if there *are* areas of consensus, they might be interesting and even significant. As Wiliam noted in 1994 and 1998:

> To the extent that the examiners *(or learners)* agree, they agree not because they derive similar meanings from the regulation, but because they already have in their minds a notion of the required standard. The consistency of such assessments depend on what Polanyi (1958) called *connoisseurship*, but perhaps might be more usefully regarded as the membership of a community of practice (Lave & Wenger, 1991). Wiliam (1994, p. 61, emphasis added)

> '… most summative assessments were interpreted not with respect to criteria (which are ambiguous) … but rather by reference to a shared *construct of quality* that exists in well defined communities of practice. (Wiliam, 1998 p. 6)

Reflect on your recent teaching and see if you can identify instances in which this would have been possible and useful.

We should perhaps see this whole judging process as one through which we facilitate learners' grappling with their emerging connoisseurship as they join us within our community of practice.

In 2008, this exact process was conducted with a class of year 10 students as part of the 'e-scape' project (Kimbell et al., 2009). At the time, we were interested in the extent to which the learners could make such judgements and how they would relate to those made by their teachers. As learners make their multiple choices about portfolio A or B, the whole class-worth of judgements can be assembled into a rank order of performance. It is worth reporting that with the 20 pieces of work being judged by the learners, their emergent rank order correlated very highly (0.88) with that generated by the e-scape marker team of teachers. There are two things to recognise about that. First, we had uncovered (quite unexpectedly) the power of this comparative judgement process as a tool for encouraging their discourse about quality. In discussion with the learners after the event, their first comments testified to the learning power that we had unwittingly exposed. 'Why didn't you show us this before we did our project … I can see how I could have made my work much better'. Naturally, I responded with the question '… what do you mean by better?' And so the discourse kicked off. Beyond the discourse, however, there is a second matter of significance. Since the learner and teacher ranks correlated so well, it is at least prima facie evidence that these year 10 learners did indeed hold a construct of quality quite close to that of their teachers. So was it so useful to provoke the discourse?

Of course, it was, not so much to *create* the construct (which was already in place), but to encourage the learners to *articulate* it. They could say that this … and that … and this … and that … are components of it. And maybe also that when you see *this* in association with *that* its especially good. We know that, in design terms, the act of expression pushes ideas forward. So too with this discourse, the act of it begins to crystallise the construct for them. It makes a vague and intangible construct into something a bit more substantive. It encourages Polanyi's *connoisseurship* to take form. As Sadler (1989) comments:

> Much more than we give credit for, students can recognize, or learn to recognize, both big picture quality and individual features that contribute to or detract from it. They can decompose judgements and provide (generally) sound reasons for them. (Sadler, 1989, p. 121)

6 Empirical Studies of Emergent 'Constructs of Quality'

Two more recent and more carefully controlled studies, one in Ireland and one in the USA, have used the same principles to explore the power of comparative judgement as a means to help learners build and articulate their personal constructs of quality.

With undergraduate construction and engineering students, Canty, Seery, Hartell, and Doyle (2017) report a study conducted at the University of Limerick.

A total of 136 undergraduates completed a design/engineering project over 12 weeks, each creating in the process a web-based portfolio of their performance and an associated end product. Using the ACJ comparative judgement tool (the online tool resulting from the e-scape project), students reviewed pairs of portfolios and were required simply to decide which (of each pair) was the stronger piece of work. The reliability coefficient for the judging session in this study was 0.98:

> The real significant point of note is that this level of reliability was achieved using holistic judgement and without providing explicit criteria to the assessors. (Canty et al., 2017, p. 5)

Looking at the impact of the holistic judging process …

> 72% of students agreed that having completed their paired judging session it broadened their perspective of capability. Having completed the ACJ assessment 79% of the students agreed that they re-evaluated their own performance in the module as a result of judging other students work. (p. 6)

Collaboration between students also provided them with the opportunity to view, discuss and appraise the quality of other students' work:

> It is hypothesised that the social dialogue that evolved as a result of the assessment approach was a factor in the consensual outcome of the peer assessment. Students were sharing interpretations of criteria and standards in a bid to establish what was of value and quality. This also provided the opportunity for the student to interpret other appraisals of their own work ….

The high levels of reliability and consensus are strong indicators that students developed an ability to judge and appraise the relevant qualities of work based on their personal construct definition:

> The holistic assessment activity broadened students' conception of quality, helping them to establish a sense of their own capability ….

An even more recent study has just been published by Bartholomew, Strimel, and Yoshikawa (2018) in the USA. A controlled study with 130 middle-school learners involved them in designing a one-page travel brochure. At the mid-point in the project, the experimental group ($n = 65$) uploaded their work as pdf files to engage in an ACJ judging session whilst the control group (also $N = 65$) printed their work for peer sharing of a more conventional kind:

> At the conclusion of the assignment, all student work (control and experimental) was assessed using ACJ and a final rank order for student products was obtained. An independent samples *t* test was conducted to investigate the difference between students which used ACJ in the midst of the design assignment (experimental) and their peers that did not (control). (Bartholomew et al., 2018, p. 12)

There are two related but separate issues of interest in this study. First, the extent to which engaging in the comparative judgement process improves the performance of the learners:

> Our analysis suggests that students who participate in ACJ in the midst of a design assignment reach significantly better levels of achievement than students who do not. While the students participating in the experimental condition varied widely in final parameter values …. they received higher parameter values overall as compared to their control-group classmates. (Bartholomew et al., 2018, p. 13)

But second – and more important for the purpose of the issues in this chapter – does the comparative judgement process in itself better enable learners to articulate and develop their personal sense of what 'quality' means?

> I learned that my peers do a lot of things differently than I do them.
>
> I learned the differences between good and bad brochures.
>
> [I learned] what things people look for while looking at a one pager.
>
> I saw patterns in the feedback and knew exactly what I needed to change.
>
> I looked at what someone and someone else said and if two people told me the same thing I looked and made changes. (Bartholomew et al., 2018, p. 14)

Clearly, it would be possible for these benefits also to arise from conventional peer discussion, but the study suggests that there is something particular about the simple and repetitive paired-judgement process that crystalises and accentuates the benefit. And a final point is worth noting. Canty's (2017) undergraduate students achieved astonishingly high correlations; the e-scape year 10 students (in 2008) achieved really good correlations; and Bartholomew's middle-schoolers were sufficiently confident to comment on '… the differences between good and bad brochures …'. We need to acknowledge that learners' personal constructs of quality, their connoisseurship, that lies at the heart of capability in design and technology take form early.

7 Summative Assessment; Flawed Practice

Awarding Organisations that certificate through project-work assessment typically publish lists of assessment criteria and provide exemplars of performance standards. A great deal of attention is focussed on these assessment criteria, for they are the levers that provide purchase on good grades and this is an arena in which both learners and teachers find themselves hugely pressured.

Nichols and Berliner (2007) remind us of Campbell's law:

> the more any quantitative social indicator is used for social decision-making, the more subject it will be to corruption pressures and the more apt it will be to distort and corrupt the social processes it is intended to monitor. (Campbell, 1979, pp. 82)

And Coe (2015) reporting to Cambridge Assessment on the 'accountability' processes at work in schools in England contributes a related pair of laws:

(i) Meddling with qualifications and accountability is irresistible to politicians,
(ii) Unintended effects are always underestimated. (Coe, 2015, slide 28)

There is worldwide evidence of these damaging coercive forces at work. Hutchins (2015) in *'Exam Factories'* provides a coruscating analysis in England; in Australia (Lobascher, 2011); and in the USA (Nichols & Berliner, 2007).

There are many dimensions to Campbell's corruption, and it strikes both at the learner and the teacher. Concerning the learner, it lies in the derailing of their striving for autonomous capability. In evidence to a Parliamentary Committee, The Vice Chancellor of Exeter University commented:

> The problem we have with A-levels is that students come very assessment-oriented: they mark-hunt; they are reluctant to take risks; they tend not to take a critical stance; and they tend not to take responsibility for their own learning. But the crucial point is the independent thinking. It is common in our institution that students go to the lecture tutor and say, "What is the right answer?" (Smith, 2008, para 129)

It also bears down hard on teachers, since league tables have rendered examination results just as important to schools and to teachers (pursuing their 'targets') as they are to learners. In countries (like England) with an increasingly fractured and privatised education system, the pursuit of a schools' competitive advantage adds ethical difficulties to the already plentiful practical difficulties of assessment. As but one example – with voucher schools in Sweden – Vlachos and Tyrefors Hinnerich (2016) and Vlachos (2018) demonstrate how grading leniency and selective examination absenteeism result in voucher schools achieving elevated placings in league tables. Where such schools are operated 'for-profit', the difficulty is obvious:

> In 2008, 64 per cent of Swedish Free Schools were run by joint-stock companies, and applications for new licenses are now predominantly from for-profits. (See, 2012, p. 4)

In the context of the arguments in this chapter, the pedagogic conundrum is clear. How can coaching to maximise the hit rate on examination criteria be squared with the development of learners' autonomous capability? And perhaps there is a way.

If learners are familiar with articulating their personal constructs, and with debating about them with their teachers and peers, then differences of emphasis and interpretation are all part of the process. Learners know that teacher X tends to value this more than that, which is different to the priorities of teacher Y. And the same holds true for differences of view within their peer group. It is not then a huge stretch to ask learners to make judgements as another person might. 'Don't judge this as yourself ... judge it as though teacher X is doing it'. This enables learners to see that the judging process depends on the understanding and the mindset of the judge, and in fact, this process can enrich learners' metacognitive grip on the constructs under discussion by forcing learners to stand outside themselves and look in *as someone else*. If then the Awarding Organisation criteria can be summarised holistically by the teacher – then this summary provides A.N.Other judging stance. So learners can then be asked quite explicitly to make judgements as though they were the Awarding Organisation. And perhaps the teacher (or better still a learner) might agree to 'play the role' (make the judging argument) for that Organisation. This enables learners to retain the integrity of their own constructs, whilst at the same time working (for extrinsic reasons) with the priorities of the Awarding Organisation. In this situation, it would be perfectly valid for the learner or the teacher to say something like ... 'whilst I think John's is a very strong piece (for these reasons), the AO would not think so as it lacks X, Y or Z'.

We have known for years that one of the strengths of design is that it empowers learners ***to think as another person***; as the user or as the manufacturer of a proposed product. My proposal here is that this ability can be extended to include assessment judgements that require learners to get inside the head of the Awarding Organisation and make judgements with ***their*** priorities. In doing this, the teacher can continue to emphasise that however well learners 'wear-the-hat' (a somewhat different interpretation of de Bono's (1971) six thinking hats) of the Awarding Organisation, it does not remove or invalidate their own personal construct of technological capability. Wearing the Awarding Organisation hat can be seen as adopting a flag of convenience – as when ships from Glasgow find themselves (for whatever reason) registered in Panama.

Reflect on your recent teaching and see if you can identify instances where getting into the head of the Awarding Organisation would be useful.

Using an approach of this kind the positive interaction of assessment with pedagogy that we observed earlier can be moved on to the end of the programme of study and operate as summative assessment for awarding purposes. And in the process, it adds a dose of metacognitive street-smart to learners' armoury.

8 Conclusion

The reconciliation of assessment with pedagogy, for examiners and teachers as well as for learners, lies in the constructs of quality that they hold. *Developing* this construct is the key for learners beginning to build their autonomous technological capability. And perversely, seeing the construct as variable between individuals, and becoming familiar with the 'changing hats' routine, has the potential actually to *strengthen* the constructs that learners hold.

So, using a comparative judgement tool, the act of engaging in the judgement process empowers learners progressively to articulate a personal view about what ***they*** mean by 'good', 'ok', 'better', 'weaker' and 'outstanding' work. In this way, they develop a rich sense of quality that enables them to ground their autonomous technological capability in their own judgements. Seen within the requirements of a capability curriculum, *the assessment process is, in itself, a learning process*. The two are inseparable.

Then there is the perspective of the Janus virtuoso. Looking one way the judgment processes undertaken by learners in the classroom provide rich opportunities to share/challenge/discuss/support the emergent connoisseurship of learners. But, looking the other way, the same comparative judgement process has powerful reach beyond individual classrooms, can link teachers into judging groups and can open the opportunity for generating local, regional and national assessments of performance.

There are so many compelling reasons for this. The resulting assessment out-turn will be reliable. And it can be operated on genuine, portfolio-based, performance assessment. And, even if they wanted to, teachers cannot 'game' the system as the assessment outcome is made up of judgements by countless other teachers in countless other schools, and 'misfit' statistics alert the administrator to any teacher whose judgements fall unacceptably wide of the consensual mark. This would be a democratic assessment model in which the construct of excellence is not held by one or two senior examiners within an Awarding Organisation but rather is held by (defined by) the teachers who are responsible for propagating it in classrooms all over the region/nation/world.

The case for comparative judgement as a means for achieving reliable school assessments was first made by Pollitt (2004) and by Pollitt and Crisp (2004), and following their work, in the e-scape project, 400+ learners from 19 schools created performance portfolios that were assessed by teachers in all the schools. Pollitt comments on the assessment out-turn of the 28 e-scape judges:

> the portfolios were measured with an uncertainty that is very small compared to the scale as a whole ....The value obtained was 0.95, which is very high in GCSE terms. (Pollitt in Kimbell et al., 2009, p. 30)

The most frequent comments by those teachers concerned the importance and the value of seeing work submitted from other schools where practice might be different. This again is assessment **as** learning, for teachers. And critically, the teachers developed views about translating the process into a national assessment system:

> Over time, this assessment technique would be likely to enhance the quality of practice of D&T teachers. Where current assessment techniques encourage a reductionist approach exemplified by algorithmic approaches to designing and a lot of pretty-but-shallow work, this technique should encourage learner collaboration; stronger independent working; more reflective ability and self-evaluation; and the ability of students to discriminate good design work from poor work. For these reasons in particular I consider that this approach has the potential to be a much improved method of assessing large numbers of students when compared with existing methods. (Teacher feedback [DP] in Kimbell et al., 2009, p. 72)

As Wiliam pointed out in 1994, to the extent that our team of teachers agreed, they agreed not because they derived similar meanings from centrally drafted assessment criteria, but because they already had in their minds a construct of what 'quality' means. It originates in assessment by comparative judgement *and it is held and used by the entire community of practice*. Assessment ***as*** learning is central to this communal enterprise for both learners and teachers.

References

Archer, B., Baynes, K., & Roberts, P. (1992) The nature of research into Design and Technology education. Loughborough: Loughborough University. Additional Information: 'Design Curriculum Matters' series. Metadata Record. https://dspace.lboro.ac.uk/2134/1687.

Bartholomew, S. R., Strimel, G. J., & Yoshikawa, E. (2018). Using adaptive comparative judgment for student formative feedback and learning during a middle school design project. *International Journal of Technology & Design Education*. Springer on-line Jan 2018. https://doi.org/10.1007/s10798-018-9442-7

Baynes, K. 1992. The role of modelling in the industrial revolution. In *Modelling, the language of designing. Occasional Paper No 1*. Loughborough University of Technology.

Black, P., & Harrison, G. (1985). *In place of confusion: Technology and science in the curriculum*. London: Nuffield-Chelsea Curriculum Trust/National Centre for School Technology.

Black, P. J., & Wiliam, D. (1998). Inside the black box: Raising standards through classroom assessment. *Phi Delta Kappan, 80*(2), 139–148.

Bray, L., & Nettleton, P. (2006). Assessor or mentor? Role confusion in professional education. *Nurse Education Today, 27*(8), 848–855. 2007 Nov; Epub 2007 Feb 12.

Bronowski, J. (1973). *The ascent of man*. London: British Broadcasting Corporation.

Butler, D., & Winne, P. (1995). Feedback and self-regulated learning: A theoretical synthesis. *Review of Educational Research, 65*(3), 245–281. American Educational Research Association.

Campbell, D. T. (1979). Assessing the impact of planned social change. *Evaluation and Program Planning, 2*(1), 67–90.

Canty, D., Seery, N., Hartell, E., & Doyle, A. (2017). Integrating peer assessment in technology education through adaptive comparative judgment. In PATT 2017. *Technology & engineering education – fostering the creativity of youth around the globe*. Philadelphia.

Coe, R. (2015). *The validity of England's Accountability data.* A presentation from the Durham University Centre for Evaluation and Monitoring (CEM) at Cambridge Assessment 1st Oct 2015.
Craft, A. (1997). *Can you teach creativity?* London: Education Now Publishing Co-operative.
De Bono, E. (1971). *Lateral thinking for management: A handbook of creativity* (p. 116). New York: American Management Association.
Department of Education & Science/Welsh Office (DES/WO) (1988) The interim report of the Design & Technology National Curriculum Working Group. DES London.
Gardner, H. (1983). *Frames of mind: A theory of multiple intelligences.* London: Heinemann.
Gorman, M., & Carlson, W. B. (1990). Interpreting invention as a cognitive process. *Science, Technology and Human Values, 15*, 2.
Hicks, G. (1983). Another step forward for design & technology. *APU Newsletter no 4.* London DES.
Hoy, W., & Tschannen-Moran, M. (1999). Five faces of trust: An empirical confirmation in urban elementary schools. *Journal of School Leadership, 9*, 184–208. (May 1999) Technomic Publishing Co. Inc.
Hutchins, M. (2015). *Exam factories: The impact of accountability measures on children and young people.* National Union of Teachers. www.teachers.org.uk
Jeffrey, R., & Woods, P. (1997). The relevance of creative teaching: Pupils views. In A. Pollard, D. Thiessen, & A. Filer (Eds.), *Children and their curriculum.* London: Falmer Press.
Jones, M., Nettleton, P., & Smith, L. (2005). The mentoring Chameleon – a critical analysis of mentors and mentees perceptions of the mentoring role in professional education and training programmes for teachers, nurses, midwives and doctors. Paper presented at the *British Educational Research Association Annual Conference.* University of Glamorgan, 14–17 September 2005.
Kimbell, R., Stables, K., & Green, R. (1996). *Understanding practice in design and technology.* Buckingham, England/Philadelphia: Open University Press.
Kimbell, R., Stables, K., Wheeler, T., Wozniak, A., & Kelly, V. (1991). *The assessment of performance in design & technology.* Schools Examination and Assessment Council. COI (D/010/B/91).
Kimbell, R., Wheeler, T., Stables, K., Sheppard, T., Martin, F., Davies, D., et al. (2009) *e-scape portfolio assessment phase 3 report.* Technology Education Research Unit. Goldsmiths University of London.
Lave, J., & Wenger, E. (1991). *Situated learning: Legitimate peripheral participation.* Cambridge, UK: Cambridge University Press.
Lobascher, S. (2011). What are the potential impacts of high-stakes testing on literacy education in Australia? *Literacy Learning: The Middle Years, 19*(2), 9–19.
Nichols, S. L., & Berliner, D. C. (2007). *Collateral damage: How high-stakes testing corrupts America's schools'.* Cambridge, MA, USA: Harvard Education Press.
Polanyi, M. (1958). *Personal knowledge.* London: Routledge & Kegan Paul.
Pollitt, A. (2004). *Let's stop marking exams.* Retrieved 2 Feb 2018 from http://www.cambridgeassessment.org.uk/images/109719-let-s-stop-marking-exams.pdf
Pollitt, A., & Crisp, V. (2004). *Could comparative judgments of script quality replace traditional marking and improve the validity of exam questions?* Paper presented at the BERA Annual Conference, UMIST Manchester, England.
Sadler, D. R. (1989). Formative assessment and the design of instructional systems. *Instructional Science, 18*(2), 119–144.
Sadler, D. R. (2009). Indeterminacy in the use of preset criteria for assessment and grading in higher education. *Assessment and Evaluation in Higher Education, 34*, 159–179.
Sadler, D. R. (2013). Opening up feedback: Teaching learners to see. In S. Merry, M. Price, D. Carless, & M. Taras (Eds.), *Reconceptualising feedback in higher education: Developing dialogue with students* (pp. 54–63). London: Routledge.

See, H. (2012). *Profit in education: Where we are now. An overview of policies and controversies from around the world.* Parliamentary Skills Group and National Skills Forum. Policy connect. http://www.policyconnect.org.uk/apgse/sites/site_apgse/files/report/340/fieldreportdownload/profitineducation-wherearewenow.pdf

Smith, S. (2008). Evidence from the Vice-Chancellor, University of Exeter. In *Children, Schools and Families Committee*, para 129. www.parliament.uk

Stephenson, J. (1992). Capability and quality in higher education. In J. Stephenson & S. Weil (Eds.), *Quality in learning: A capability approach in higher education* (p. 208). London: Kogan Page.

Twietmeyer, G. (2012). What is Kinesiology? Historical and philosophical insights. *Quest, 64*(1), 4–23.

Vlachos, J. (2018). *The difficult task of evaluating performance*. A presentation at Research Ed: 10th March 2018. Sweden: Haninge.

Vlachos, J., & Tyrefors Hinnerich, B. (2016). *The impact of upper-secondary voucher school attendance on student achievement. Swedish evidence using external and internal evaluations.* Institute for Labour Market and Education Policy. Working paper 2016:9.

Wiliam, D. (1994). Assessing authentic tasks: Alternatives to mark-schemes. *Nordic Studies in Mathematics Education, 2*(1), 48–68.

Wiliam, D. (1998, July). *The validity of teachers' assessments*. Paper presented to Working Group 6 (Research on the Psychology of Mathematics Teacher Development) of the 22nd Annual Conference of the International Group for the Psychology of Mathematics Education. Stellenbosch, South Africa.

Technology Education Pedagogy: Enhancing STEM Learning

John G. Wells and Didier Van de Velde

Abstract Depending on the educational context, technology education can be addressed in curricula as a separate or cross-curricular subject, and one that is either compulsory or an elective. Regardless, the nature of technology is clearly integrative, given that technological systems must incorporate and contextualize knowledge and practices from a broad range of disciplines. Furthermore, the scope of research and implementation strategies observed in authentic educational contexts embraces the breadth of integrative possibilities. Technology educators, therefore, have an intrinsic opportunity to investigate the potential of multiple approaches in technology education pedagogies. Furthermore, technology educators might also find themselves confronted with misconceptions about the nature of technology and the versatility of its research, design, and production processes, especially during curricular and interdisciplinary STEM team collaborations. In addition, in an educational context where technology, at one end of the spectrum, might well be misperceived as an applied activity, the technology educator is capable of introducing a richer and more accurate picture of the integrative nature of technology. At the other extreme, when only procedural or declarative aspects are emphasized, or when conceptual knowledge is handled only in the abstract, the technology educator is able to strengthen relationships with math and science by bringing in conceptual knowledge through the design of technological systems and demonstrating the authentic utility of scientific knowledge in context.

Set within the context of STEM education reform, this chapter presents an exploration into the pedagogical continuum for teaching technology education through

J. G. Wells (✉)
Virginia Tech, Blacksburg, VA, USA
e-mail: jgwells@vt.edu

D. V. de Velde
Catholic Education Flanders, Catholic School Network, Brussels, Belgium
e-mail: Didier.VandeVelde@katholiekonderwijs.vlaanderen

P. J. Williams, D. Barlex (eds.), *Pedagogy for Technology Education in Secondary Schools*, Contemporary Issues in Technology Education,
https://doi.org/10.1007/978-3-030-41548-8_12

integrative approaches with the intent to enhance STEM learning. The chapter is framed by the pedagogical premise underpinning technology education and describes, from that perspective, how a balanced curricular approach can be achieved through a range of integrative pedagogies when properly supported through curricular and instructional professional development.

1 Context of STEM Education

STEM is an acronym for science, technology, engineering, and mathematics. It is not a discipline, not a meta-discipline, not a field of study, not a curriculum, nor is it a single school subject to be taught. STEM is a concept intended to promote integrative approaches to teaching and learning. A concept meant to go beyond the traditional siloed, mono-disciplinary approach with an experiential learning approach where students integrate disciplines within authentic, relevant learning scenarios.

1.1 Educational Reform

The pedagogical foundations established by early philosophers provided a framework around which education developed globally. As an example, the USA in the late 1800s used it as the structure for identifying the core STEM subjects for secondary education that would best align with those of higher education (NEA, 1894; Ravitch, 2000). Though initially conceived as an integrative educational approach, by the mid-nineteenth century, the US pedagogical trend for teaching core subjects gradually began de-emphasizing integration of STEM content and practices, and steadily increasing the preference for silo pedagogical practices; those where core subjects were taught independently and in isolation from one another. Confronted by decades of perceived threats to national security and economies in the latter part of the century, the declining U.S. global prominence was blamed on the educational system for not adequately preparing students in mathematics and science (Berube & Berube, 2007). As a result, the national education reform agenda called for curricula emphasizing instruction that focused on the silo approach for teaching core content knowledge over that of acquiring general STEM literacy. Specifically, core subjects, such as science and mathematics, were deemed critical in educating students, while those such as technology education were viewed as ancillary and relegated to electives.

In the last decades of the twentieth century, education was trending toward developing breadth of understanding through pedagogies that prepared learners with a more comprehensive level of STEM literacy. This perspective is evident within the international discourse surrounding education and economic policy where STEM education is generally viewed as serving multiple objectives that, while distinguishable, are also closely intertwined. Since the launch of Sputnik in 1957, most countries have considered the development of subjects such as science and mathematics crucial for strategic national security and economic development (Banks & Barlex, 2014). Economic considerations play a particularly important role in influencing educational reform and often lead to policies attempting to coordinate a country's educational system in ways designed to support the labor market. Such was the case for the U.S. in the late 1980s, where new national science curricula were developed specifically calling for the integration of science, technology, and mathematics (American Association for the Advancement of Science [AAAS], 1989), and more recently in the current Next Generation Science Standards (NGSS Lead States, 2013) which are explicit in using engineering design as the instructional vehicle for teaching science. Such standards revisions reflect an educational reform intent not only on improving STEM education, but also on meeting national workforce needs (Institute of Education Services [IES], 2017). In spite of these efforts, many countries around the world remain concerned about the continued decline in student interest for science and technology studies and careers, even as the demand for such graduates grows (Organization for Economic Co-Operation and Development [OECD], 2008). The OECD concluded (p. 14) that the provision of accurate information made available to students, parents, and the education community is necessary for increasing the attractiveness of science and technology studies and careers. Furthermore, a growing number of researchers are advocating for implementing a technology/engineering design based learning (DBL) approach to STEM education given the mounting evidence demonstrating it to be efficacious for promoting critical thinking while increasing student motivation and interest in STEM subjects (Barak & Assal, 2016; Kelley & Knowles, 2016; Wells, 2016b, 2017).

2 Pedagogical Commons: Strategies for STEM Integration

Many European countries describe STEM education as a priority area often linking it to the socio-economic aspects of science (Kearney, 2016, p. 5). Consequently, recommendations for curricular reform advocated for pedagogies that illustrate the relevance of each STEM discipline to the learner's immediate sociocultural environment, as well as connecting to their expanding global community. As a result, curricula were revised to reduce the targeting of subject content in favor of those redesigned to be more cross-curricular and with an increased focus on the use of knowledge, critical thinking skills, and problem-solving abilities within authentic, real-life contexts. The pedagogy of technology education is one wherein learners engage with their technological world to better understand the nature of the technologies and their accompanying design and research practices that sustain it. As an experiential pedagogy, it embraces an integrative view of learning where question posing is used to encourage critical thinking and using instructional strategies that refer to authentic practices. This integrative approach to STEM education is a signature pedagogy unique to technology education that aligns with and supports the pedagogies of other inherent disciplines both in theory and in practice.

2.1 Integrative Approaches

Unique to technology education is instruction deliberate in teaching the breadth of inherent disciplinary content and practices as imposed on the learner by the design of a technological solution, or in understanding the design of technological systems. Specifically, the pedagogy of technology education embraces and capitalizes on an integration of multiple disciplinary content and practices demanded of the learner as they work toward a plausible design solution with a reference towards authentic practices and contexts. Moreover, as opposed to the traditional silo approach, the integration of content and practices in technology education is not contrived. To the contrary, it is predicated on authentic problem scenarios that impose on a learner the "need-to-know" and requiring them to use their resident knowledge of STEM disciplines to formulate questions and/or seek answers for that which they do not yet know. In this way, technology education has the potential to reflect the nature of technology through integration of STEM content and practices. Conceptually, based on the grade levels and curricular goals, integration is a move away from the traditional mono-disciplinary approach toward a progression of integrative strategies from multidisciplinary, interdisciplinary, and ultimately to a transdisciplinary approach (Drake & Burns, 2004). When implemented through an integrative technology education pedagogy, interdisciplinary and transdisciplinary are found to be those which best enhance STEM learning.

The *interdisciplinary* approach to integration softens the focus on discrete disciplines, where they are less prominent as a specific subject and addressed more as embedded common learnings. The recognition of the pedagogical commons (Wells, 2019b) is an important motivational element for encouraging teachers to engage in interdisciplinary STEM teams: common themes, common pedagogies, common practices, meaningful contexts, and cross-cutting concepts. For instance, one of these common practices is the heuristic nature of problem solving in E, T, and M, designing in T and E, and inquiry in S, T, E and M. The "E" in STEM offers opportunities for strengthening teacher recognition of the value of technology education in supporting authentic learning through its quantitative approaches, the design process and relationships with natural science and mathematics. One potential point of friction is found within the tension between the practical use of scientific knowledge by T and E educators versus S educators who tend to focus on conceptual understanding within ideal environments, typically with minimal use of real-world experiences that are effective in dispelling misconceptions often resulting from learning in the abstract. Hence, the interdisciplinary approach requires teachers to make a conscious effort to achieve a pedagogical balance between experiential learning and the careful, gradual construction of conceptual scientific knowledge.

The *transdisciplinary* approach is one which exemplifies the authentic integrative nature of technology. Often characterized as a blending of project-based, problem-based, and place-based learning, the design based learning (DBL) model of integration builds student understanding of disciplinary connections requisite for making informed, strategic decisions that best meet design criteria. A transdisciplinary approach does not attempt to directly or didactically teach discrete subject matter knowledge. Instead, educators employ strategies such as question posing, guided discovery, and predictive analysis to teach targeted disciplinary content and/or practice based on the need-to-know as dictated by the design scenario.

Among the inter and transdisciplinary pedagogical approaches to integration, it is important to recognize that what distinguishes one from the other is the perceived degree of separation between subject areas. Equally important is that implemented through an integrative technology education pedagogy, these approaches provide a platform for progressively de-emphasizing the stereotypical siloed perception of the STEM subjects among collaborating teachers and instead emphasize the pedagogical commons reflective of genuine disciplinary integration (Wells, 2019b).

Design Based Learning Fundamental to the teaching of technology education is a pedagogy that capitalizes on the integrative nature of technological design based learning as an instructional strategy for teaching about technology. Technological design based learning aligns with the pedagogical framework of Integrative STEM Education (Wells, 2013), where learners construct knowledge through engagement in the phases of technological design. The PIRPOSAL model (Fig. 1; Wells, 2016a) captures the fluid nature of technological design and depicts it as *phases of engagement* the learner encounters when working toward a design solution. Phases are not steps students commit to memory and follow ritualistically (McCormick, 2004), but

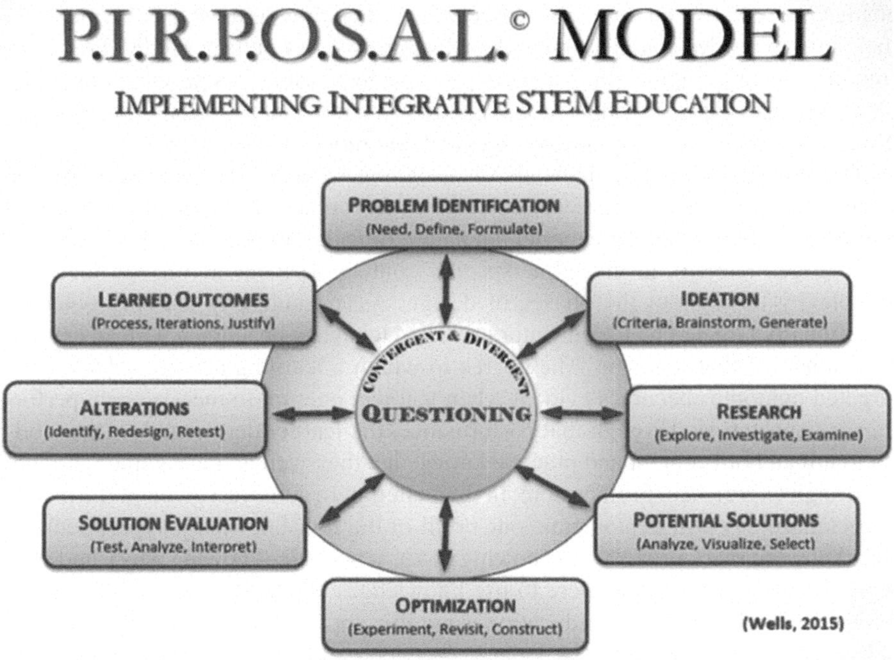

Fig. 1 PIRPOSAL model

rather reflect the fluidity in their designerly thinking as dictated by the questions confronting them regarding what they need to know. The process of design is, therefore, guided by the questions imposed on the learner at any given point throughout the phases of design. Whether it be a technological question (function, behavior, structure) or one of the contents (bioprocessing, liquid/gas flow rates, chemical conversion), in all phases of design, students are confronted by designerly questions that generate a genuine need-to-know. In responding to designerly questions, learners oscillate seamlessly between convergent thinking (resident knowledge – what I know) and divergent thinking (new knowledge – what I need to know) in building the body of knowledge necessary for producing a viable design solution. These rapid, continuous transitions between what I know (knowledge domain) and what I need to know (concept domain) lead to informed design decisions that foster habits of mind and higher order thinking skills characteristic of problem solvers.

The value of integration through technological design is realized by the opportunities it provides students to utilize their knowledge and skills, recognize content connections, develop blended disciplinary perspectives, achieve depth and breadth of understanding, develop positive attitudes toward learning, and more thoroughly explore the curriculum (Huber & Hutchings, 2005; Lipson, Valencia, Wixson, & Peters, 1993). For example, in determining the best materials for use as a bridge truss, a student would conduct experimentation (scientific inquiry) to collect and analyze data on various truss materials, and then make a strategic decision based on the results. Similarly, students exploring the ideal geometry for truss design will engage in predictive analysis and mathematical modeling as a means of comparing their prototype designs. The natural inclusion of cross-cutting STEM heuristics (inquiry, design, engineering, modeling) in technological design encourages the blending of disciplinary-specific pedagogies and movement beyond siloed disciplines toward the more realistic pedagogical commons (Wells, 2008).

The imposed cognitive demands of designing serve as the glue for engaging learners in multiple STEM subjects from an authentic experiential learning approach. In particular, the experiential nature of technological design based learning engages learners in discipline-specific content/practices at varying levels of complexity throughout the phases of design. As a result, any learning outcome intentionally targeted by the educator or STEM team can be aligned with an explicit assessment of that outcome. The degree to which a learning activity achieves the targeted outcomes becomes evident when learners attempt to meet a given performance expectation. Through that performance, the learner demonstrates their ability to utilize both content and practice knowledge they were to have gained through the design experience. Furthermore, in their attempt to achieve a given performance expectation they will demonstrate one or all of the four knowledge types: declarative (knowing that), procedural (knowing how), schematic (knowing why), and strategic (knowing when and where to use knowledge) (Li & Shavelson, 2001; Ryle, 1949). Teaching technology education using this approach to learning is pedagogically challenging, requiring a balance between didactic and student centered instruction, and applying both formative and summative assessments to ensure effective implementation of strategies for STEM practices. One way teachers can be encouraged to explore this approach is to ease the challenges by providing exemplars demonstrating how integrative approaches in technology education can enhance the learning of STEM subjects.

2.2 Pedagogical Approaches in Belgium

In the recently reformed Flemisch secondary education (Flemish Government, 2018), we find the acronym STEM appointing a group of tracks including pre-academic natural and industrial science tracks as well as vocational education tracks. In middle school, besides compulsory math, science and technology, we find integrative STEM as an elective subject that allows young people to orientate themselves in a field of further secondary study.

Besides this, the acronym STEM is also used in government policies that want to encourage young people to consider a study track in higher secondary and post-secondary education that is in line with one or more components of STEM. In this context, STEM stands for a wide range of study tracks in which science and/or technology and/or engineering and/or mathematics play an important role (Flemish Government, 2012). These developments are sometimes combined with a pedagogical discourse that connects STEM with more active and integrative (and, therefore, more meaningful) learning. This approach aims to make the "STEM subjects" more relevant and motivating. (Flemish Government, 2015). An important number of secondary schools seize "STEM initiatives" as a path to implement more integrative pedagogies.

At the level of the Flemish secondary curriculum, the standards are no longer appointed strictly to the traditional school subjects. The curriculum standards are grouped according to the European key competences (European Parliament, 2006). This not only gives rise to more coherence in the goals for mathematics, science, and technology, but also indicates how these goals are related to one another. There are also a group of common curriculum standards that define "STEM-practices" for all pupils, incorporating inquiry, design, problem solving and modeling.

For several decades, attempts have been made to give technological literacy a place not only in the middle school but also in higher secondary education. Due to a lack of dedicated technology teachers and a fear of too much fragmentation in the curriculum, in recent years, many schools have introduced elective integrative STEM courses (some form of pre-university engineering) within pre-university science tracks.

At this point, we can observe signs that some aspects of technological literacy, probably under the influence of STEM trends, are becoming adopted by other subjects. For instance, Mathematics responds to digitization by focusing more on logic and discrete mathematics (such as graph theory). In Geography, there is more attention for sustainable development in relation to technology use. In the interplay between Mathematics and Physics, technical systems are often articulated in an experimental context. In their pedagogies, more attention is paid to contexts, problem solving, modeling and systems thinking.

3 Integrative Exemplars: Enhancing STEM Learning

All teachers strive to meet the educational needs of their students, but often find themselves constrained in doing so as a result of their initial mono-disciplinary preparation. Educators attempting to move beyond their traditional pedagogical preparation and adapt new, integrative pedagogies are faced with the challenge of knowing how to best design and deliver their instructions differently. The following technology education exemplars are provided to describe and demonstrate how such challenges can be addressed when implementing integrative STEM education.

3.1 Prototype Vehicle Project

Central to the Flemish STEM@school research project (Thibaut et al., 2018) was the collaborative development of three interdisciplinary STEM projects for grade 9 students by teacher design teams and researchers. Each project was designed to take about 10 weeks to complete, with 1 hour of physics and 3 hours of engineering (STEM@school, 2018) delivered each week throughout the school year. Mathematics lessons were incorporated within both the physics lessons and engineering design on a "just-in-time" basis when and where needed. One of the projects tasked students with designing and programming a prototype vehicle able to travel along a cascading array of green lights, the green wave, without having to stop at intersections. Completing the design task required learners to develop knowledge about kinematics, linear functions, and Arduino programming in the process of designing a mechanical prototype with drives and transmissions. To calibrate their system, students used a graphical representation of a function to map the controllers' digital output value (representing the voltage on a controllable DC motor) to the speed of the vehicle. Because the vehicle travels the green wave in a straight line within a run-on zone, both the steering and the non-linear acceleration phases need not be considered. To intentionally address the sociocultural engineering parameters associated with such autonomous transportation systems, students were required to conduct a case study as part of the project. In doing so, they investigated the pros and cons of autonomous vehicles, starting first with an analysis of their own mobility choices. Through reflections on a wide range of variables from traffic congestion to pollution, students generated transportation solutions designed to provide more sustainable forms of mobility. In this way, the project aligned with technology education content and practices authentically situated within the broader sociocultural context of engineering design.

Teachers from across the 30 participating schools implementing the STEM projects reported that the interdisciplinary approach, where inquiry is embedded within the design process, resulted in students far more motivated to learn compared to traditional siloed science instruction. Furthermore, participating math teachers, who at first were skeptical of the interdisciplinary approach and feared the potential loss of critical deductive thinking, soon were convinced of the motivational value of having students gather their own data for use in math courses. Specifically, math teachers realized that their students were better able to recognize the importance of different representations (table, graph, formula) and the possibilities for modeling and predicting system behaviors (Van de Velde, Van Boven, Dehaene, Knipprath, & De Cock, 2016). To ensure student recognition of such disciplinary interactions, it is important that technology, math, and science teachers intentionally point out similarities and differences in the use of symbols and units, e.g., for slope and intercept. As observed in this example, interdisciplinary STEM projects naturally impose design issues and/or failures on teachers and students alike, requiring them to address the limitations of their prototyping and modeling in predicating system functions; e.g., energy level of individual batteries was a critical variable, students realized they must account for when accurately calibrating vehicle speed. Unanticipated factors such as batteries were difficult to predict, but resulted in significant learning as students worked to solve problems imposed by design failure.

The majority of schools involved in the project focused on teaching the conceptual elements embedded within inquiry and prototype design. However, some schools were structured as traditional industrial science and pre-engineering programs. As a result, teachers at these schools were often more concerned with the production of high-quality industrial prototypes and, in some cases, fell prey to "the tyranny of product outcomes" (McCormick & Davidson, 1996) where the undue attention given to high-quality production diverts attention away from the more conceptual aspects of research and design. In addition, because project tasks involved interdisciplinary subject matter, participating teachers debated the best way textbooks should be used for integrating the content. They resolved the issue by using instructional materials from both the problem description (design brief) and separate subject-specific texts. Such pedagogical adjustments required to implement the STEM projects were imparting a new instructional paradigm on teachers where their concept of learning progression evolved from teaching math first, using that acquired knowledge in science, followed by application in technology, to a more integrative view where interdisciplinary interactions occur in a more natural way.

Adoption of new pedagogical practices is not without its challenges. As members of interdisciplinary teams, STEM@school teachers using new integrative STEM education approaches in the context of a pré-university interdisciplinary STEM program recognized that there are several critical factors that affect successful implementation (Van de Velde et al., 2016). A recent literature review on the effects of subject integration confirms many critical points that could also be observed in the project (Wilschut & Pijls, 2018). Most important is having a shared STEM mission among collaborating teachers and school administration, providing an educationally supportive environment for all subject teachers involved. It is important that teachers and school leaders reflect on what they want to achieve and choose the appropriate resources. Involvement and enthusiasm of teachers should be based on thorough knowledge about subject integration. If a team is not convinced, it is better not to start with experiments. Teachers involved must have cooperation skills. Other critical elements include sufficient time for teachers to meet and co-plan the integration of inquiry and design based learning approaches and a spirit of collaboration among all teachers involved. Teams having a strong collaborative spirit report better educational outcomes which creates mutual dependencies among teachers and the potential to challenge established pedagogical beliefs among those involved. Also critical for successful STEM initiatives is motivating teachers to become involved and collaborate in integrative STEM education. As such, involvement of all school subjects becomes a relevant factor for building broad interdisciplinary school support. Ultimately, the innovative power of a school becomes inextricably connected to the intensity of collaboration within that school (Van der Bolt, Studulski, Van der Vegt, & Bontje, 2006).

It is important for school management to consider whether strong subject-oriented teacher sections are desirable in relation to the intended integration. On the one hand, they can provide strong professional expertise and subject-specific coherence, while on the other hand, they can act as inhibitors defending subject-centered approaches (Wilschut & Pijls, 2018). Moreover, it is also important to implement strategies for effectiveness and teacher clarity in pedagogies for integrative STEM education. These recommendations can support the coaching behavior of the teachers involved.

Teachers of any subject that are involved in an interdisciplinary STEM team attempting to use T/E DBL approaches meet important challenges (Van de Velde et al., 2016). Science teachers have to deal with skills such as designing, programming, working with electronic kits, with which they are often not very familiar. Math and science teachers can struggle with disciplinary perspectives of learning progressions designed to avoid presenting misconceptions and facilitating critical deductive thinking. For example, when designing integrative STEM instruction, math teachers are often expected to change the sequence of topics. In doing so, deductive aspects of the learning progressions must then be revised and replaced with a new balance in planning for a pedagogically desirable variation of topics to be covered. As a result, these teachers are challenged to combine the deductive thinking used in mathematics with the experimental and problem-solving thinking used in science and technology. However, experimental thinking has historically played an important role in the growth of mathematical knowledge and is also supported by important work in the context of "inquiry-based mathematics teaching" (Winslow, 2017). In light of this, it would be important to revalue these connections between math and science thinking among members of integrative STEM education teams, as they redesign their traditional mono-disciplinary curricula.

Technology and engineering teachers are expected to pay more attention to connections between technology, science and mathematics. These teachers report that attitudes such as being innovative, flexible, dealing with failure, collaborative, problem-oriented are very important for teaching when using design based learning approaches. Moreover, they feel having sound horizontal and vertical curricular knowledge about the STEM curriculum is an asset in their collaborative interdisciplinary work. Some important concerns during the introduction of integrative STEM education initiatives are topics such as assessment of STEM competences, using effective pedagogies, and practical concerns regarding the setup of experiments and design tasks. To implement the prototype vehicle project, the teachers involved had the opportunity to share their experiences and concerns in a teacher network providing a platform for professional development and encouragement.

3.2 Design Based Biotechnical Learning

The contexts within which all technological activities occur have been said to reside within three mutually interdependent domains: physical, informational, and biological (International Technology Education Association [ITEA], 1996). Although the biological domain is arguably the most inclusive of all STEM disciplines, the integration of science and/or mathematics within technology education continues to be approached primarily through only the physical and informational domains. Presenting students with such a singular technological domain perspective limits their recognition of genuine connections among STEM disciplines within the design of technological systems and also limits the educator in demonstrating the true integrative nature of technology education. Designing technological solutions requiring inclusion of all three technological domains presents a powerful learning approach that helps students of all ages recognize the natural connections among the entire spectrum of disciplines: STEM and others. Many countries, recognizing the value of the design based biotechnical learning (DBBL) approach, have incorporated it into their national standards and curricula (Jones, 1997; Smith, 1988; Korean Institute of Curriculum and Instruction, 2002; Ferguson, 2009; ITEA, 2000). As a result, technology educators across the globe implementing the DBBL approach have shown that it naturally engages the learner in need to know the full spectrum of disciplinary content and practices (Robertson, n.d.; Dunham, Wells, & White, 2002; France, 1997, 2007, 2015). As an integrative STEM education approach, DBBL naturally and seamlessly immerses the learner in all three domains of technological activity, where the learning of STEM content and practices results from that which is inherently imposed by the design of a biotechnological solution.

DBBL Problem Scenarios Regardless of grade level (elementary, middle, or high school), the most effective approach for preparing teachers to integrate STEM content and practices is engaging the teacher, as a student themselves, in the very design challenges they plan to use in their classrooms. This is the preparatory approach followed in the Design Based Biotechnical Learning (DBBL) graduate course taught at Virginia Tech where K-12 teacher–students experience STEM integration directly through immersion in biotechnical design challenges. The design challenges used in the DBBL course are drawn from more than 40 Problem Scenarios (ProbScens) included in the DBBL Teaching Guide (Wells, 2019a), all of which are developed to engage learners at any grade level in challenges addressing engineering issues destined to confront humanity in the twenty-first century (National Academy of Engineering [NAE], 2009). DBBL ProbScens are open-ended design challenges that place learners in real-life scenarios tasking them with designing biotechnical solutions within a given set of constraints and parameters. In every ProbScen, the overarching system requirement is the support and use of living organisms (or parts thereof) as the biological tool needed for resolving a human need.

As explained earlier in the PIRPOSAL model, student designers will progress through the eight phases of design in response to questions imposed on them by the challenge. These imposed cognitive demands require students to recognize what they know (convergent thinking) and then determine what they need to know (divergent thinking) about the content and practices involved, e.g., what they know and need to know about the technology (materials, properties, systems), science (biology, physics, chemistry) and mathematics (computation, predictive analysis, modeling). As with any lesson, the teacher determines what specific STEM learning outcomes students will be expected to achieve, as well as how students are expected to demonstrate that achievement through learning products that can range from simple drawings and design concepts, to more complex models or working prototypes. Materials used for constructing prototypes vary depending on the design solution, but typically include common classroom or household supplies such as plastic bottles, flexible tubing, balsa, cardboard, and the like.

Photobioreactor Design Challenge ProbScen 3D (Algal Photobioreactor, Fig. 2) is the biotechnical exemplar selected from the DBBL graduate course to illustrate the integrative STEM education approach. When implementing ProbScen 3D numerous learning outcomes are targeted for intentionally teaching specific STEM content and practices throughout the phases of design. Described in the following sections are several instructional strategies used in teaching some of the learning outcomes for ProbScen 3D. Specifically, the outcomes selected to illustrate the teaching and assessment of targeted STEM content and practices determine the volume and flow rate of water within the system (math), designing an airlift pump to circulate water through the system without harming the algae (technology), and the basic metabolic processes algae use to remediate phosphate-contaminated water (science). Artifacts generated by the K-12 teacher-students enrolled in the graduate course will be used to highlight the demand for different knowledge types, along comments to provide insight from the educator's perspective on the value of having their students design to understand authentic connections among disciplines.

The context of ProbScen 3D is the contamination of local stream water from the overuse of fertilizers. The challenge asks students to design and develop a biotechnical (biological/technological) system that will grow green algae and use these living organisms to remediate the contaminated streams (remove excess phosphates). In the DBBL course, the authenticity of the design challenge scenario is established by addressing this environmental impact condition as one that can be found anywhere fertilizers are used and connecting the issue directly to the student. Student awareness about this environmental issue is raised by using available local/global resources (daily news, YouTube videos, etc.) to locate current information regarding the impact on clean water resulting from the overuse of chemical

Plant & Animal Applications

ProbScen 3D

Student ProbScen

AGRICULTURE

Algae Photobioreactor

Context:

Algae are organisms that range in size from microscopic to large seaweeds. Many uses have been found for these organisms. For example, large brown seaweeds are being used to extract alginates, which can be made into varnishes and gelatins. Micro-algae have been recognized for uses ranging from animal, plant, and human foods, to antibiotics, vitamins, and food colorings. Using micro-algae as a food source for humans, animals, and plants has taken on greater importance in recent years. The nutrients and environmental conditions needed for algae to grow are very simple, which makes it appealing as a food crop for people in developing countries. Micro-algae also appeal to US farmers as an excellent alternative to chemical fertilizers, and are easily produced using various types of photobioreactors. A photobioreactor is an easily constructed technological system used in the growing and harvesting of micro-algae. *Spirulina* and *Chlorella* are freshwater algae easily grown in plastic tubing under warm, sunny conditions.

Challenge:

Because of your knowledge of photobioreactors, you have been hired by a local farmer to design a system to grow and harvest *Spirulina* or *Chlorella* for use as an organic fertilizer (bio-fertilizer) instead of the chemical fertilizers normally used. Your system must use clear plastic tubing for the growth chamber, an air lift pump mechanism for circulating the algae so as not to injure them, and a method that will determine the best time to harvest the algae crop. The farmer also wants the system to be an energy efficient system that will deliver the bio-fertilizer directly to the field!

 209

Fig. 2 ProbScen 3D – context and challenge

fertilizers. Emphasizing that overuse is not something done only by farmers but, by ordinary families such as their own when caring for lawns in residential areas, connects students with the issue. What students come to realize is that a significant amount of fertilizer is regularly carried away from both farming fields and residential areas in rain water runoff which pollutes their local waterways with too much phosphate. The excess phosphate in the stream leads to algal blooms that will eventually affect the health of all other aquatic organisms. This opening discussion is meant to gain student attention and, in this case, directly connect them to the human need for clean water as a way of getting them ready to engage in the design challenge (Gagne, Wager, Golas, & Keller, 2004). It is also an opportunity for the teacher to make connections between key STEM subjects and other curricular areas (information searches, scientific experiments, math calculations, etc.); students will need to draw on when designing their solution. Ensuring students see connections between subject areas can easily be done by having them write short descriptions in their Interactive Engineering Journals (IEJ) (Fig. 3) of what they understand is the human need, what problem they see needs to be solved, and explaining why a biotechnical system is the most appropriate solution. Doing so helps the learner recognize the problem to solve, the design requirements that must be met, and what technological and biological functions their system must perform.

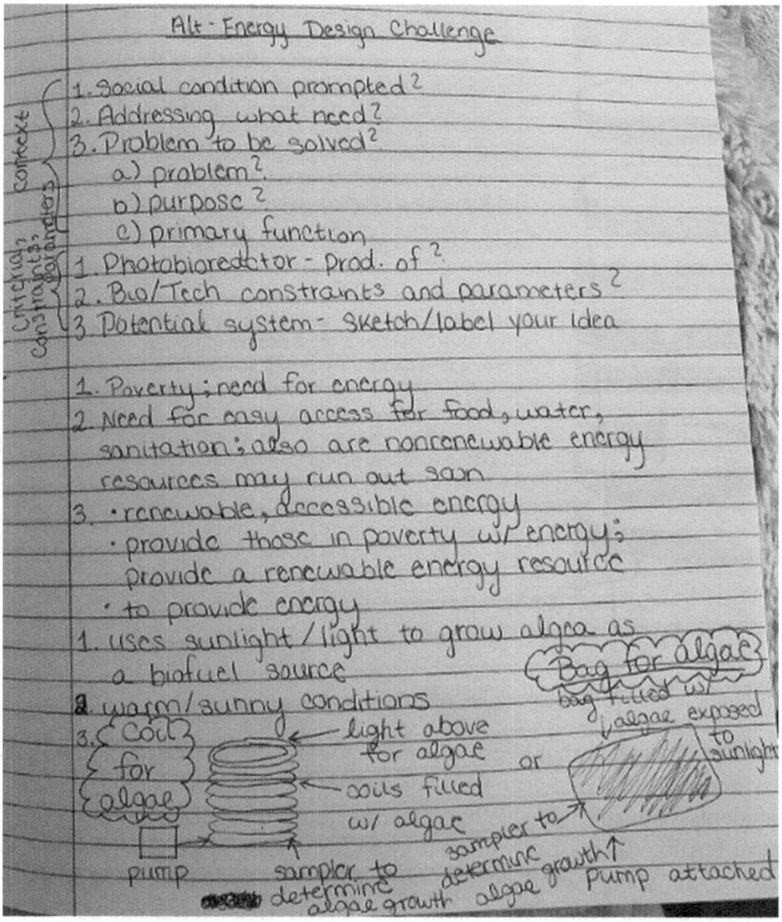

Fig. 3 Student notes: problem identifcattion

Teaching Strategy The photobioreactor challenge tasks the learner, in this case teachers in the DBBL course, with designing and building a system prototype as proof of concept (a feasible solution) for using algae as a tool for removing excess phosphate from stream water. Pedagogically ProbScen 3D provides the ideal educational context for using technological design to connect student learning of science, mathematics, and language arts directly with the issue of environmental sustainability. To enhance retention and promote critical thinking (Gagne et al., 2004), ProbScen 3D can be implemented as an extension lesson (building on prior instruction) modeling a transdisciplinary approach to STEM integration. Specifically, an extension lesson is meant to guide students through the design phases with the explicit goal of intentionally supplementing and extending their prior learning of specific subjects (STEM and other) by connecting those subjects on a need-to-know basis within the design challenge.

When the photobioreactor challenge is first introduced to students they are confronted with the need to know the general concept of a reactor – definition, components, function, etc. The general reactor concept is then refined by including characteristics specific to a "bio" reactor (one that must include a biological component) and finally to recognize the unique characteristics of a "photo" "bio" reactor (one using organisms that require light). These reactor concepts can easily be conveyed to learners at any grade level using images either provided by the teacher or located through student internet searches. In their searches for photobioreactor images, students will quickly find that the two main types are the coil and fence closed-system designs (Fig. 4). The *photo*bioreactor introduces a plant science criterion that imposes on students the need to consider their prior knowledge about how plants utilize sunlight (photosynthesis), the importance of nutrients to plant growth (cellular processes), and the type of plants suitable for a typical coil or fence closed-system design (unicellular algae). Whether designing the coil or fence

Fig. 4 Coil and fence photobioreactors

system, students will engage in technological decision-making regarding such things as reactor construction materials, pumping mechanisms, sensors, adhesives, and tools. Their design will also need to consider how contaminated stream water might be introduced into the system, how it will circulate throughout the system without damaging the living algae, and then finally returned as clean water to the stream. Such questions confront students with the need to regularly revisit certain technological, scientific, and mathematical concepts they have previously been taught.

As teachers in the DBBL course engaged in the Ideation, Research, and Potential Solutions phases of design, one of the technological problems confronting them was how to circulate algae throughout the system without damaging the plants. Solving this problem required teachers to research the various types of pumps (impeller, piston, etc.), which eventually led them to an airlift pump (Fig. 5) as a potential solution that serves this purpose. With the circulation problem resolved, teachers turned their attention to determining such things as the total amount of liquid (volume) their system can accommodate, deciding how fast algae should move through the system (flow rates), calculating the ideal number of algae (plants) to grow within that volume (quadrat sampling), etc. In doing so, teachers are quick to recognize that students, building on previous math instruction for determining surface area and volume, could use the diameter and length of the tubing to calculate the total volume of liquid the system can accommodate. It is at this point that teachers begin to realize how ProbScen 3D not only teaches students at any grade level the technological concepts involved in designing biotechnical systems, but also naturally connects those concepts to their grade-appropriate math and science knowledge.

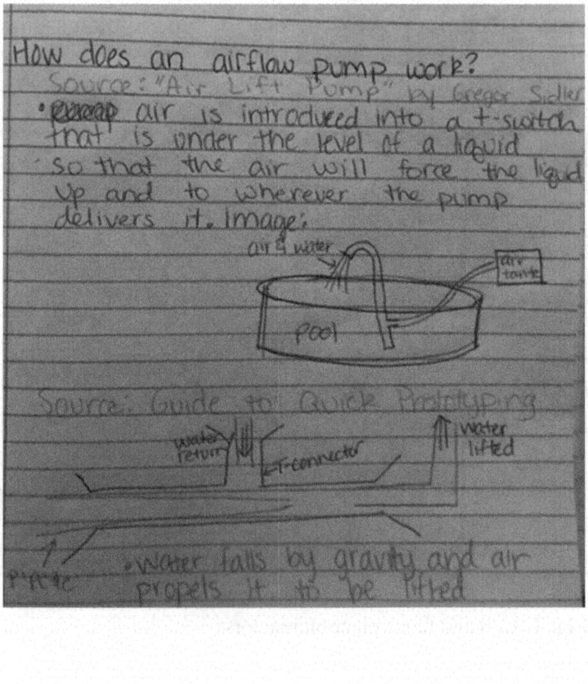

Fig. 5 Airlift pump design

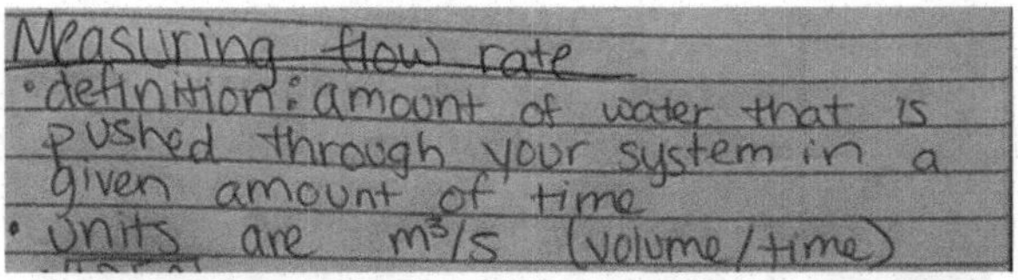

Fig. 6 Measuring flow rate

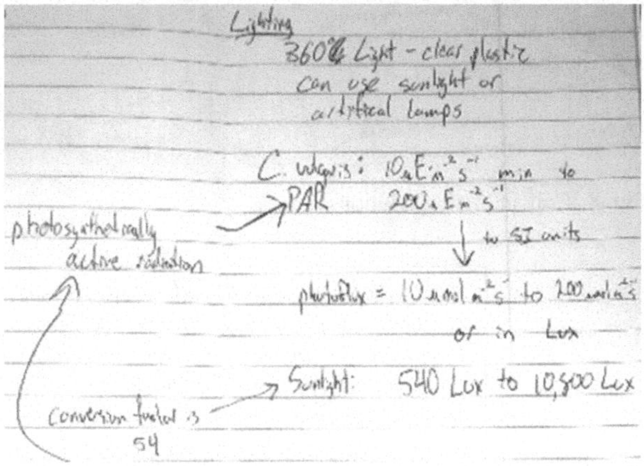

Fig. 7 Calculating lux requirements

Specifically, using math is to predict volume and flow rates (Fig. 6) or to calculate the lux requirements (Fig. 7) plants need from sunlight. And using the scientific method to determine what nutrient (phosphates) plants get from the water, and a spectrometer to determine if remediated water from their system is free of contaminants (between 0.01 and 0.03 mg/L). In this extension lesson exemplar, the necessary grade-appropriate science, math, and language arts concepts and skills needed to design the prototype are those that students were previously taught through the curriculum. Specifically, this example presents the photobioreactor design challenge as an extension lesson to intentionally connect previously taught technology concepts concurrently with those of science, math, language arts, and many others within a meaningful technological context. Furthermore, it is important to note that for students to make these connections, the systems they design need not be sophisticated. In preparing their proof-of-concept solutions, low-tech prototypes (Fig. 8) constructed with common classroom materials can serve well to demonstrate (and teachers assess) a student's practical and conceptual understanding. The educational value in using DBBL (ProbScen 3D) to teach STEM content and practices was summed up nicely (Fig. 9) by one of the teachers in the DBBL course. In addition, reflections teachers wrote in their IEJ (Fig. 10) captured their own learning processes, along with recognition of connections among STEM disciplines being a learning outcome inherent to T/E DBL.

Designerly Ways of Knowing At all grade levels, from elementary to high school, the photobioreactor design challenge can be used to teach technological design while at the same time being intentional in the teaching of any other STEM content

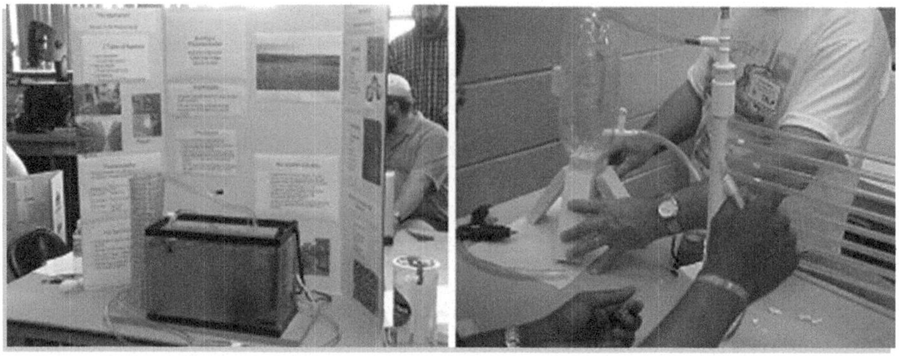

Fig. 8 Coil and fence solutions

A design based teaching/learning strategy allows for multiple solution pathways as well as opportunities to iterate and redesign. In each iteration students gain a deeper understanding of both the scientific and technological components needed to meet the design criteria. The T/E DBL process allows students to proceed in development of a prototype similar to real world applications. The non-linear approach to T/E design - the ability to move through different design phases at any juncture in the process - generates questions that aid not only in refinement of the prototype, but also in students' understanding of all the STEM concepts involved. Students are guided in the design process through carefully crafted questions posed on them as they engage in each phase of design. These questions are designed to lead students to particular learning outcomes that they would not be able to achieve on their own. With the achievement of the learning outcomes, students are able to successfully develop habits of both mind and hand.

Fig. 9 Teacher perspective on DBBL

or practice inherent to the designed solution. Design based learning, arguably the signature pedagogy of technology education, clearly shows that having students design to understand is the key educational goal. Furthermore, equally important is that this method of knowledge acquisition, characterized as "designerly ways of knowing," is unique to technology education and distinct from methods used in other disciplines. Designerly ways of knowing is a method of acquiring knowledge used during the design of a technological solution that is patently distinct from any of the other more commonly recognized methods such as scientific inquiry. At the elementary and secondary education levels, designerly ways of knowing is singularly

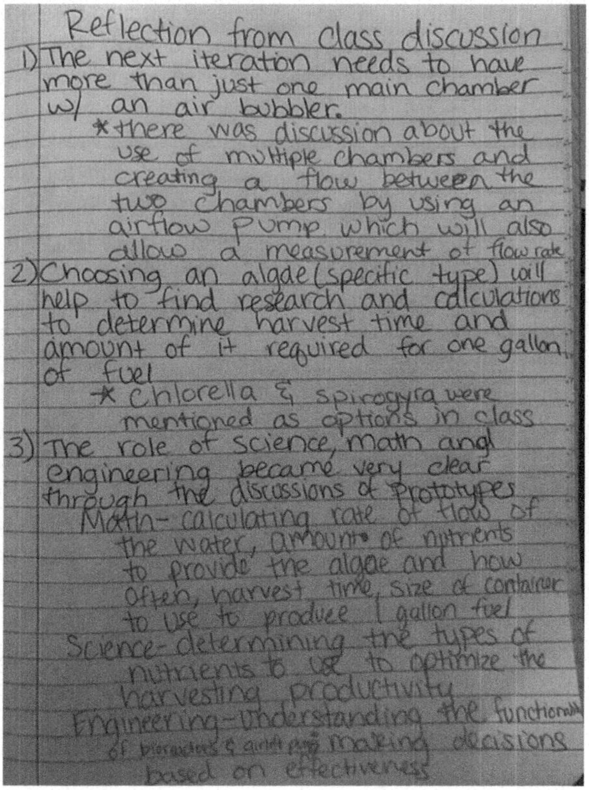

Fig. 10 IEJ teacher reflection

unique to the field of technology education, where novice designers gain knowledge and achieve understanding by grappling with ill-defined problems, being solution-focused, employing convergent and divergent modes of constructive thinking, using graphic codes (sketches, diagrams, and drawings) to translate abstract requirements into concrete objects, and then reading and writing in object languages based on those codes. This approach to developing a human cognitive ability is unique in its preparation of those individuals who will be needed to solve the complex, real-world problems of the twenty-first century.

3.3 Pedagogical Implications

As characterized previously in the Flemish project example, the interdisciplinary approach emphasizes the teaching of discrete STEM disciplines, but softens the focus by presenting them as embedded common learnings through well-timed, as-needed parallel lessons to teach the subject matter (i.e., physics and math) essential to a given design phase. As used in the Flemish project, the interdisciplinary approach induces a motivational problem- and inquiry-based context towards not only the technological design process but also the embedded science and math lessons. The transdisciplinary approach described in the photobioreactor exemplar is one where STEM disciplines are used extensively without overtly teaching them. As a transdisciplinary example, the photobioreactor demonstrates the intentional teaching of targeted STEM content on a need-to-know basis through experimentation on sub-systems students are considering for use in their design. On an as-needed basis, the teacher introduces experimentation (science inquiry) on individual sub-systems as a lesson to intentionally teach targeted subject matter such as physics and math. In the photobioreactor, a sub-system is needed for circulating water, nutrients, and algae, but in such a way so as not to damage the living organism. Several technology options are possible, one being an airlift mechanism where bubbles rising in a vertical column are used to carry/push/draw algae through the system. Discrete physics and math concepts and formulas can be taught by having students conduct experiments to determine the optimal bubble size based on algae size, the best rate at which bubbles should be produced to create an adequate flow of water, the correct amount and rate to add nutrients, and the ideal tubing diameter for growing the type of algae they selected. Subject-specific content and practice are overtly taught/learned through these parallel lessons, which is then used by students when they re-engage in a given phase of design.

When engaging upper-level high school students (or even pre and in-service teachers) in technological/engineering design based learning, the educator recognizes that students at this level already possess significant breadth and depth of resident STEM knowledge. Building on a high level of prior knowledge, the integrative STEM education approach promotes critical thinking (reasoning/sense making) by having students use their understanding of disciplinary interdependence in making strategic decisions appropriate for meeting the design criteria. An immersive teaching strategy engages the learner in the design challenge by providing not much more than the context, challenge, and parameters presented by the Problem Scenario. This is truly an experiential learning approach where teaching occurs through question posing and guided discovery, and only as-needed at any time while students are engaged in the various phases of technological design.

4 Durable Changes in Teacher Practice

Imparting change in participant understanding, practice, and beliefs toward an expressed end is a long-recognized overarching goal of professional development (Griffin, 1983). With that end in mind, the goal of integrative STEM education professional development for technology educators would be no different: "The ultimate goal of all professional development is improved student achievement" (Mundry & Loucks-Horsley, 1999, p. 3). In the past two decades, results from the research investigating the characteristics of successful professional development indicate that they all incorporate a similar set of key factors (Assis-Cezarino, 2004; Garet, Porter, Desimonre, Birman, & Yoon, 2001; Guskey, 2000, 2002, 2003;

Kennedy, 1999; Loucks-Horsley, Hewson, Love, & Stiles, 1996; Wells, 2007). Top among those factors is professional development that is teacher-centered, focuses attention on pedagogies found effective in integrating content and practice, and provides teachers with first-hand engagement in disciplinary integration as a precursor to curricular redesign (Wells, 2007, p. 113). Professional development leading to durable changes in teacher practices requires rethinking of both our curricula and teaching strategies in the context of these factors.

Given what research in cognitive science has revealed about connecting teaching with student learning, the growing mandate for K-12 teachers to use integrative STEM education practices is clearly challenging teachers currently in the classroom, those newly prepared as teachers, and our models of teacher preparation. Adding to this challenge is the increasing demand to use T/E DBL as a pedagogical vehicle and curricular focus for implementing STEM integration (Denson & Lammi, 2014; Lewis, 2006, 2007; NGSS Lead States, 2013; Schunn, 2008). There is growing evidence that technological/engineering design based learning better engages students in the learning process, helps promote understanding of disciplinary connections, and fosters critical thinking (Kelley, 2008, 2011; Wells, 2016b, 2017). However, in spite of such evidence, the question remains whether current and preservice educators are receiving the preparation needed to effectively implement this type of integrative STEM education.

Teachers of any subjects that are involved in an interdisciplinary STEM team attempting to use T/E DBL approaches meet important challenges (Van de Velde et al., 2016). Science teachers have to deal with skills such as designing, programming, and working with electronic kits, with which they are often not very familiar. Math and science teachers can struggle with disciplinary perspectives of learning progressions designed to avoid presenting misconceptions and facilitating critical deductive thinking. For example, when designing integrative STEM instruction, math teachers are often expected to change the sequence of topics. In doing so, deductive aspects of the learning progressions must then be revised and replaced with a new balance in planning for a pedagogically desirable variation of topics to be covered. As a result, these teachers are challenged to combine the deductive thinking used in mathematics with the experimental and problem-solving thinking used in science and technology. However, experimental thinking has historically played an important role in the growth of mathematical knowledge and is also supported by important work in the context of "inquiry-based mathematics teaching" (Winslow, 2017). In light of this, it would be important to revalue these connections between math and science thinking among members of integrative STEM education teams as they redesign their traditional mono-disciplinary curricula.

Technology and engineering teachers are expected to pay more attention to connections between technology, science and mathematics. These teachers report that attitudes such as being innovative, flexible, dealing with failure, collaborative, problem-oriented are very important for teaching when using design based approaches. Moreover, they feel having sound horizontal and vertical curricular knowledge about the STEM curriculum is an asset in their collaborative interdisciplinary work. Some important concerns during the introduction of integrative STEM initiatives are topics such as assessment of STEM competences, using effective pedagogies, and practical concerns regarding the setup of experiments and design tasks.

Teacher networks can offer opportunities to share these concerns and discuss lessons learned. In this way, they contribute to professional development. Policy makers in several European countries have launched STEM knowledge networks that are trying to generate synergies among STEM stakeholders in combining bottom-up and top-down initiatives. In most cases, such initiatives are bolstered by evidence from higher education and research. As a result, policy makers often support school projects and professional development programs for teachers and school administration.

Many of today's teachers traditionally prepared to teach a mono-discipline are seriously challenged to conceive, design, and implement STEM integration in general, no less to use the technological design based learning approaches as a vehicle for STEM integration. Meeting this challenge requires preparing teachers more as "adaptive experts" (Hammerness et al., 2005) capable of responding to the demands of open-ended, design based instruction within ill-structured classroom environments. Preparing teachers as adaptive experts means moving them beyond their established beliefs and preconceived notions of traditional classroom practice so they teach differently, overcoming their "problem of enactment" (Kennedy, 1999, p. 70) to implement what they know using many different ways of teaching simultaneously, and responding effectively within complex classroom environments by developing adaptive teaching routines (teaching habits of mind) to guide instructional decisions they are confident will support student learning. These are well-documented challenges that must be overcome when learning to teach, but which become significantly magnified when attempting to prepare teachers capable of using technology education as a means for integrating STEM subjects.

In education, the primary goal of any professional development is to affect some degree of change in teacher practice that ultimately improves student learning (Darling-Hammond, 2000; Guskey, 2003). Change theory (Rogers, 2003) recognizes that there are stages of concern teachers progress through as they are confronted with innovations in education. Implementing STEM integration, and in particular through technological/engineering design based learning, is an instructional innovation that raises real concerns for the educator regarding their content, instructional design, and pedagogical preparedness. Any one of these are formidable challenges that create barriers to change in practice. Rogers (2003) identified five factors known to affect teacher receptiveness to instructional innovations: relative advantage, compatibility, complexity, trialability, and observability. To be open to a new innovation, the educator must perceive that it will be as good or better than what they already do, feel that it aligns well with their current educational values, believes that it will not be too difficult for them to implement, has a chance to try out the innovation, and is given the opportunity to observe how others are using it in their classrooms. Durable professional development addresses all of these factors by providing teachers with evidence that the new practice will positively impact student learning outcomes, and affording them multiple opportunities to practice the new techniques and strategies prior to implementing it in their classroom (Wells, 2007). To successfully incorporate integrative STEM education at the classroom level, such professional development will be needed to mitigate the challenges and barriers teachers face when adopting new integrative practices.

References

American Association for the Advancement of Science [AAAS]. (1989). *Science for all Americans*. Washington, D.C.: Author.

Assis-Cezarino, K. (2004). *Strategies for sustainable professional development programs to promote effective pedagogical use of instructional Technology in Teaching*. Unpublished doctoral dissertation. Morgantown, WV: West Virginia University.

Banks, F., & Barlex, D. (2014). *Teaching STEM in the secondary school: Helping teachers meet the challenge*. London: Routledge. Internet: https://dandtfordandt.wordpress.com/

Barak, M., & Assal, M. (2016). Robotics and STEM learning: Students' achievements in assignments according to the P3 Task Taxonomy – Practice, problem solving, and projects. *International Journal of Technology and Design Education, 28*(1), 121–144.

Berube, M., & Berube, C. (2007). *The end of school reform*. Landham, MD: Rowman and Littlefield.

Darling-Hammond, L. (2000). Teacher quality and student achievement: A review of state policy evidence. *Educational Policy Analysis Archives, 8*(1). Retrieved April 7, 2018 from https://epaa.asu.edu/ojs/article/view/392/515

Denson, C., & Lammi, M. (2014). Building a framework for engineering design experiences in high school. *Journal of Technology Education, 26*, 75–87.

Drake, S., & Burns, R. (2004). *Meeting standards through integrated curriculum*. Alexandria, VA: Association for Supervision and Curriculum Development.

Dunham, P., Wells, J., & White, K. (2002). Biotechnology education: A multiple instructional strategies approach to biotechnology education. *Journal of Technology Education, 14*(1), 65–81.

European Parliament, Council of the European Union. (2006). *Recommendation on key competences for lifelong learning*. Internet: https://eur-lex.europa.eu/legal-content/EN/TXT/?uri=celex%3A32006H0962

Ferguson, D. (2009). *Development of technology education in New Zealand schools 1985–2008*. Research paper published by the New Zealand Ministry of Education. Available at https://technology.tki.org.nz/Teacher-education/Archived-papers/Development-of-technology-education-in-NZ-1985-2008

Flemish Government (2012). Actieplan voor het stimuleren van loopbanen in wiskunde, exacte wetenschappen en techniek 2012-2020. Stuk 1478– Nr. 1. 8 februari 2012 (2011-2012) (publication in Dutch). Internet: https://onderwijs.vlaanderen.be/sites/default/files/atoms/files/STEM-actieplan.pdf

Flemish Government (2015). STEM-kader voor het Vlaamse Onderwijs. Brussel. Departement Onderwijs en Vorming (publication in Dutch). Internet: https://www.vlaanderen.be/publicaties/stem-kader-voor-het-vlaamse-onderwijs

Flemish Government (2018). Decreet tot wijziging van de Codex Secundair Onderwijs van 17 december 2010, wat betreft de modernisering van de structuur en de organisatie van het secundair onderwijs. Stuk 1469 Nr. 3 van 28 maart (publication in Dutch). Internet: https://www.vlaamsparlement.be/parlementaire-documenten/parlementaire-initiatieven/1245244

France, B. (1997). *Realising the technology curriculum: Professional development in biotechnology education* (Unpublished D. Phil. thesis). Centre for Science, Mathematics and Technology Education Research, University of Waikato, Hamilton, New Zealand.

France, B. (2007). Location, location, location: Positioning biotechnology education for the 21st century. *Studies in Science Education, 43*(1), 88–122.

France, B. (2015). Technological literacy: A realizable goal or a chimera? ACE papers, issue 5: Issues in educational professional development, paper 3. Available: https://researchspace.auckland.ac.nz/bitstream/handle/2292/25056/ACE_Paper_3_Issue_5.pdf?sequence=1

Gagne, R., Wager, W., Golas, K., & Keller, J. (2004). *Principles of instructional design* (5th ed.). Independence, KY: Cengage Learning.

Garet, M., Porter, A., Desimonre, L. Birman, B., & Yoon, K. S. (2001). What makes professional development effective? Results from a national sample of teachers. *American Educational Research Journal, 38*(4).

Griffin, G. A. (Ed.). (1983). *Staff development. Eighty-second yearbook of the National Society for the study of education.* Chicago: University of Chicago Press.

Guskey, T. R. (2000). *Evaluating professional development.* Thousand Oaks, CA: Corwin Press.

Guskey, T. R. (2002). Does it make a difference? Evaluating professional development. In *Educational Leadership* (pp. 45–51).

Guskey, T. R. (2003). Analyzing lists of the characteristics of effective professional development to promote visionary leadership. *National Association of Secondary School Principals, 87*(637), 38–54.

Hammerness, K., Darling-Hammond, L., Bransford, J., Berliner, D., Cochran-Smith, M., McDonald, M., et al. (2005). How teachers learn and develop. In L. Darling-Hammond & J. Bransford (Eds.), *Preparing teachers for a changing world: What teachers should learn and be able to do* (pp. 358–389). San Francisco: Jossey Bass.

Huber, M. T., & Hutchings, P. (2005). *Integrative learning: Mapping the Terrain.* Washington, D.C.: Association of American Colleges and Universities.

Institute of Education Services [IES] (2017). *Technical working group on career and technical education.* Summary report. Washington, D.C. Available: https://ies.ed.gov/ncer/whatsnew/techworkinggroup/pdf/CTE_TWG.pdf

International Technology Education Association [ITEA]. (1996). *Technology for all Americans: A rationale and structure for the study of technology.* Reston, VA: ITEA.

International Technology Education Association (ITEA) (2000). Standards for technological literacy: Content for the study of technology, Reston VA: Author.

Jones, A. (1997). Technology education in the New Zealand curriculum. In J. Burns (Ed.), *Technology in the New Zealand curriculum – perspectives on practice* (pp. 46–59). Palmerston North, New Zealand: Dunmore Press.

Kearney, C. (2016). *Efforts to increase students' interest in pursuing mathematics, science, technology studies and careers.* National measures taken by 30 countries – 2015 report. Brussels, Belgium: European Schoolnet.

Kelley, T. R. (2008). Cognitive processes of students participating in engineering-focused design instruction. *Journal of Technology Education, 19*(2), 50–64.

Kelley, T. R. (2011). Engineer's notebook – a design assessment tool. *Technology and Engineering Teacher, 70*(7), 30–35.

Kelley, T. R. & Knowles, J.G. (2016). A conceptual framework for integrated STEM education. *International Journal of STEM Education.* Internet: https://stemeducationjournal.springeropen.com/articles/10.1186/s40594-016-0046-z

Kennedy, M. (1999). The role of pre-service teacher education. In L. Darling-Hammond & G. Sykes (Eds.), *Teaching as the learning profession: Handbook of teaching and policy* (pp. 54–86). San Francisco: Jossey Bass.

Korea Institute of Curriculum and Instruction (2002). *A study of the systemization of objectives and contents of school 'Practical arts' and 'Technology & Hoe Economics'.* Republic of Korea, Seoul, Korea.

Lewis, T. (2006). Design and inquiry: Basis for accommodation between science and technology education in the curriculum? *Journal of Research in Science Teaching, 43*(3), 255–281.

Lewis, T. (2007). Engineering education in schools. *International Journal of Engineering Education, 23*, 843–852.
Li, M., & Shavelson, R.J. (2001). *Examining the linkage between science achievement and assessment.* Paper presented at the American Educational Research Association Annual Meeting, Seattle.
Lipson, M., Valencia, S., Wixson, K., & Peters, C. (1993). Integration and thematic teaching: Integration to improve teaching and learning. *Language Arts, 70*(4), 252–264.
Loucks-Horsley, S., Hewson, P., Love, N., & Stiles, K. (1996). *Designing professional development for teachers of science and mathematics*. Thousand Oaks, CA: Corwin Press.
McCormick, R. (2004). Issues of learning and knowledge in technology education. *International Journal of Technology and Design Education, 14*(2), 21–44.
McCormick, R., & Davidson, M. (1996). Problem solving and the tyranny of product outcomes. *Journal of Design & Technology Education, 1*(3), 230–241.
Mundry, S., & Loucks-Horsley, S. (1999). *Designing professional development for science and mathematics teachers: Decision points and dilemmas* (NISE Brief 3(1)). Madison, WI: University of Wisconsin.
NAE (National Academy of Engineering). (2009). *Grand challenges for engineering*. http://www.engineeringchallenges.org/challenges.aspx. Accessed 19 Mar 2011
National Education Association [NEA]. (1894). *Report of the Committee of Ten on Secondary School Studies: With the reports of the conferences arranged by the committee*. New York: American Book Company.
NGSS Lead States. (2013). *Next generation science standards; for states, by states*. Washington, D.C.: The National Academies Press.
Organization for Economic Co-Operation and Development [OECD]. (2008). *Encouraging student interest in science and technology studies*. ISBN 978-92-64-04069-4. Available: https://read.oecd-ilibrary.org/science-and-technology/encouraging-student-interest-in-science-and-technology-studies_9789264040892-en#.WVpUdoTyjIU#page5
Ravitch, D. (2000). *Left back: A century of failed school reforms*. New York: Simon and Schuster.
Robertson, A. (n.d.). *Tech systems for year 3–4 students: Hydroponics*. Available: http://technology.tki.org.nz/Resources/Case-studies/Classroom-practice-case-studies/Food-and-biotechnology/Tech-systems-for-year-3-4-students-Hydroponics
Rogers, E. (2003). *Diffusion of innovations*. New York: Macmillan Publishing.
Ryle, G. (1949). *The concept of mind*. New York: Barnes & Noble.
Schunn, C. (2008). Engineering educational design. *Educational Designer, 1*, 1–23.
Smith, J. E. (1988). Biotechnology. In *New studies in biology* (2nd ed.). London: Edward Arnold.
STEM@school (2018). *Learning modules for 9th grade: Autonomous driving car*. Internet: http://www.stematschool.be/en/our-learning-modules
Thibaut, L., Ceuppens, S., De Loof, H., De Meester, J., Goovaerts, L., Struyf, A., et al. (2018). Integrated STEM education: Conceptualizing an instructional approach for secondary education. *European Journal of STEM Education, 3*(1), 02. https://doi.org/10.20897/ejsteme/85525
Van de Velde, D., Van Boven, H., Dehaene, W., Knipprath, H., & De Cock, M. (2016). *Pré-university STEM education: how do teachers perceive the implementation? Publication in Dutch: Doorstroomgericht STEM-onderwijs: hoe beleven en percipiëren leraren in de tweede graad van het secundair onderwijs de opstart?* Brussels, Belgium: Impuls – Acco. Internet: https://www.acco.be/nl-be/items/TAB000086/Impuls-Jaargang-2016-2017%2D%2D-enkel-abonnement.
Van der Bolt, L., Studulski, F., Van der Vegt, A., & Bontje, D. (2006). *De betrokkenheid van de leraar bij onderwijsinnovaties Een verkenning op basis van literatuur*. Utrecht, Netherlands: Sardes.
Wells, J. G. (2007). Key design factors in durable instructional technology professional development. *Journal of Technology and Teacher Education, 15*(1), 101–118.
Wells, J. G. (2008). *STEM education: The potential of technology education*. Invited paper presented at the 95th annual conference of the Mississippi Valley technology teacher education conference. St. Louis, MO. Available: http://www.mississippivalley.org/archives-2

Wells, J. G. (2013). Integrative STEM Education at Virginia Tech: Graduate preparation for tomorrow's leaders. *Technology and Engineering Teacher, 72*(5), 28–35. Reston, VA.

Wells, J. G. (2016a). PIRPOSAL model of integrative STEM education: Conceptual and pedagogical framework for classroom implementation. *Technology and Engineering Teacher, 75*(6), 12–19.

Wells, J. G. (2016b). Efficacy of the technological/engineering design approach: Imposed cognitive demands within design based biotechnology instruction. *Journal of Technology Education, 27*(2), 4–20.

Wells, J. G. (2017). Design to understand: Promoting higher order thinking through T/E design based learning. *Proceedings of the technology education New Zealand and International conference on technology education-Asia Pacific*, 325–339. New Zealand: TEMS Education Research Center, University of Waikato.

Wells, J. G. (2019a). *Design based biotechnology literacy: Teaching guide* (13th ed.). Blacksburg, VA: Technology Education Biotechnology Curriculum Project; Biosens, (self-published).

Wells, J. G. (2019b). STEM education: The potential of technology education. In M. Daugherty & V. Carter (Eds.), *The most influential papers presented at the Mississippi Valley technology teacher education conference. Council on technology and engineering teacher education, 62nd yearbook* (pp. 195–229). Muncie, IN: Ball State University. Available: http://ctete.org/wp-content/uploads/2019/05/2019-CTETE-Yearbook-Mississippi-Valley-Conference-21st-Century.pdf

Wilschut, A. Pijls, M. (2018). *Effecten van vakkenintegratie: een literatuurstudie.* Kenniscentrum Onderwijs en Opvoeding Hogeschool van Amsterdam.

Winslow, C. (2017). *Meria practical guide to inquiry based mathematics teaching*. Available: http://www.meria-project.eu

Teaching Problem-Solving in the Digital Era

Moshe Barak

Abstract This chapter addresses four unique dimensions of fostering technological problem-solving in the digital era. First, in an era of rapid technological changes, the traditional term problem-solving has acquired a supplementary aspect: inventing new products and services that create new needs that people had not thought of before, for example, the iPhone. The second aspect refers to adopting methods for Systematic Inventive Thinking (SIT) by carrying out systematic manipulations with the attributes or components in a system to solve a problem or create a new product. The third part of the chapter relates to fostering pupils' computational thinking (CT), which relates to solving problems, designing systems, and understanding human behavior by drawing on concepts fundamental to computer science. The fourth part of the paper suggests a more rationalized view of applying the project-based learning methodology in the technological class. The chapter stresses that applying PBL in schools could be more effective after students have gained some basic knowledge and working skills in learning a new subject.

1 Introduction

In the last half century, we have experienced the 'digital revolution' and 'information revolution,' which are dramatically affecting almost every aspect of our lives, for example, the economy, society, workplace, leisure time, and education. Youngsters today think, shop, spend leisure time, and learn differently compared to 20 or 30 years ago. As technology educators, we are required to re-examine what we teach in school, and how we teach it, in order to make technology studies as relevant as possible to students' daily lives and interests, and provide them with the knowledge and tools they need to integrate successfully into today's dynamic society, economy, and workplace.

Let us look, for example, at the changes that took place in the process of driving by car into the big city and parking. In the past, a driver had to navigate his way

M. Barak (✉)
Ben-Gurion University of the Negev, Beer Sheva, Israel
e-mail: mbarak@bgu.ac.il

P. J. Williams, D. Barlex (eds.), *Pedagogy for Technology Education in Secondary Schools*, Contemporary Issues in Technology Education,
https://doi.org/10.1007/978-3-030-41548-8_13

through the city streets, look for a parking place on the street or in a parking lot, and pay for parking by inserting coins into a machine or buying a parking ticket from a nearby machine or shop. New technologies are now helping a driver navigate into the city, find a parking place, and pay for parking by smartphone. We are rapidly approaching the point where we will use an autonomous car that navigates itself, finds parking, and overcomes obstacles without the help of a human driver. On the other hand, the new digital technologies might collect information about people's movements on the roads, and violate peoples' rights for privacy, which is an ethical issue.

These examples of how new technologies impact our lives demonstrate that in order to make technology education relevant in light of the today's technologies, teaching technological problem-solving nowadays must engage students in understanding and developing digital innovative products and services, beyond the design and construction of small electro-mechanical devices, as was common in schools for many years. Today, more than ever, teaching technological problem-solving is not just about dealing with peoples' specific needs or technical issues, but also about preparing students to integrate into the sophisticated technological world in which new innovative products and services appear rapidly.

To elaborate on this challenge, I would like to address four central aspects in this chapter related to teaching problem-solving in the digital era. The first topic in our dissection will relate to a new view of problem-solving and product development, with a focus on *developing new products and services that create new needs and push the market forward*, rather than investigating people's explicit needs. The second subject deals with solving problems and developing new products by the method entitled *Systematic Inventive Solving (SIT)*, which includes attribute analysis and doing systematic manipulations with components and attributes in a given system. The third topic for discussion has to do with fostering *computational thinking* in technology education, with an effort to shed light on this term and demonstrate its application within the context of technological problem-solving. The last section addresses the advantages and limitations of applying *project-based learning* as a pedagogy for teaching technological problem-solving, and suggests an approach for the gradual implementation of this instructional approach in school. A summative discussion and conclusion section will close this chapter.

2 Innovative View of Technological Problem-Solving: From Answering Needs to the Invention of Innovative Products and Services

Historically, problem-solving had to do with answering human needs and volitions. In an era of rapid technological changes, companies often invent new products, technologies, and services that stimulate new needs people have not thought of before. In other words, the old paradigm 'products and services follow needs' is being replaced by the paradigm 'new products and services are designed to create new needs.' A good example of this process is the invention and development of the smartphone. If people would be asked what kind of telephone they would like 20 years ago, no one would have said that he/she needs a small telephone to carry in their pocket enabling them to connect to everyone, everywhere. No one would also have said that the phone should include a camera, Internet, Email, Google, Facebook, navigational software, or WhatsApp applications. In fact, the digital revolution is changing the way we communicate with each other, listen to music, watch movies, shop, book a vacation, handle our bank account, or learn in school and outside of it. All these services have been developed thanks to the imagination and initiatives of scientists and technologists who understand people's 'hidden needs' or know how to produce new needs.

An important aspect of teaching technological problem-solving today also refers to marketing a new product or service. Students who deal with the invention of new technological products must learn basic concepts about marketing, for example, developing marketing channels, expanding sales through new uses for the product, targeting buyers and non-buyers, evaluating if the product is profitable, and closing the gap between the early market and the mainstream. Many innovative products, services, and marketing methods are developed by carrying out systematic manipulations on existing products, for example, changing components or variables in a system, as will be discussed in the following section.

The message to technology educators is that we must provide our students with effective tools for inventing new products and services that no one would have thought of before in the hope that people will ask later "how had we managed without them until now?" The new vision of problem-solving in technology education is not how to find solutions to technical problems or answer given needs, but how to invent new products and services that create new needs and enrich people's lives. Within the context of engaging students in developing innovative technological products and services, there is also room to address market and business development aspects.

3 Teaching Methodologies for Inventive Thinking and Problem-Solving

Naturally, our discussion comes to the question of how to develop creative thinking in technology education. One must admit that despite the huge amount of literature on fostering creativity in science, technology, engineering, or management and trade, the issue has remained rather vague. For example, creativity is often associated with terms such as 'thinking out of the box,' lateral thinking, or brainstorming, which actually deal with a random search for new or original ideas. These methods sometimes work, but often help only little to achieve the goal (Chamorro-Premuzic, 2015; Litchfield, 2009).

In this chapter, I would like to shed light on an alternative approach according to which inventive solutions to problems or the development of innovative products are achieved by conducting systematic manipulations or alterations of the attributes and components existing naturally in a system.

3.1 Attribute Analysis

The literature on product development, marketing, and trade shows that many new products and services are developed as evolutions of older products (Payson, 1997). For example, it is easy to identify that many Apple products such as computers and smartphones in the year 2019 are evolutions of the company's products from 25 years ago. From time to time, however, Apple comes up with brand new products, such as the Apple Watch.

One way of developing new innovative products from existing ones is to perform an attribute analysis of the components of a known product and produce a new product in which some parameters of the existing product have been changed. For example, let us list the components of a conventional chair:

- Seat
- Legs
- Back

The next step is listing the attributes (not functions) of each component:

- Material.
- Shape
- Size
- Color
- Hardness/softness
- Waterproof

Now, we can try carrying out different manipulations with the attributes of the chair components. For example, if we eliminate the back, the chair becomes a stool. To this end, it is useful to use the SCAMPER method described in the next section.

3.2 Scamper

Alex Osborn (1963), a pioneering teacher of creativity, identified nine principal ways of manipulating a subject, entitled SCAMPER, as listed below:

- S – Substitute
- C – Combine
- A – Adapt

Fig. 1 A beach chair as an evolution of a conventional chair

- M – Modify/Magnify
- P – Put to Other Uses
- E – Eliminate
- R – Rearrange/Reverse

Concerning the chair example mentioned above, we could think about *combining* the seat and the back and *eliminating* the legs to create a beach chair, as seen in Fig. 1.

Barak (2007) showed an example of how eliminating a component or function from a device could be an effective tool. In 1990, when mobile phone calls were very expensive, an Israeli cellular company came out with the Mango mobile phone in which the dialing function was eliminated. The user could receive calls but dial only one number by pressing the digit 1. Companies provided this device to drivers or service technicians so that they could receive calls when out of the office, but call back only to the office. The Systematic Inventive Thinking (SIT) method described in the following section has some parallel lines with SCAMPER, but includes even more powerful problem-solving and new product development tools.

3.3 Systematic Inventive Thinking (SIT)

Systematic Inventive Thinking (SIT) is a method of finding solutions to problems by making systematic alterations or manipulations with a system's components and attributes, rather than searching randomly for ideas using methods such as brainstorming. The SIT method (Boyd, 2007; Boyd & Goldenberg, 2013; Goldenberg & Mazursky, 2002; Barak, 2004, 2010) was derived from the TRIZ theory of inventive problem-solving (Altshuller, 1988).

Among the principles or tools learned in the SIT course are:

- *Unification*: solving a problem by assigning a new use or role to an existing object
- *Multiplication*: solving a problem by introducing a slightly modified copy of an existing object into the current system
- *Division*: solving a problem by dividing or cutting an object or subsystem and reorganizing its parts
- *Change relationships between variables (attribute dependency)*: solving a problem by adding, removing, or altering relationships between variables

- *Removal*: solving a problem by removing an object (with its main function) from the system
- *Inversion*: solving a problem by inverting the structure or functions of components in a system

Following is an example of use of the SIT method by students in a final exam during an inventive problem-solving course held at Ben-Gurion University of the Negev (Barak & Albert, 2017). One of the exam questions was "Suggest a method of how to encourage youngsters to use special night buses for going out on weekends instead of driving a car (especially for those who drink…)."

Examples of conventional solutions to this question are providing free buses or having more police on the roads. To find an inventive solution using the SIT method, we first make a list of the components in the world of the problem: *youngsters, cars, police, buses, pubs, alcohol, music, etc.* We then try to find a solution by carrying out systematic manipulations in the system's components according to SIT principles. One of the SIT principles or 'tools' is *unification*: assigning a new use or role to an existing object in the system. A solution some students suggested was *serving drinks and playing music on a bus*. In terms of the SIT method, this solution assigns the role of a pub to a bus, as illustrated in Fig. 2.

In their book entitled *Creativity in Product Innovation* (2002), Goldenberg and Mazursky use the term 'Creativity templates' to describe the SIT and other systematic methods for thinking along inventive routes in order to target creative thoughts. These methods are often manifested in the sequence of inventing a new product or a new configuration for a product, and then inferring the benefits or market values of the product. These authors define this perspective as 'Function Follows Form'

Fig. 2 Pub-bus: assigning the role of a pub to a bus

(p. 41), in contrast to the conventional approach of 'Form Follows Function.' The notion of 'Function Follows Form' has been associated with some of Apple's revolutionary products in which people were first surprised by a new product design but quickly learned to use and enjoy the product. Goldenberg and Mazursky (1999) suggested the concept of 'The voice of the product' to describe the process in which new innovative products create a market, in contrast to the conventional approaches of 'The voice of the customer' or 'The voice of the market' in developing new products.

The concept of 'Function Follows Form' in contrast to 'Form Follows Function' has been discussed in the context of architectural design. For example, when an architect pursues designing a unique or 'iconic' building for a shopping center or a library, the 'form' of the building might significantly affect its function. According to famous American architect and writer Frank Lloyd Wright (1908), "Form follows function – that has been misunderstood. Form and function should be one, joined in a spiritual union."

Another example of applying SIT concepts for problem-solving and new product development is the case of developing a 13-km fast lane on Road no. 1 from Jerusalem and Ben Gurion Airport to Tel Aviv. Drivers who want to enter Tel Aviv quickly are required to pay a toll ranging from 7 to 105 ILS (Israeli shekels) (£1.5–22.5), depending on traffic on the regular road, as illustrated in Figs. 3 and 4. To take advantage of the fast lane for the benefit of the entire population and not just rich people, public buses, taxis, and private cars carrying 1 + 3 or more passengers can travel on this route for free.

Fig. 3 A fast lane on the road from Jerusalem to Tel Aviv

Fig. 4 Solving a problem by applying attribute dependency: the toll for traveling in the fast lane escalates as traffic on the regular road increases

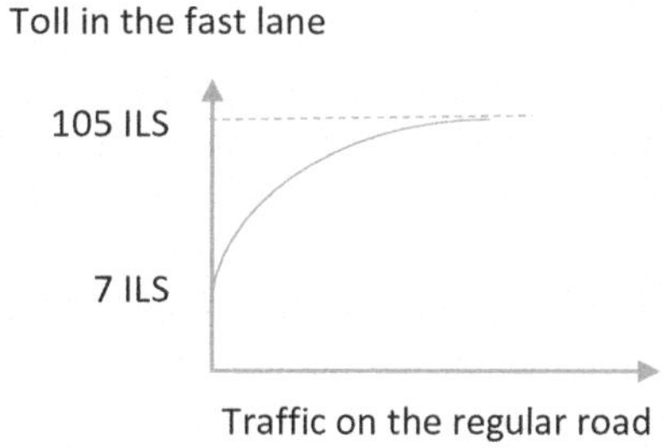

The system described in Figs. 3 and 4 is unique in that the toll for driving in the fast lane might change from minute to minute, and drivers see the actual toll as they approach the entrance to the fast lane. The project also includes a free parking area and rapid shuttle to the city center, not described in the current paper. This system is computer-controlled and has been working very successfully for several years.

A Question for Readers

A woman booked two expensive flight tickets online for herself and her daughter. By mistake, she wrote the daughter's first name on both tickets. The flight company absolutely refused to change the name on the tickets or give a refund for the ticket and suggested that the consumer purchase an additional ticket.

Suggest a solution to the woman based on the SIT method.

Turner (2009) presents the 'Advanced Systematic Inventive Thinking' (ASIT) system as an eco-design strategy in the context of technology education. Moon, Ha and Yang (2012) show the use of ASIT for structured idea creation to improve the value of construction design.

4 Fostering Computational Thinking (CT) in Technology Education

In the Introduction to this chapter, we have seen that teaching technology education and technological problem-solving today must address the digital and information revolution affecting almost every aspect of our lives. In the past, educators often associated the term computer literacy with teaching computer science, and programming in particular, to students of all ages. Recently, educators and stakeholders have recognized that computer science is not a required subject area for all students, and is not always prioritized by school boards, partially because of the lack of qualified teachers and other essential resources for teaching this subject as a core subject area (Yadav, Hong, & Stephenson, 2016). Recently, however, attention has focused on the term computational thinking (CT), which refers to a variety of aspects related to understanding and using computer systems.

Yadav, Stephenson, and Hong (2017) point out that CT offers an encompassing approach that exposes students to computing ideas and principles within the context of the subject areas they are already learning, for example, mathematics, science, technology, or the humanities. Wing (2006) defined computational thinking as "solving problems, designing systems, and understanding human behavior by drawing on concepts fundamental to computer sciences." According to this author, CT involves three key concepts:

- Algorithm – a step-by-step series in instruction
- Abstraction – generalizing and transferring the problem-solving process to similar problems
- Automation – using digital and simulation tools to mechanize the problem's solution

Several researchers (Barr, Harrison, & Conery, 2011; Barr & Stephenson, 2011) describe computational thinking as a problem-solving process that includes:

- Formulating problems in a way that enables us to use a computer and other tools to help solve them
- Logically organizing and analyzing data
- Representing data through abstractions, such as models and simulations
- Automating solutions through algorithmic thinking (a series of ordered steps)
- Identifying, analyzing, and implementing possible solutions with the goal of achieving the most efficient and effective combination of steps and resources
- Generalizing and transferring this problem-solving process to a wide variety of problems

Yadav et al. (2017) quote a report by the US National Council for Research (NRC) according to which "computational thinking is a cognitive skill the average person is expected to possess, involving, for example, the use of heuristics, a problem-solving approach that involves applying a general rule of thumb or strategy that may lead to a solution." This definition indicates that CT has some parallel lines with the concept of SIT discussed in the previous section, which also suggest that problem-solving often requires using rich strategies and heuristics, rather than carrying out a random search or following a linear process.

In the following sections, I will present examples of how technological problem-solving, design, and new product development are a natural platform for fostering computational thinking.

4.1 Example A: Using a Computer for Audio Analysis in Learning About Sound, Waves, and Communication Systems

At Ben-Gurion University of the Negev, we developed a STEM-oriented program (30 h) for junior high schools in which the students learn about sound, waves, and communication systems (Awad & Barak, 2016). The course aims at providing junior high-school students with: (1) scientific concepts, such as transitive wave, longitude wave, period (T), frequency (f), wavelength (λ), amplitude (A), sound velocity (v), and sound propagation on different materials or states of matter; and (2) technological concepts, such as sound system, microphone, speaker, amplifier, amplification process, analog-to-digital conversion process, digital sound. Class sessions combine the teacher's presentations, hands-on lab work, using simulation, and project-based learning. In this chapter, I present only one example of students' activities in

Fig. 5 Students use the Audacity software to observe the signal obtained from a microphone connected to a computer's audio input

the class – use of the Audacity professional software for audio signal recording and analysis. Figure 5 illustrates the case of students connecting a microphone to a computer's audio input and observing the signal obtained on the computer screen.

Figure 6 shows the signals obtained from the microphone in three cases: a single knock on the table, three slow knocks on the table, and three quick knocks on the table. One can see that the signals presented in Fig. 6a show the frequency of the sound obtained from knocking on the table, while Fig. 6b, c demonstrates frequencies of the slow and fast knocks on the table.

The example described in Figs. 5 and 6 demonstrates how learning technology deals with fostering students' computational thinking (CT). After the students learn the basic scientific, technological, and computing aspects of sound, it is the educator's role to encourage students to suggest innovative applications of these tools, such as devices, control systems, or services, based on sound analysis and recognition. For example, the students can apply this system to explore the velocity of sound in different materials such as air and water, measure the distance between two objects in robotics, or build a control system activated by sound recognition. As mentioned earlier in this chapter, the challenge for the teacher and the students is to develop new products and services that people have not thought of before.

4.2 *Example B: Learning Concepts of Digital Image Processing*

Image processing is one of the most fascinating subjects in the field of computer science and technological applications. Although professional image processing uses advanced mathematics and programming, we can teach basic concepts of image processing to young children (for example, junior high school students), as was observed in a course developed at Ben-Gurion University of the Negev (Barak & Asad, 2012).

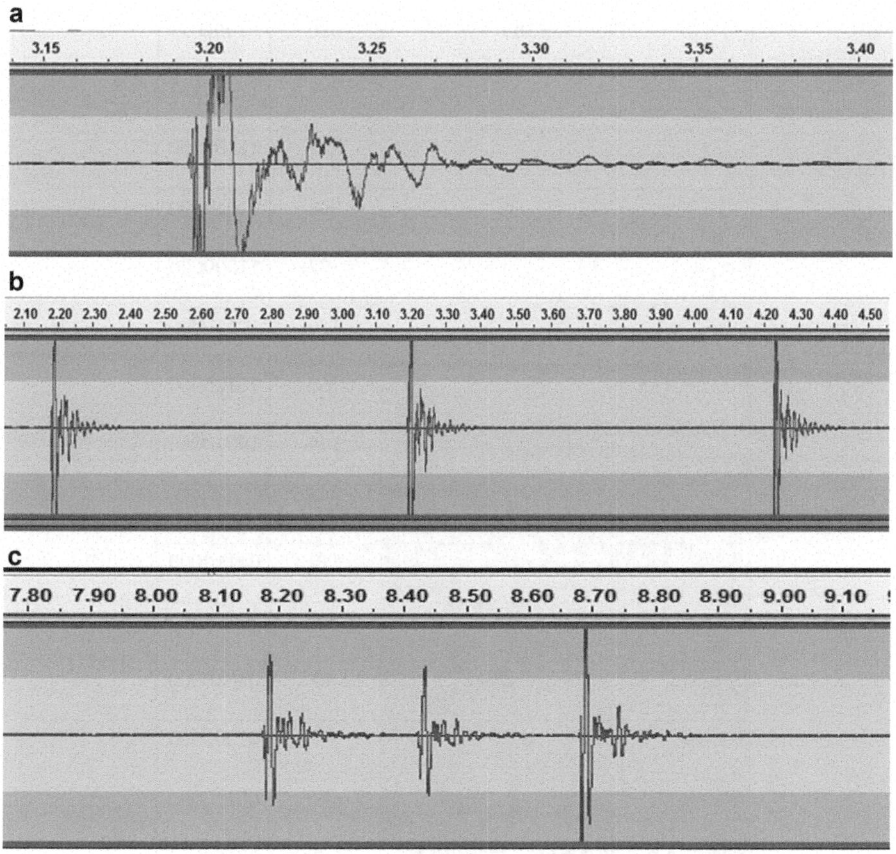

Fig. 6 Analyzing the sound obtained from a microphone in the cases of one knock, three slow knocks, and three quick knocks on a table. (**a**) A single knock on the table recorded by Audacity (time in seconds). (**b**) Three slow knocks recorded by Audacity (time in seconds). (**c**) Three quick knocks recorded by Audacity (time in seconds)

Figure 7 shows that in presenting a black-and-white image, which is actually composed of gray levels, we can use a format of 8-bit digital data, which distinguishes between 256 brightening levels from 0_{10} (full black) to 255_{10} (full white).

Figure 8 shows an example of 'digitizing' a picture to 12 × 16 pixels, assigning a number (brightness level) to each pixel, and finally presenting the entire picture in a matrix of 12 × 16 numbers. It is worth mentioning that in real life, for example, on a computer screen or on a digital camera, we often use much higher resolution, such as 1440 × 1080 pixels.

So far, we have seen the method of digitizing a black-and-white picture. How is a color picture digitized? In a color picture, each pixel gets a different color. From physics, we know that any color could be created by a composition of the three basic colors of red, green, and blue (RGB). Figure 9 illustrates an example of creating an orange pixel through the composition of red = 255, green = 178, and blue = 102

Color Intensity	DEC	BIN
	255	11111111
	224	11100000
	208	11010000
	192	11000000
	176	10110000
	160	10100000
	144	10010000
	128	10000000
	112	01110000
	96	01100000
	80	01010000
	64	01000000
	48	00110000
	32	00100000
	16	00010000
	0	00000000

Fig. 7 Representation of 256 brightness levels, from black (0_{10}) to white (255_{10})

(for details, search RGB orange in Google). A colored picture of n x m pixels is created by a composition of three matrixes of the same dimension representing the R, G, and B colors.

In summary, when we take a picture with our digital camera, we are actually saving a file of numbers, and computers can handle numbers very well. For example, if we want to increase the brightness of an entire image, we will add a number, for example, 20, to the brightness level of each pixel.

Question for Readers

- Check the resolution of pictures on your computer screen, digital camera, or mobile phone.
- Suggest a method to erase a section of a picture.
- Read about the objectives and methods of file compression.
- Calculate the file size of a colored picture of 1024 × 768 pixels (8 bit/pixel) without compression.

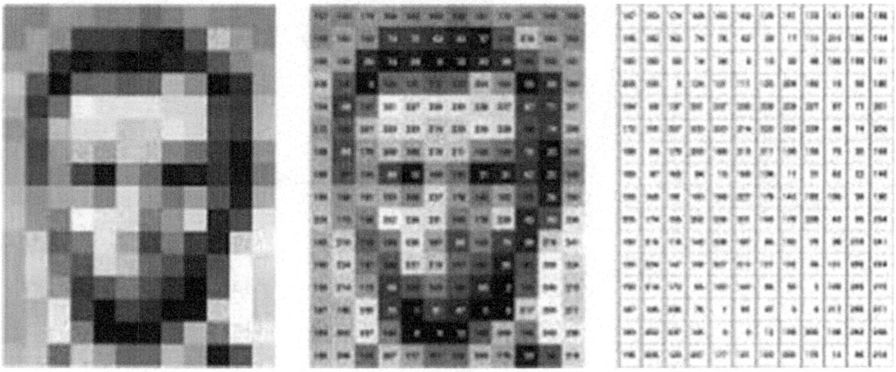

Fig. 8 Image representation by 12 × 16 pixels at brightness levels 0–255

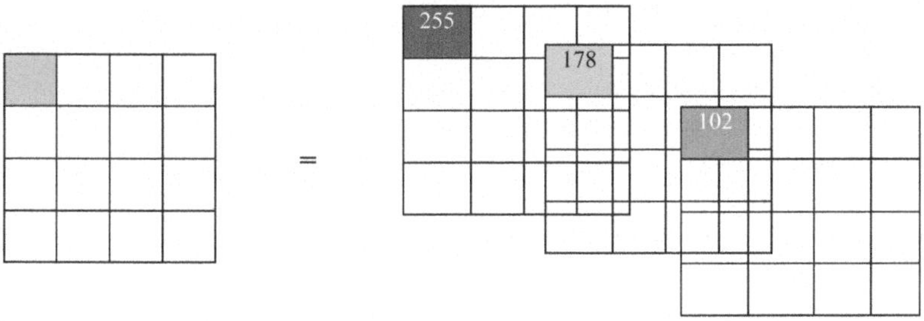

Fig. 9 An orange pixel created by the red, green, and blue (RGB) colors

4.3 *Facial Recognition*

Facial recognition is one of the most interesting image processing applications (Midrak, 2018). However, how can we use a digital image to identify a person's face? Following is a simple method for facial recognition that was learned successfully by junior high school students (Barak & Asad, 2012). Figure 10 shows that for facial recognition, we can measure eight parameters from the picture of a face:

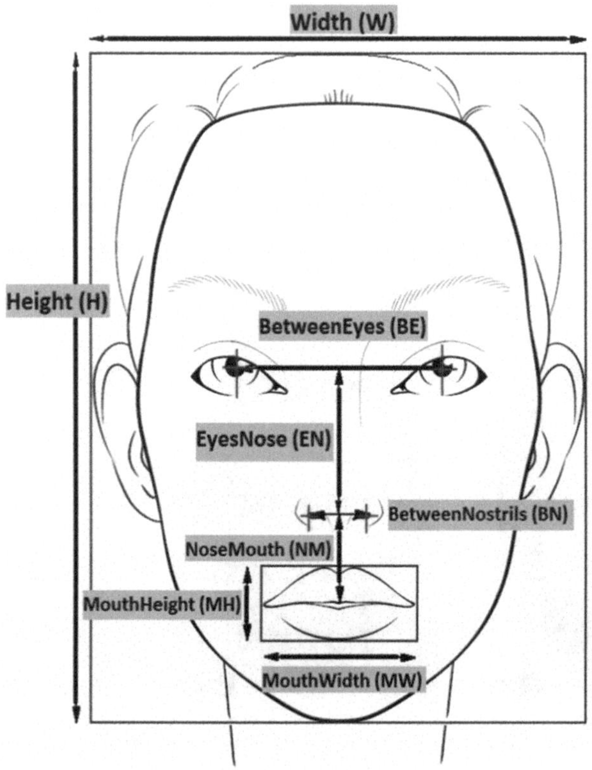

Fig. 10 Eight parameters used for facial recognition

- BE – distance between eyes
- BN – distance between nostrils
- MW – mouth width
- EN – distance between eyes and nose
- NM – distance between nose and mouth
- MH – mouth width
- W – total width
- H – total height

Let us assume that we have values of the eight parameters for two faces:

$$\text{Face } 1 = \{\text{BE1, BN1, MW1, EN1, NM1, MH1, W1, H1}\} \tag{1}$$

$$\text{Face2} = \{\text{BE2, BN2, MW2, EN2, NM2, MH2, W2, H2}\} \tag{2}$$

Since the faces could be of different sizes, we normalize the six parameters B1 to MH by dividing each one by total width w or total height h, respectively. We get six normalized parameters a, b, c, d, e, and f for each picture, as seen in Eqs. 3 and 4.

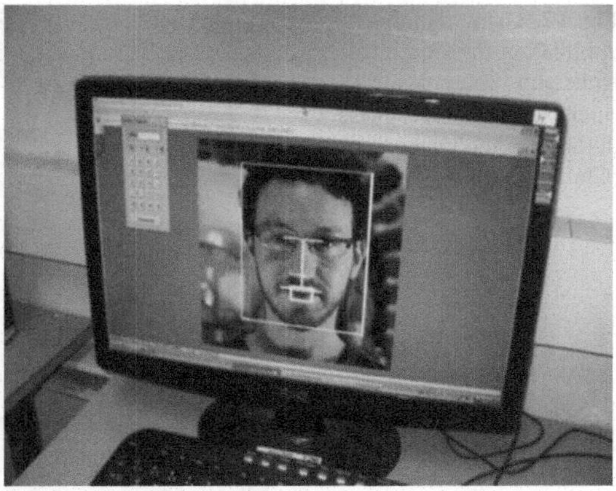

Fig. 11 Measuring the eight facial parameters

$$\begin{aligned} a1 &= BE1/W1 \quad b1 = BN1/W1 \quad c1 = MW1/W1 \\ d1 &= WN1/H1 \quad e1 = NM1/H1 \quad f1 = MH1/H1 \end{aligned} \tag{3}$$

$$\begin{aligned} a2 &= BE2/W2 \quad b2 = BN2/W2 \quad c2 = MW2/W2 \\ d2 &= WN2/H2 \quad e2 = NM2/H2 \quad f2 = MH2/H2 \end{aligned} \tag{4}$$

In this process, a vector of parameters a–f represents each facial picture.

The last step is calculating the difference D between the vectors of two faces by the formula presented in Eq. 5.

$$D = \sqrt{\begin{aligned} &(a1-a2)^2 + (b1-b2)^2 + (c1-c2)^2 + (d1-d2)^2 + (e1-e2)^2 \\ &+ (f1-f2)^2 \end{aligned}} \tag{5}$$

Equation 5 shows the calculation of the 'Euclidean distance' between two vectors of six dimensions each, which can be explained as an extension of the Pythagoras theorem for calculating the diagonal in a straight triangle. Students can perform the process described above using a spreadsheet.

In class, the students worked in pairs, took photos of themselves, and manually measured the eight parameters BE, BN, MW, EN, NM, MH, W, H of their photos, as illustrated in Fig. 11. They entered the data into a simple computer program that created a database of vectors for all students' faces in the class.

To identify an 'unknown' student, the program calculates the difference D between his/her vector of facial parameters and the vector of each student included in the class database. The case with the smallest D would indicate the face of the

Fig. 12 Using a mobile application for measuring fruits and vegetables ripeness quality, freshness, durability, and taste (Clairfruit, 2018)

unknown student. An interesting question was to check the rate of the system's success or failure in facial recognition using this method.

A Question for Readers

Suggest a method for number recognition, namely, recognition of digits 0–9.
Propose practical applications for this tool.

In the technological class, the students could raise their own examples of using digital technologies for systems design and problem-solving as either class exercises or final projects. For example, in one of the middle classes in which the course on image processing (Barak & Asad, 2012) was taught, the students were interested in using their digital cameras to measure the height of the tower in the mosque from which the muezzin calls people to come to prayer. They took pictures of the high tower, a lower building, and a one-meter-high pillar. By observing the number of pixels at the height of each image and finding the scale ratio, they calculated the height of the mosque and the building examined.

Professional facial recognition systems use more sophisticated algorithms than the method described above. However, this program demonstrates a way of fostering students' computational thinking, including many of the aspects described by Barr and Stephenson (2011), as mentioned above. After learning this unit, students could understand some image processing concepts and use professional applications wisely to develop their new projects on the computer or smartphone. For example, the Israeli company Clairfruit (2018) developed a mobile application for measuring ripeness quality, freshness, and durability of fruits and vegetables, as seen in Fig. 12. It is the challenge for technological education to develop students' aptitudes and thinking patterns for inventing such products on their own.

5 Applying Project-Based Learning in School: The Need for Careful Design of Students' Assignments

Since teaching problem-solving is a central subject in technology education, it is essential to look at the general literature on problem-based learning and project-based learning, which have become among the most common methods for applying student-oriented instruction. Problem- or project-based learning is aimed at engaging students in solving reality-based problems, and encouraging them to become active, independent, and collaborative learners. The digital revolution that has been affecting all aspects of our lives, including education, has pushed forward efforts to introduce PBL into traditional schooling because students today have access to tremendous resources and tools on the network for investigating scientific and technological issues, suggesting solutions to problems and designing innovative technological systems. However, despite the wide consensus in the literature about the advantages of PBL over traditional schooling, educators are increasingly aware of the limitations of applying these methods within the regular school context. Kirschner, Sweller and Clark (2006) write about the failure of constructivist-oriented instructional methods such as discovery and problem-based and inquiry-based learning, because the notion of minimal guidance during learning does not work. Minimally guided instruction is less effective and efficient than instructional approaches, which place strong emphasis on guiding the student learning process (Hushman & Marley, 2015). Some supporters of PBL (Hmelo-Silver, 2004; Hmelo-Silver, Duncan, & Chinn, 2007; Savery, 2006) address the limitations of this method, and mention that it is important to tailor the scope and complexity level of assignments to students' prior knowledge and skills, and provide instruction and support in order to reduce the cognitive load and enable students to learn in a complex domain. Dolman et al. (Dolmans, de Grave, Wolfhagen, & van der Vleuten, 2005) also write that PBL curricula should consist more of tutor guidance at the beginning through shared guidance of both the students and the tutor, and move to more student guidance at the end.

To address this issue, we developed the P3 Task Taxonomy, which distinguishes between three levels of student assignments:

- *Practice*: exercises and closed-ended tasks in which learners know the final solution in advance and can check if they arrived at the correct answer
- *Problem-solving*: small-scale, open-ended tasks in which students might use different solution methods and arrive at different answers
- *Projects*: challenging open-ended tasks in which the students take part in defining the problem, setting objectives, identifying constraints, and choosing the solution method

For example, let us look at the case of building an automated watering system for a home garden that contains flowers, plants, and grass.

- At the practice level, the gardener receives the full design of the irrigation system and has to install pipes, sprinklers, etc., set up the electronic controller, and check the system's operation including troubleshooting.
- At the problem-solving level, the gardener receives the design of the garden, must learn the irrigation required for each part or component in the garden and design the optimal irrigation system to achieve this end while saving water.
- At the project level, the gardener's task is to design the entire garden. He/she must propose a garden that suits the owners' expectations, on the one hand, and comply with local soil and climate conditions, on the other hand.

This example demonstrates that only an experienced designer can cope with a task at the project level. The P3 Task Taxonomy helped in designing students' tasks in the course on science, wave, and communications systems (SWCS) mentioned above (Awad & Barak, 2016) and a robotics course for junior high schools (Barak & Assal, 2018). An earlier version of this taxonomy guided the design of students' tasks in a project for developing a system for computer engineering studies at the university level (Kastelan et al., 2014).

To conclude, project-based learning is certainly an excellent approach for developing students' problem-solving creativity competencies. However, this method should be applied gradually, while adapting the tasks assigned to the students' prior knowledge and skills. In the technological lab, in particular, students might want to use sophisticated, expensive, or dangerous equipment and tools such as 3D printers, electronic micro-controllers, or machines for metal or wood works. Tutor guidance is essential, and learners can only work with such equipment independently after acquiring the appropriate basic knowledge and experience.

6 Summary and Conclusions

This chapter casts light on four unique characteristics of fostering technological problem-solving in the digital era, as illustrated in Fig. 13:

Following is a brief summary of the four distinctive aspects of technological problem-solving discussed in this work:

- In the modern area, inventing new products and services is often designed to create new uses and needs in ways people had not thought of before. This is an innovative approach, beyond the traditional view of problem-solving as identifying and answering people's needs and desires.
- Using methods for 'Systematic Inventing Thinking' means carrying out systematic manipulations with a system's attributes or components to solve a problem or create a new product. This is an opposite, and complementary, approach to the conventional method of randomly searching for new ideas by methods such as brainstorming.
- Fostering students' computational thinking (CT), which is an important target of technology education today, relates to solving problems, designing systems, and

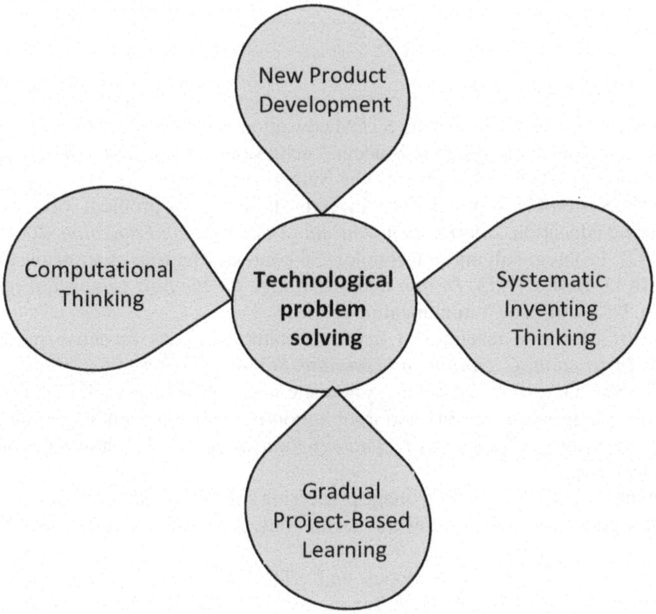

Fig. 13 Aspects of fostering technological problem-solving in the digital era

understanding human behavior by drawing on concepts fundamental to computer science. This may include, for example, data collection, analysis and representation, problem decomposition or abstraction, and using algorithms, procedures, automation, and simulation for problem-solving.

- A rationalized approach to applying the project-based learning (PBL) methodology in the technological class takes into account that many students could benefit from PBL only after they have gained some basic knowledge and working skills in learning a new subject. The P3 scale – practice, small-scale problem-solving, and broad open-ended projects – could help educators in designing effective curricula and project-based learning in the technological class.

6.1 Concluding Comment

The challenge for technology teachers and students today is to design products and services for the future, and to use new and emerging technologies in their design proposals, as suggested by Barlex and Trebell (2008). The new technologies are often virtual, global, and dynamic, and they rapidly produce new products and services extensively affecting individuals' lives, society, and the economy. The hope is that the ideas discussed in this chapter will help educators achieve this goal.

References

Altshuller, G. S. (1988). *Creativity as an exact science*. New York: Gordon and Breach.

Awad, N., & Barak, M. (2016). Exploring STEM education in the context of learning about sound, waves and communication systems: Students' achievements and motivation. *Pupils' attitude towards technology* (PATT-32). Utrecht, The Netherlands, August 23–26.

Barak, M. (2004). Systematic approaches for inventive thinking and problem solving: Implications for engineering education. *International Journal of Engineering Education, 20*(4), 612–618.

Barak, M. (2007). Problem solving in technological context: The role of strategies, schemes and heuristics. In D. Barlex (Ed.), *Design and technology for the next generation* (pp. 154–169). Whitchurch, UK: Cliffeco Communications.

Barak, M. (2010). How the teaching of heuristic methods affects inventive problem solving. *Technology, Instruction, Cognition and Learning, 8*(3–4), 273–296.

Barak, M. &, Albert, D. (2017). Fostering systematic inventive thinking (SIT) and self-regulated learning (SRL) in problem-solving and troubleshooting processes among engineering experts in industry. *Australasian Journal of Technology Education, 4*(1–14). http://ajte.org/index.php/AJTE/article/view/45

Barak, M., & Asad, K. (2012). Teaching image processing concepts in junior high school: Boys' and girls' achievements and attitudes towards technology. *Research in Science and Technological Education, 30*(1), 81–105.

Barak, M., & Assal, M. (2018). Robotics and STEM learning: Students' achievements in assignments according to the P3 Task Taxonomy – Practice, Problem solving, and Projects. *International Journal of Technology and Design Education, 28*(1), 121–144. https://doi.org/10.1007/s10798-016-9385-9

Barlex, D. M., & Trebell, D. (2008). Design-without-make: Challenging the conventional approach to teaching and learning in a design and technology classroom. *International Journal of Technology and Design Education, 18*(2), 119–138. https://doi.org/10.1007/s10798-007-9025-5

Barr, D., Harrison, J., & Conery, L. (2011). Computational thinking: A digital age skill for everyone. *Learning & Leading with Technology, 38*(6), 20–23. https://www.iste.org/docs/learning-and-leading-docs/march-2011-computational-thinking-ll386.pdf

Barr, V., & Stephenson, C. (2011). Bringing computational thinking to K-12: What is involved and what is the role of the computer science education community? *ACM Inroads, 2*(1), 48. https://doi.org/10.1145/1929887.1929905. https://c.ymcdn.com/sites/www.csteachers.org/resource/resmgr/BarrStephensonInroadsArticle.pdf

Boyd, D. (2007). A structured, facilitated team approach to innovation. *Organization Development Journal, 25*(3), 119–122.

Boyd, D., & Goldenberg, J. (2013). *Inside the box: A proven system of creativity for breakthrough results*. New York: Simon & Schuster.

Chamorro-Premuzic, T. (2015). Why group brainstorming is a waste of time. *Harvard Business Review*, 25.

Clairfruit. (2018). *A mobile application for measuring fruits quality and taste*. https://www.clarifruit.com/

Dolmans, D. M., de Grave, W., Wolfhagen, I. P., & van der Vleuten, C. M. (2005). Current perspectives: Problem-based learning: Future challenges for educational practice and research. *Medical Education, 39*(7), 732–741.

Goldenberg, J. & Mazursky, D. (1999). The voice of the product: Templates of new product Emergence. *Creativity and Innovation Management*, 157–164.

Goldenberg, J., & Mazursky, D. (2002). *Creativity in product innovation*. Cambridge, UK: Cambridge University. https://doi.org/10.1017/CBO9780511674464

Hmelo-Silver, C. E. (2004). Problem-based learning: What and how do students learn? *Educational Psychology Review, 16*(3), 235–266. https://dandtfordandt.wordpress.com/2016/09/11/dt-for-the-next-generation/

Hmelo-Silver, C. E., Duncan, R. G., & Chinn, C. A. (2007). Scaffolding and achievement in problem-based and inquiry learning: A response to Kirschner, Sweller, and Clark 2006. *Educational Psychologist, 42*(2), 99–107.

Hushman, C. J., & Marley, S. C. (2015). Guided instruction improves elementary student learning and self-efficacy in science. *Journal of Educational Research, 108*(5), 371–381.

Kastelan, I., Lopez, B. J. R., Artetxe, G. E., Piwinski, J., Barak, M., & Temerinac, M. (2014). E2LP: A unified embedded engineering learning platform. *Journal of Microprocessors and Microsystems, 38*(8), 933–946.

Kirschner, P. A., Sweller, J., & Clark, R. E. (2006). Why minimal guidance during instruction does not work: An analysis of the failure of constructivist, discovery, problem-based, experiential, and inquiry-based teaching. *Educational Psychologist, 41*(2), 75–86.

Litchfield, R. (2009). Brainstorming rules as assigned goals: Does brainstorming really improve idea quantity? *Motivation & Emotion, 33*(1), 25–31.

Midrak, R. L. (2018). What is facial recognition? *Lifewire, 3*. https://www.lifewire.com/how-does-a-computer-recognize-your-face-4154178

Moon, S., Ha, C., & Yang, J. (2012). Structured idea creation for improving the value of construction design. *Journal of Construction Engineering & Management, 138*(7), 841–853.

Osborn, A. F. (1963). *Applied imagination: Principles and procedures of creative problem solving* (3rd ed.). New York: Charles Scribner's Sons.

Payson, S. (1997). Product evolution and the classification of business interest in scientific advances. *Knowledge & Policy, 9*(4), 3–26.

Savery, J. R. (2006). Overview of problem-based learning: Definitions and distinctions. *Interdisciplinary Journal of Problem-Based Learning, 1*(1), 9–20.

Systematic Inventive Thinking [SIT]. (2018). From Wikipedia. https://en.wikipedia.org/wiki/Systematic_inventive_thinking

Turner, S. (2009). ASIT—A problem solving strategy for education and eco-friendly sustainable design. *International Journal of Technology & Design Education, 19*(2), 221–235.

Wing, J. M. (2006). Computational thinking. *Communications of the ACM, 49*(3), 33–35.

Wright, F. L. (1908). Quotations page. http://www.quotationspage.com/quotes/Frank_Lloyd_Wright/.

Yadav, A., Hong, H., & Stephenson, C. (2016). Computational thinking for all: Pedagogical approaches to embedding 21st century problem solving in K-12 classrooms. *Tech Trends, 60*(6), 565–568. https://doi.org/10.1007/s11528-016-0087-7

Yadav, A., Stephenson, C., & Hong, H. (2017). Computational thinking for teacher education. *Communications of the ACM, 60*(4), 55–62.

Pedagogical Approaches to Vocational Education

P. John Williams

Abstract The thesis of this chapter is that a disconnect exists between the goals of vocational education and the pedagogies most effective in the achievement of those goals. The chapter will describe the current situation, propose frameworks within which a critique of the current situation can be made and then make some conclusions about the elements of appropriate pedagogical practice.

The context for this chapter is the situation in Australia. It is acknowledged that the place of vocational education in different countries varies, but it is hoped that the discussion about the relationship between the goals of education and the pedagogies to achieve those goals will be applicable to a range of contexts.

1 Introduction

The vocational education system in many countries, and in particular the Australian system with which the author has familiarity, has significant industry involvement in the determination of the knowledge and skills to be mastered in preparation for students to be placed in each particular industry. This knowledge and these skills are organized to form a developmental instructional sequence, and then stated in the language of competencies. Students practice the competencies until they are judged to have achieved mastery, and then they move on to the next sequence of competencies. The resultant list of competencies to be achieved forms the basis of a mastery instructional program and represents the measures of external standards. This mechanism for institutional and government control has the effect of rendering vocational training instrumental and disjointed, but administratively accountable. The

P. J. Williams (✉)
Curtin University, Perth, WA, Australia
e-mail: pjohn.williams@curtin.edu.au

P. J. Williams, D. Barlex (eds.), *Pedagogy for Technology Education in Secondary Schools*, Contemporary Issues in Technology Education,
https://doi.org/10.1007/978-3-030-41548-8_14

instructional approach in this context most commonly involves the demonstration of the application of knowledge to a skill, followed by the student's practice until mastery is achieved. The curriculum consists of a sequence of tasks, sometimes related, and the pedagogy applied is skill demonstration followed by individual practice and support.

On the other hand, many employers report the need for new employees who have developed a range of transversal competencies and adaptable occupational knowledge, such as thinking creatively, problem solving and collaboration, which can best be achieved through a focus on learning processes. They argue that specific competencies related to machines and manipulative skills can be taught on the job, and these change within short periods of time anyway, whereas the transversal competencies are required to enable employees to be versatile and flexible within an organization.

The pedagogical approach required to address these transversal competencies would tend to be more student centred and project or problem based, involving a focus on a design process (Billett, 2016), rather than teacher centred with a focus on competencies. This would enable the students to have a degree of creative input into their studies, which would in turn facilitate the development of the transversal competencies.

2 Status of Vocational Education

The status of vocational education is an essential contextual consideration in this discussion. It has a low status, which stems from the enduring philosophy that vocational education is somehow more simple than other forms of education encompassing the arts, humanities or the sciences, and so the curriculum and the pedagogies associated with it can also be more simple. These social values have been perpetuated by social elites who have little understanding of the essence of the system, and certainly have not participated in it themselves. The belief that vocational education is of low value, involves routine tasks and lower order thinking, is reflected in the instrumentalist approach taken to the curriculum and its delivery. This low status (socially and intellectually) has mitigated against permitting those in the system playing a more influential role in its design.

The low status of vocational education represents the confluence of a number of factors and in turn generates attitudes which reinforce this status. It is:

- Positioned as pragmatic and instrumental
- Held to be easy to learn and straightforward
- Ranked low on qualifications frameworks
- Restricted to low academic achievers
- Creates barriers for community and parent engagement
- Assessed binarially (competent or not)

As someone who received a PhD in political philosophy and then went even further to become a motor cycle mechanic, Matthew Crawford (with the support of others such as Barbieri-Low, 2007; Epstein, 1998) makes the case for the intellectual sophistication of trade work through his experience of finding manual work more engaging intellectually compared with his work as Director of a Washington think tank, where he could not see the rationale for being paid. "Given the intrinsic richness of manual work – cognitively, socially, and in its broader psychic appeal – the questions becomes why it has suffered such a devaluation as a component of education" (Crawford, 2009, p. 27).

The significance of this situation is summarized well by Billett (2016):

> Probably the greatest challenge for vocational education is the perennial issue of its standing. It is difficult to summon interest, engage parents, and engage industry and enterprises in productive ventures, let alone garner governmental support, when the knowledge required for effective occupational practice is downplayed and marginalised, and the shortest and most cost-efficient models of education are selected for implementation. (p. 211)

3 History

The structure of vocational education in Australia began to take shape in the late 1980s and development continued into the 1990s, as one of the outcomes of the neoliberal reforms that were being implemented in many countries around this time (Harvey, 2007). The instrumentalist perspective of policy makers of the time resulted in a vocational education focus on the skill requirements of employers, which were represented by competencies. These were standardized into measures of observable performance, and thereby used to hold educators accountable through a centralized bureaucracy. This competency-based form of vocational education is globally pervasive (Hodge, Atkins, & Simons, 2016); therefore, in Australia, and in many developed and developing countries, the vocational education curriculum has been organized this way for many decades now.

As early as 1991 the influential Finn Report (Australian Education Council, 1991) proposed that worker flexibility should be an important characteristic of vocational training. This was mostly unheeded, and the competency-based training system has become embedded, even though continuing critique has generally judged it as having failed to provide for flexibility because of the system of achievement of specific and often discrete competencies (Smith & Keating, 2003).

In Australia in 1997, vocational competencies were grouped together into training packages which represented particular industrial groupings. In the 20 years since this innovation, the system has been remarkably successful by many measures. An OECD report on the system in 2008 estimated that 80% of occupations in Australia were covered by the system (Hoekel, Field, Justesen, & Kim, 2008). In 2015 there were 18,151 nationally-endorsed competency documents with another 2993 competencies endorsed at the state level. These competencies were gathered into 82 Training Packages, a figure indicating the number of discrete industries served by the system. In terms of the number of students in these competency-based programs, the National Centre for Vocational Education Research (NCVER, 2016) estimated that in 2015 there were 4.5 million enrolments. Considered in relation to the population, this number of enrolments represents about 26.8% of all Australians aged 15–64 years (Hodge, 2016, p. 172).

Government promotion of such statistics is often used to indicate the success of the vocational education system. The approach has enabled an instrumentalist bureaucratic system which seemingly can guarantee a competent workforce based on centralized control (Billett, 2011) of a national uniform system. The official discourse which links competency education with responsible government provision of a competent workforce has supported this system.

The promotion of flexible delivery of vocational education courses in the mid-1990s in Australia provided the opportunity for the implementation of a more flexible range of pedagogies. Robertson (2009) analysed this development through the lens of Bernstein's theoretical framework (2000). This framework incorporates the notion of 'framing' which refers to the nature and control of elements of pedagogy (content, pace, sequence, control, etc.). Where there are explicit rules governing pedagogical practice, the framing is strong, where the pedagogue has more control and so can be flexible, the framing is weak.

Robertson concluded that the new policies that supported flexible delivery of vocational education courses strengthened the role of government and industry and weakened teacher pedagogical autonomy in the delivery of vocational competency based training. In the industry–teacher relationship, traditionally dominated by the industry, the power of industry was strengthened, and in the teacher–learner relationship, the learners became more dominant. So, "the legitimised discourses of flexible delivery and flexible learning represent a general weakening of the influence of the teacher in both regulative and instructional discourses" (Robertson, 2009, p. 8).

It seems in some ways counterintuitive that the option of flexible delivery would not provide more autonomy for the teacher in utilizing a range of pedagogies in delivering content. However, it has not been possible to separate the outcomes, in this case competency-based training packages, from the process of delivering the outcomes (Wheelahan, 2008). The forms of evaluation determine the pedagogic practice by providing the required criteria (Bernstein, 2000).

An additional opportunity for variability in the pedagogical approaches of traditional vocational institutes in Australia was presented by the granting to them of authority to offer bachelor's degrees in 2010. Such degrees have been situated as "…courses that develop discipline skills and employability skills with the emphasis less on 'employ' and more on 'ability'". The emphasis is on developing critical, reflective abilities, with a view to empowering and enhancing the learner (Harvey, 2007, p. 3). This is an approach that embeds employability into the core teaching and learning practices and is opposed to more traditional [university] courses where employability might be addressed through placements and internships (Pollard, Vincent, & Wilson, 2016). In some instances, this broader educational brief has provided opportunities for the use of more diverse and learner-centred pedagogies, but these pedagogies have generally not influenced the traditional pedagogical approaches related to the delivery of competencies.

So the initial conceptualization of the vocational education curriculum as competency based, and the teacher-centred pedagogies accompanying competency development in students, has endured. The approach has been supported by a bureaucracy that promotes the system as successful because it is large, has little internal variation, is manageable and is easily reportable.

4 Critique of Competency-Based Education

It has been clear for some time that a competency-based curriculum is not the best mechanism to achieve the kind of adaptable skills that are recognized as being necessary for a twenty-first-century trade (Jackson, 1993; Stevenson, 1992). The vocational education learning model is

> … based on behaviourist notions of knowledge and skill, embedded in a fragmented model of qualifications and a fragmented VET [Vocational Education and Training] system, all designed to serve a fragmented market based on exchanges between putatively ontologically distinct rational self-maximising individuals. (Wheelahan, 2008, p. 4)

There remains a focus on skills at the expense of other less observable, higher level attributes and more conceptual forms of knowledge which are important for employees to possess (Ashworth & Saxton, 1990) and which represent significant elements of the employment community of practice into which employees are expected to be inculcated.

When the workforce skills are broken down into specific competencies and this becomes the focus of education and assessment, there may be no opportunity for trainees to develop an integrated approach to their work (Buchanan, Yu, Marginson, & Wheelahan, 2009), which entails integrating competencies together in complex tasks that require a broad, overview approach rather than an atomized competency approach.

Some posit that the consequences of this focus on specific skill competencies prevent the participants of this form of education from participating effectively in the 'knowledge economy' (Wheelahan, 2008). A critical view would here interpret the maintenance of a neoliberal agenda by preventing the particular social classes who are aligned with vocational education from participating fully in the contemporary economy: teaching working class kids for working class jobs. This also has the effect of limiting effective movement from vocational education into other forms of higher education in which a certain level of generalizable knowledge is assumed.

The commonly accepted notion of situated learning (Lave & Wenger, 1991) implies that competency based vocational education is problematic. Situated learning reflects the notion that the learning required in a particular workplace setting can only be acquired through participation in that workplace. The learner progressively acquires the necessary knowledge through participation in the tasks and processes of the workplace, guided by those experienced in the workplace. Eventually they become a recognized practitioner. This notion of learning is clearly at odds with a context where workplace skills are broken down into specific and discrete competencies which are taught and achieved in a training institution for later transfer (hopefully) into a workplace environment.

There seem to be two incongruities in relationship to the continued dominance of competency-based education systems for vocational education. One is the notion that complex learning can take place simply through the mastery of dozens of specific competencies, which contradicts contemporary thinking about learning, teaching and transfer. Secondly, despite that fact that representative industry bodies developed the competency statements on which the curriculum is built, which constitute training for that industry, industry also states that what they need in new employees is not technical competency achievement, but a range of other 'softer' skills which they deem more important for successful employment.

Despite a range of robust critiques of the competency-based approach in the literature over a long period of time, which indicate its narrow focus and inflexible structure are inappropriate, little has changed in the vocational education system generally and in the pedagogies employed in particular.

5 Pedagogical Considerations

The consideration here is a forward looking one. For the delivery of a narrow instrumental and competency-based form of vocational education, pedagogies are predetermined to also be narrow, limited and focussed on the student practice of skills until they are mastered. However, if the conceptions of vocational education are broadened to encompass a more robust curriculum, which is also focussed on the types of skills that employees need for the future work environment, and which is considerate of learner's interest and readiness, then the consideration of a broader range of pedagogies is possible.

The discourse around good teaching in vocational education is no different to the general discourse: teachers are facilitators of learning, learners play an active role in the construction of knowledge, higher order thinking skills are incorporated and learning is situated and social (Gamble, 2013; Smith & Blake, 2006).

Billett (2016) succinctly outlines the challenges for vocational education to develop in students:

1. An understanding about their selected occupation,
2. The canonical knowledge of the occupation,
3. Occupational principles and practices that can be adapted to particular work settings and tasks, and
4. The broad range of capacities required to achieve these goals (p. 205).

The 21C skills (or core skills, generic competencies, employability skills, transversal skills and so on) have come to prominence recently in terms of general education, but these are not new concepts in vocational education and have been promoted as important since the early 1990s (Australian Education Council, 1991). These skills need to be embedded within a context, in this case, of occupational practice. There seems to be a consensus that general competencies are meaningless unless they are embedded in a context of domain-specific knowledge – this provides a context for their deployment, gives them meaning and provides an opportunity for them to be evaluated. Pedagogies can be used as opportunities for students to develop and apply their generic skills. Situating opportunities for practice in problem scenarios are likely to be more effective than focussing on directed instruction.

Commonly there are two communities of practice within which students become integrated through developing understandings of the nature of the communities – one is in an education setting and the other is in a workplace setting. The balance between the time spent in these two communities during the period of education varies, but the goal is the same: to develop the support structures necessary for each student to understand the norms and practices which enables an induction into the community. One of the roles of the educator is also to frame the educational community practice as authentically as possible so that it aligns with the workplace community.

Rather than a focus on the development of specific and often isolated competencies, the sequencing of experiences which provide for the application of knowledge which clearly integrates theory and practice would be more appropriate. As students mediate these experiences, they construct the knowledge necessary to understand the workplace, develop necessary competencies, apply higher order thinking skills to particular situations and adapt to different contexts.

This pedagogical approach is situated across traditional boundaries, aiming for the development of a professional identity, "in contrast to the usual transmission-style of teaching in higher education or the watch-me-then-do competency-based style of teaching in vocational education" (p. 8). The goal is to develop a professional identity through the development of appropriate disciplinary habits, while at the same time critiquing and challenging the professional culture. In the case studies described by Pollard et al. (2016), this was achieved through studio-based practice, and creative practice with the discipline. The studio-based practice largely involved a project-based pedagogy of students working in teams guided by the lecturer; complexity was carefully scaffolded, and teamwork skills were explicitly taught. The practice with the discipline was achieved through visits by industry experts and students undertaking work-integrated learning through an industry placement.

After an examination of issues related to pedagogies in vocational education, through the lenses supplied by Dewey, Schon and Usher, and in the context of training vocational education teachers in a university program, Arden, Danaher and Tyler (2005) concluded that there are four core elements to consider when conceptualizing appropriate pedagogies:

1. The curriculum needs to be truly learner-centred, holistic and interconnected – that is, founded in genuine and authentic experience in which learners are engaged and making the link between institution-based and work-based learning (mutual enhancement).
2. The pedagogy needs to be both *experiential and experimental,* connecting learning with real life/work situations, concerned with the interplay between subject matter and the learner's own experience and encouraging personal growth through reflection on existing as well as new experience (experimentation).
3. Key learning and assessment activities need to be learner-centred, growth enhancing, situation focused and problem-based, facilitating a learning environment where learners reflect as a social group or learning community on their practice – their theory-in-action – moving through a process of identifying, reflecting on, teasing out and theory-testing practice problems and dilemmas in order to 're-map' their future courses of action.
4. As reflective and reflexive practitioners themselves, the teachers need to 'walk the talk' and use the above strategies and principles as a foundation for development as well as the regular review and evaluation of their curriculum and pedagogy to ensure the continued rigour, coherence and relevance to the needs of learners of that curriculum and pedagogy (p. 43)

Lave and Wenger (1991, p. 34) suggest that "the generality of any form of knowledge always lies in the power to negotiate meaning of the past and future in constituting the meanings of present circumstances".

6 Frameworks for Reconceptualizing Vocational Pedagogy

There are many critiques of vocational education from a range of perspectives. In this chapter, the goal of the analysis is related to pedagogy, and as such the theoretical construct within which the analysis is framed should also be related to pedagogy and derive from how learning takes place when the learning is about work. Following is a discussion of a number of concepts, selected because they are conducive to framing an appropriate vocational education pedagogy.

6.1 Social Practice

A theory of social practice (Ollman, 1976) consists of two fundamental elements:

1. It is only through practice that knowledge and understanding develop. People are the sum of, or are formed by, their social relationships, which involve action, reaction and practice. Practice forms knowledge and understanding (not the reverse), and there are no solitary learners.

2. All the practices in which individuals engage are social. Humans are essentially social; the political, economic, cultural and social structures are both shaped by people and in turn shape people as participants. No person is solitary.

Within this theory of social practice, Martire and Lave (2016) elaborate on the place of competency-based teaching and institutional-based learning, which assumes:

1. Teaching is a prerequisite for learning. Teachers teach novices. A teacher is 'someone who knows', teaching someone who doesn't. So without teachers teaching, novices won't/can't learn, and can't learn from each other.
2. Learning is the mental exercise of internalizing knowledge, thus it can only be done by 'an individual.'
3. To learn is to *acquire*, internalize and accumulate knowledge/skill transmitted from senior knowers. When something is learned, the learning is done, the next step is to transport it to where it can be applied.
4. Teaching is understood as a matter of delivering a curriculum (produced by others beforehand and elsewhere) that guides and specifies processes of transmitting information or knowledge.
5. Schools are designed primarily to separate learners from the other contexts that compose their everyday lives, contexts in which they are expected eventually to apply what they've learned (p. 260).

Social practice theory suggests that people learn about work through what they are doing, through something like legitimate peripheral participation in the heterogeneous multiple practices, with co-participants, in ways of being peripherally part of other related practices that configure 'jobs'. The point is that there is much to changing participation in ongoing practice that only comes about in relations with things, people, technologies, other workers, processes of producing, etc., in situ (Martire & Lave, 2016, p. 263).

6.2 Discourse Community

Moraitis, Carr, and Daddow (2012) have used the concepts of discourse community and discoursal identity to construct a pedagogy that enables students: (1) to learn the 'knowledge' and 'language' of their course, both for work and further study; and (2) to begin to develop a critical perspective on the discourse community into which they are being inducted (p. 58).

Their starting point was based on critiques of competency (Wheelahan, 2010), which advocated centralizing knowledge within the curriculum, and critiques of progressivist pedagogy (Martin & Rose, 2008), which related to the language of the knowledge, and the explicit instruction aligned with that goal. Their project set up collaboration between language teachers and discipline teachers in order to initiate students into the community through an integrated pedagogy.

6.3 Signature Pedagogies

A pedagogical framework which is gaining some currency in vocational education is related to signature pedagogies. This concept, developed by Schulman (2005), relates the 'signature' ways a particular workplace community thinks and acts, to teaching for participation in that community.

> Signature pedagogies make a difference. They form habits of the mind, habits of the hand and habits of the heart.... they prefigure the culture of professional work and provide the early socialisation into the practices and values of a field. Whether in a lecture hall or a lab, in a design studio or a clinical setting, the way we teach will shape how professionals behave. (Shulman, 2005)

Throughout history, the classic signature pedagogy of vocational education was an apprenticeship model. The socialization of new community members into the ways the community thinks and acts through participation in the community provided a thoroughly integrated (habits of the mind, habits of the hand and habits of the heart) form of education. The establishment of a formal system of vocational education to enable regulation and accreditation, representing as it does a false dichotomy between theory and practice, put an end to the apprenticeship as a signature pedagogy.

Whether all the attributes required in any specific vocation can be captured as a 'signature' is debatable, as increasingly there are calls for vocational education to address a wide range of capabilities. A 2014 international forum of 197 vocational educators from 65 countries proposed the following set of capabilities:

1. Routine expertise (being skilful)
2. Resourcefulness (stopping to think to deal with the non-routine)
3. Functional literacies (communication and the functional skills of literacy, numeracy and ICT)
4. Craftsmanship (vocational sensibility, aspiration to do a good job, pride in a job well done)
5. Business-like attitudes (commercial or entrepreneurial – financial or social – sense)
6. Wider skills (for employability and lifelong learning) (Lucas, 2014)

It is difficult to conceive of a replacement for the signature pedagogy of apprenticeship, which would encompass the manifold expectations of vocational education.

6.4 Practice Architectures

The theory of 'practice architectures' was used by Kemmis and Green (2013) in their research about vocational education teachers conceptions of their pedagogy, developing from the notion of 'learning architectures' proposed by Wenger (1998). This theory proposes that the architecture of an organization determines the practices of that organization. It is described by Kemmis and Grootenboer (2008) in this way:

> Organizations, institutions and settings, and the people in them, create practice architectures, which prefigure practices, enabling and constraining particular kinds of sayings, doings and relatings among people within them, and in relation to others outside them. The way these practice architectures are constructed shapes practice in its cultural-discursive, social-political and material-economic dimensions, giving substance and form to what is and can be actually said and done, by, with and for whom. (p. 57)

Practice architectures relate to ways of saying and thinking, ways of doing things and ways of relating to people. When moving from one architecture to another (a workplace to an institution) or in attempting to change elements of an architecture (introducing new pedagogies into a traditional environment), tensions arise. Understanding the conflicting elements of different architectures assists in enabling change.

There are a number of commonalities which arise from these conceptual frameworks which will be discussed in the next section.

7 Discussion of Appropriate Pedagogies

A process approach to curriculum, instruction and assessment is held to more closely align with achieving goals associated with adaptability, work innovation and meeting particular workplace requirements than one based on pre-specified statements of measurable competence, standardized content and education processes focused on such outcomes (Billett, 2016, p. 199). As indicated, vocational learning needs to be much broader than the mastery of specific competencies, and include skills such as collaboration, metacognition and self-management.

An appropriate pedagogical approach to vocational education is situated across traditional boundaries, aiming for the development of a professional identity, through the development of appropriate disciplinary habits, while at the same time critiquing and challenging the professional culture. In the case studies described by Pollard et al. (2016), this was achieved through studio-based practice, and creative practice with the discipline.

Lucas, Spencer, and Claxton (2012) examined ten critical issues, and developed a spectrum related to each. The usefulness of this way of thinking derives from the location on the spectrum informing decisions about vocational pedagogy (Fig. 1).

To supplement these ten issues, Lucas et al. (2012, p. 115) developed, and provided tentative answers to, questions which inform vocational education curriculum and pedagogy:

1. *What is the goal and so the desired outcomes of vocational education today?*

   They suggest that the goal of vocational education is working competence in a chosen vocational area, and that there are six desired outcomes in all vocational education – routine expertise, resourcefulness, functional literacies, craftsmanship, business-like attitudes and wider skills.
2. *Can different kinds of vocational education be usefully categorized in order to make it easier to decide how best to teach them?*

Fig. 1 Vocational pedagogy – ten dimensions of decision making

They suggested that vocational education could be categorized into three elements – learning about working with physical materials, learning about working with people and learning about working with symbols (words, numbers and images).

3. *Which learning and teaching methods are best suited to delivering the desired outcomes in a specific vocational subject?*

   They suggested that effective vocational teaching requires a blend of hands-on or first-hand learning with critical reflection, collaboration and feedback in the context of strong relationship between teacher and taught.

4. *How is the choice of teaching methods influenced by context – the characteristics of vocational learners, the skills of vocational teachers and the settings in which the learning takes place?*

The settings are highly complex, and the teaching workforce is generally underqualified and undertrained.

An innovative training program developed by Sefton, Waterhouse, and Deakin (1994) was developed by grouping competency standards into larger clusters with a knowledge-skill mix, described as 'holistic competencies'. The mix included literacy, numeracy, self-confidence and work competencies. This was accompanied by a rejection of a deficit model of learning, and built on learner's strengths and abilities. However, it was found that the pedagogical skills of the instructors were inadequate to adapt to this form of integrated training – working in teams and focusing on general workplace needs rather than a range of specific competencies.

Compounding this issue further, vocational pedagogues are often theorists who work in non-vocational institutions of higher learning; few vocational instructors have education qualifications, and the training (or the opportunity) necessary to research their own practice. In addition, there is a dichotomy between institutional educators and workplace educators, resulting in separate rather than blended learning, so the relationship between workplace and institutional learning environments is often not seamless, and the institutional environment's capacity for meaningful situated learning is limited. While it is appropriate that industry have input into this process, it is educators who have the expertise to devise appropriate curriculum and pedagogies.

Smith and Blake (2006) saw teachers as facilitators of learning, with the learner playing an active role in construction of knowledge. Eight characteristics of facilitative teaching were noted: an emphasis on the workplace as a meaningful learning context; interactive approaches to cognitive and performative aspects of learning; work-ready learning outcomes; learner collaboration in determining learning and assessment processes; learners as co-producers of knowledge; recognition of prior learning; flexible teaching strategies for different learning styles; and social interaction as integral to the learning process.

In summary, Posner (1982) proposes that an appropriate vocational pedagogy needs to provide for the following:

- Placement of the learner at the centre of the educational experiences
- A broad range of experiences
- Opportunities for students to extend what they have learnt in different situations
- Engagement in non-routine activities
- Exposure to problem situations – selective application of knowledge and skills – development from novice to expert
- Problems need to be presented in a scaffolded way – support with possible heuristics, present 'half-solved' problems – means less cognitive load, and a heightened chance of novice success
- Problems structured and presented to focus on a particular purposeful aspect of learning, not just general experiences of problem solving
- Students placed in the role of practitioners – encourage them to think and act as though they were actors in their vocation
- Emphasis on engagement, level of engagement is correlated with the quality of the learning outcomes

Apart from the inculcation into a vocational community of practice, the characteristics in this list reflect the characteristics of good pedagogy in general education. This confirms to some extent the confluence of the pedagogies of vocational education and general education. For some time now (Williams, 1998), and further reinforced in the literature cited in this chapter, there has been a recognition that many of the goals of vocational education are similar to the goals of general education, particularly in terms of the development of transversal skills. It therefore follows that the pedagogies to achieve those goals will be similar.

8 Conclusion

The vocational education enterprise is complex and significant and, over many years, has proved to be resilient to change. The fundamental difficulty is that those who have the power to make changes prioritize criteria which does not indicate a need to change: large enrolments, manageable reporting structures and compliant constituents. Conversely, those who see the need for change (educators, teachers and students) act in a stigmatized social context and thereby lack the power to be effectual.

From a correlated perspective, what needs changing involves complex variables: social stigma, instructor's skills, industrial-institutional relationships, reporting achievement mechanisms. So even if the will for change resided with those who had the power to do so, significant complexity remains.

However, critique and theory, supported by research, provides indicators as to what pedagogical principles would be more appropriate than those currently utilized. These are those which are student centred, implemented in authentic contexts, social, and cross-disciplinary, really just sound pedagogical principles of general education. So it would seem that in order for vocational education to fulfil its potential, better satisfy the needs of employers and develop students to their potential, there is no need to develop unique pedagogies but to subscribe to those which are proposed as suitable for sound general education.

In a postscript to this chapter, the Australian Government announced on November 28, 2018, a review of the vocational education sector (Prime Minister of Australia, 2018). The terms of reference include the following statement:

> The Australian VET [Vocational Education and Training] system is complex, delivered through Commonwealth and state/territory funding, policy and regulation, and a network of public, private and industry providers. It provides services to 4.2 million students (24.1 per cent of the Australian population aged 15–64 years) and encompasses 4200 registered training providers. Against this backdrop, significant shifts have occurred in Australian industry and its workforce, with demand for skills shifting from manufacturing to the services sector and emerging industries such as advanced manufacturing and ICT. It is timely to consider how the system can better deliver for Australian job-seekers and employers now and into the future. As well as ensuring the system can better respond to skills shortages, there is a need to capitalise on available data to drive improvements in quality, results and employment outcomes. The system should be positioned to meet the information and training needs of school leavers to support them to secure employment, and to enable people to update their skills at any point during their working life. **Industry feedback suggests that vocational education must not only focus on specific employment skills, but build up foundational life, literacy and numeracy capability.**

References

Arden, C., Danaher, P., & Tyler, M. (2005). Struggling for purchase? What shape does a vocational education and training agenda take within a contemporary university education faculty? *International Journal of Pedagogies and Learning, 1*(1), 32–47.

Ashworth, P., & Saxton, J. (1990). On 'competence'. *Journal of Further and Higher Education, 14*(2), 3–25.

Australian Education Council. (1991). *Young people's participation in postcompulsory education*. Canberra, Australia: Australian Government.
Barbieri-Low, A. (2007). *Artisans in early imperial China*. Seattle, WA: University of Washington.
Bernstein, B. (2000). *Pedagogy, symbolic control and identity: Theory, research, critique (revised)*. Lanham, MD: Rowman & Littlefield.
Billett, S. (2011). *Vocational education: Purposes, traditions and prospects*. Dordrecht, The Netherlands: Springer.
Billett, S. (2016). Beyond competence: An essay on a process approach to organising and enacting vocational education. *International Journal of Training Research, 14*(3), 197–214. https://doi.org/10.1080/14480220.2016.1254365
Buchanan, J., Yu, S., Marginson, S., & Wheelahan, L. (2009). *Education, work and economic renewal*. Melbourne, VIC: Australian Education Union.
Crawford, M. (2009). *Shop class as soul craft*. New York: Penguin.
Epstein, S. (1998). Craft guilds, apprenticeship, and technological change in preindustrial Europe. *The Journal of Economic History, 58*, 684–713.
Gamble, J. (2013). Why improved formal teaching and learning are important in technical and vocational education and training (TVET). In UNESCO-UNEVOC (Ed.), *Revisiting global trends in TVET: Reflections on theory and practice* (pp. 204–238). Bonn, Germany: UNESCO.
Harvey, D. (2007). *A brief history of neoliberalism*. Oxford: Oxford University Press.
Hodge, S. (2016). After competency-based training: Deepening critique, imagining alternatives. *International Journal of Training Research, 14*(3), 171–179. https://doi.org/10.1080/14480220.2016.1261432
Hodge, S., Atkins, L., & Simons, M. (2016). Towards an epistemically neutral curriculum model for vocational education: From competencies to threshold concepts and practices. *International Journal of Training Research, 14*(3), 230–243. https://doi.org/10.1080/14480220.2016.1256895
Hoekel, K., Field, S., Justesen, T. R., & Kim, M. (2008). *Learning for jobs. OECD reviews of vocational education and training: Australia*. Paris, France: Organisation for Economic Cooperation and Development.
Jackson, N. (1993). If competence is the answer what is the question? *Australian and New Zealand Journal of Vocational Education Research, 1*, 46–60.
Kemmis, R., & Green, A. (2013). Vocational education and training teachers' conceptions of their pedagogy. *International Journal of Training Research, 11*(2), 101–121. https://doi.org/10.5172/ijtr.2013.11.2.101
Kemmis, S., & Grootenboer, P. (2008). Situating praxis in practice: Practice architectures and the cultural, social and material conditions for practice. In S. Kemmis & T. J. Smith (Eds.), *Enabling praxis: Challenges for education* (pp. 37–62). Rotterdam, The Netherlands: Sense.
Lave, J., & Wenger, E. (1991). *Situated learning: Legitimate peripheral participation*. Cambridge, UK: Cambridge University Press.
Lucas, B. (2014). *Vocational pedagogy*.. Report of the UNESCO-UNEVOC virtual conference, May 12–26. Bonn, Germany: UNESCO-UNEVOC International Centre for Technical and Vocational Education and Training.
Lucas, G., Spencer, S., & Claxton, G. (2012). *How to teach vocational education: A theory of vocational pedagogy*. London: City and Guilds Centre for Skills Development.
Martin, J., & Rose, D. (2008). *Genre relations: Mapping culture*. London: Equinox.
Martire, A., & Lave, J. (2016). Afterword: Social practice theory and learning work. *International Journal of Training Research, 14*(3), 256–266. https://doi.org/10.1080/14480220.2016.1267833
Moraitis, P., Carr, A., & Daddow, A. (2012). Developing and sustaining new pedagogies: A case for embedding language, literacy and academic skills in vocational education curriculum. *International Journal of Training Research, 10*(1), 58–72. https://doi.org/10.5172/ijtr.2012.10.1.58
NCVER. (2016). *Total VET students and courses 2015*. Adelaide, SA: National Centre for Vocational Education Research.

Ollman, B. (1976). *Alienation: Marx's conception of man in capitalist society* (2nd ed.). Cambridge, UK: Cambridge University Press.

Pollard, V., Vincent, A., & Wilson, E. (2016). Learning-to-be in two vocationally-oriented higher education degrees. *On the Horizon, 23*(1), 7–15. https://doi.org/10.1108/OTH-06-2014-0021

Posner, G. (1982). A cognitive science conception of curriculum and instruction. *Journal of Curriculum Studies, 14*, 343–351.

Prime Minister of Australia. (2018). *Media release*. https://www.pm.gov.au/media/delivering-skilled-workers-stronger-economy. Retrieved 30 Nov 2018.

Robertson, I. (2009). Developments in pedagogic nomenclature in Australian vocational education: Evolution or revolution? *International Journal of Training Research, 7*(2), 106–121. https://doi.org/10.5172/ijtr.7.2.106

Sefton, R., Waterhouse, P., & Deakin. (1994). *Breathing life into training: A model of integrated training*. Canberra, Australia: Australian Department of Employment, Education and Training. ERIC ED 398 441.

Shulman, L. (2005). Signature pedagogies in the professions. *Daedelus, 134*, 52–59.

Smith, E., & Keating, J. (2003). *From training reform to training packages*. Tuggerah, NSW: Social Sciences Press.

Smith, P., & Blake, D. (2006). *Facilitating learning through effective teaching: At a glance*. Melbourne, Australia: National Centre for Vocational Education Research (NCVER). http://www.ncver.edu.au/publications/1660.html. Retrieved 23 Feb 2018.

Stevenson, J. (1992). *Contradictions, complications and competencies: Who cares?* Paper presented at the student services – bridge to success – a TAFE national conference on student services. Brisbane, QLD.

Wenger, E. (1998). *Communities of practice*. Cambridge, UK: Cambridge University Press.

Wheelahan, L. (2008) *Can learning outcomes be divorced from process of learning? Or why training packages make very bad curriculum*. Paper presented at the AVETRA 2008 conference. VET in context. Retrieved from www.avetra.org.au/annual_conference/papers.shtml

Wheelahan, L. (2010). *Why knowledge matters in curriculum: A social realist argument*. New York: Routledge.

Williams, P. J. (1998). The confluence of the goals of technology education and the needs of industry: An Australian case study with international application. *International Journal of Technology and Design Education, 8*, 1–13.

Teaching Technology in "Poorly Resourced" Contexts

Mishack T. Gumbo

Abstract There is a widely held perception that teaching Technology in "poorly resourced" contexts is nearly unfeasible. In this chapter, I present a counter view to conventional understandings of "poorly resourced". Communities inhabiting such contexts engage localised technologies on daily basis to sustain their livelihoods. Teachers can use such resources to teach Technology too. This chapter contributes a paradigm shift from disregarding these contexts and the transformative and creative approaches to the teaching of Technology.

1 Introduction

If we consider rural or indigenous environments (commonly perceived) as "poorly resourced" contexts, it should be noted that "rural communities have achieved hard-won methods of managing their affairs, and each rural community has developed sophisticated social networks and cultural practices" (Gardener, 2008, p. 9). Resources that enable these practices have been inadequately explored for teaching Technology. Mawson (1999) and Lee (2011) find it awkward that the cultural practices of communities in these contexts suffer a lack of consideration despite the expectation that they follow the countries' curricula. For example, the Curriculum and Assessment Policy Statement (CAPS) for Technology Grades R – 12 in South Africa includes indigenous technology (Department of Basic Education, 2011). However, teachers have a tendency to disregard it in their teaching, a practice which is also motivated by a disregard of this aspect by facilitators of the teachers'

M. T. Gumbo (✉)
University of South Africa (UNISA), Pretoria, South Africa
e-mail: Gumbomt@unisa.ac.za

P. J. Williams, D. Barlex (eds.), *Pedagogy for Technology Education in Secondary Schools*, Contemporary Issues in Technology Education,
https://doi.org/10.1007/978-3-030-41548-8_15

professional development workshops. The CAPS's use of the phrases "wherever possible", "made aware" or "become aware" also treats indigenous technology shallowly.

This chapter focuses on teaching Technology in "poorly resourced" contexts. Students who attend in rural schools can be considered educationally disadvantaged compared to their urban peers (Alloway, Gilbert, Gilbert, & Muspratt, 2004). This chapter, therefore, critiques "poorly resourced", and emphasises the richness that resides in contexts perceived as "poorly resourced" – tools, equipment, consumable materials, and curriculum materials. In other words, this chapter introduces the notion of "resource sensitive teaching" as it outlines a "culturally responsive teaching" approach. The chapter encourages Technology teachers to be more resourceful and in effect, adopt resource sensitive teaching. It is important to advance this line of thought in order to make Technology teachable in such contexts. A teaching framework, some teaching ideas, content, and an example of a lesson are offered ultimately. These contexts may be erroneously perceived as poor, but it is a matter of thinking how to tap into locally available resources in such contexts to make the teaching of Technology possible, vibrant, and relevant. Hence, this chapter is transformative as it can help Technology teachers, stakeholders in education and scholars to think differently about teaching Technology in "poorly resourced" contexts.

2 Contextualising the Phenomenon

Technology teachers who teach in rural contexts can ill afford to sit arms-folded waiting for the Ministry of Education to supply resources. In contexts such as South Africa, teachers often complain about lack of resources (Gumbo, 2016), thus allege that they are unable to teach Technology as a result. While it is a fact that the Ministry of Education has a responsibility to resource schools, waiting on this expectation handicaps teachers as they fail to take advantage of the richness of local resources. The problem is that teachers perceive resources from a conventional (high-tech) perspective (Sade, Moreland, & Jones, 2007), denying themselves opportunities to contextualise the teaching of Technology in line with the resources daily used by communities. In addition, often when ministries supply resources, these are divorced from local contexts. Studies by scholars such as Seemann (2000), Ockenden (2014) and Gumbo (2016) raise issues about contextual relevance, and offer ideas to draw on local, available resources. This calls for some pondering: Who lives there? How do they view reality? What are their ways of knowing? What defines activities of their livelihoods? What resources do they engage? Grafting the teaching of Technology on these contexts can create opportunities to utilise the locally available resources.

3 Defining Concepts

3.1 *Poor*

There is a tendency to define "poor societies" externally especially from the perspective of urban establishments with their high-tech amenities. There are four perspectives on poverty: income poverty/consumption poverty, material lack such as shelter, clothing and furniture, and capability deprivation (Chambers, 2006); Ludi & Bird, 2007). But Chambers (2006) cautions:

> What poverty is taken to mean depends on who asks the question, how it is understood, and who responds. Our common meanings have all been constructed by us, non-poor people. They reflect our power to make definitions according to our perceptions. Whose reality counts? Ours, as we construct it with our mindsets and for our purposes? Or theirs as we enable them to analyse and express it? (p. 1).

Chambers' definition discourages underrating of societies that are defined as poor. That plays down local knowledge, skills, practices, methods, technics, processes and activities. Hence, I propose this definition:

Poverty means the extent of lack in certain areas (resources, services, capabilities, etc.) judged or measured against what is available locally, not against external especially urban establishment, techniques and processes.

We should look at the resource rich side of "poorly resourced" contexts as a starting point – people in that part of the world engage available resources in their daily activities; they cannot be utterly lacking. They thrive on apprenticeship (Eady, Harrington, & Jones, 2010) which is technological in nature. Thus, teaching Technology should be in consonant with local problems and resources to ensure relevance and sustainable development ultimately.

3.2 *Rural*

The Fiji Ministry of Education (2011) defines a rural school as one that is 10–20 km from a town boundary, equal to or greater than 20 km from a town boundary, and very remote. Lack of streetlights and electricity in the house, 4- to 22-wheeled engine power transport modes, schools, professionals, etc. are defining characteristics of rural areas against urban areas. "So today there is the view that those areas where black people live in the country are poor and decaying, full of disposed people and old-fashioned culture, and that they have become places which are a trap for the old and the young" (Gardener, 2008, p. 9). Such areas are alternative referred to as remote areas in terms of them being cut out from urban contexts. Hence, rural development efforts are an attempt to make rural environments catch up with urban environments instead of acknowledging local resources and building thereon. Rural schools are alleged as lacking classrooms, access to services such as water and electricity, landline telephones and internet access, libraries, and the like (Gardener, 2008). There is no regard for open classrooms where the young are taught everywhere and anytime, and the elders being the live libraries, for example. Gardener (2008, p. 9) argues that "educational and other research today finds that it is no longer appropriate or useful to define urban in terms of rural or the other way around". To still do so "is to create a competitive relationship between them, to the disadvantage of rural areas" (Gardener, 2008, p. 9). Lee (2011) raises an element of culture pertaining to indigenous communities, acknowledging the interdependence between design and culture as critical in the teaching of Technology. The inclusion of historical, societal, cultural and environmental aspects will enable this criticality and advantage students to understand technology much better (Lee, 2011). Culture means the "relationship between a given group of people and their environment. It includes patterns of production and consumption and the beliefs, values and structures that maintain these patterns" (Kokko & Dillon, 2010 in Lee, 2011, p. 43).

3.3 Resourcefulness

Shade (2001, p. 89) defines resourcefulness as physical tools to scientific knowledge, wherein material objects function as resources when used to secure desired ends. According to Shade (2001), resources also include "spiritual" or human means such as habits of persistence. It is irrefutable that rural/indigenous people have manipulated the natural environment by engaging their knowledge and skills to satisfy their needs and wants. Spirituality, which is a compass of their lives, has assisted them in this venture. The San people of Southern Africa, for example, are guided by their spirituality in how to treat the animal, eland. It is, however, concerning that these knowledge and resources have not yet been used in teaching and learning. In South Africa,

> researchers from the CEPD and from the Universities of Fort Hare and the Witwatersrand have together established how little schools draw on the many sources of expertise and numerous possible forms of support to be found in all communities, no matter how poor. (Gardener, 2008, p. 11)

Knowledge and resources such as this need to be amassed for educational purposes.

4 A Framework for Teaching in "Poorly Resourced" Schools

This chapter is framed in Ladson-Billings' (2000) culturally relevant teaching for its suitability to can support teaching Technology in "poorly resourced" contexts. According to Marchant (2009), students come to class with knowledge of technological practices in their communities that can facilitate learning new concepts and easily understanding the tasks/projects assigned. Teachers should critically examine their own views about knowledge, human nature, values, and society if they mean to adopt a culturally relevant teaching. They should open up to students' technological worldviews. The logic behind this is that children are first taught by their community, a teaching role that is especially done by the elders.

Furthermore, blacksmiths, architects, creative industrialists, food design and processing specialists, agriculturalists, expert poets, musicians, dancers, historians, cultural interpreters, and people skilled in traditional forms of knowledge (Gardener, 2008) are local human resources that can be consulted to enrich the teaching and learning of Technology. Their use has the potential to help teachers to overcome past inappropriate teaching practices. Teachers can invite them as para-teachers to explain certain concepts and processes to the students. Gardener goes on to claim that rural communities also have material resources that can provide the much-needed assistance to local schools. So, Technology teachers should primarily look at surroundings for resources.

Effective teachers explore ways to integrate traditional and contemporary culture in their teaching – through textbooks, artefacts, stories, etc. that represent a pluralistic view; such a view has a space for indigenous contexts as well. Walker (2008, p. 63) defines effective teachers as those who have "been the most successful in helping" students to learn. According to Walker (2008), students characterise them as prepared, positive, creative, fair, compassionate, hold high expectations, display a personal touch, cultivate a sense of belonging, have a sense of humour, respect students, forgive and admit mistakes. "The most effective teachers are resourceful and inventive in how they teach their classes" (Walker, 2008, p. 65). Certainly, teachers are the most resourceful of professionals, and Technology teachers arguably the most resourceful of all.

Effective teachers in indigenous contexts, therefore, possess several qualities that are aligned with ubuntu. Ubuntu is "an African value system that means humanness or being human, a worldview characterised by such values as caring, sharing, compassion, communalism, communocracy and related predispositions" (Khoza, 2005, p. 269). Though Khoza (2005) defines the concept from an African perspective, he opines that its philosophy can be applied universally. For instance, the Aboriginal world view is framed on the unity and coherence of people, nature and time (Christie in Harris, 1992). Caring, sharing, compassion, communalism and communocracy could be used to promote cooperative learning in knowledge co-construction, co-investigation of existing solutions or designs, co-design new solutions, and so forth. Unity and coherence of people, nature, and time can be used in the students' design projects, i.e. how they organise themselves as teams, show responsibility towards nature for sustainable goals, and manage time. This way a teacher becomes a caring project manager and advisor.

Culturally relevant teaching accommodates students' home languages to better understand what they are taught. Teachers who use language interaction patterns that approximate the students' home cultural patterns are more successful in boosting student academic performance (Ladson-Billings, 1995).

Conclusively, a culturally relevant teaching makes education to be available, accessible, acceptable and adaptable to students. This has a transformative function, i.e. to change the mindset from perceiving contexts as "poorly resourced" to richly resourced.

5 An Overview of a Few Studies and Their Findings

According to Ockenden (2014), integrating indigenous knowledge and perspectives in the curriculum has proven to be the key to engaging indigenous students in learning and associated with better results. Furthermore, this enables indigenous and non-indigenous students to appreciate indigenous cultures, address racism and contribute towards a more positive school culture and affirm indigenous self-identities so that indigenous students can develop a sense of belonging at school (Ockenden, 2014, p. 14).

Lingam and Lingam (2013) conducted a mixed methods case study on school resources vis-à-vis rural primary schools in Fiji. They sampled 52 teachers from 13 rural schools to survey their perceptions about their level of satisfaction regarding resources availability. School resources included the building, classrooms, staff room, library, teaching aids, curriculum materials, reference material, furniture, gardening tools, science equipment, sports equipment, toilet facilities, and multimedia. The main findings revealed that children who attend the school were mostly from low-income families some of whom depended on the sale of copra for their livelihood. Generally, the people in those villages relied on subsistence farming. However, teaching in the school conformed to the Ministry of Education's prescriptions and guidelines. One of the participants' words were captured thus from the interview: "No priority, as the community around the school is not rich enough to provide financial help to the school… even through fundraising we cannot raise enough money" (Lingam & Lingam, 2013, p. 2165). Another participant claimed: "They cannot afford school resources as they are subsistence farmers" (Lingam & Lingam, 2013, p. 2166). These responses represent views that looked to external help for resources instead of local resources – a disempowering thinking. A Technology teacher, for example, could assign students activities based on the technology and designs around the agricultural activities and copra harvesting.

Australian Human Rights Commission (2008) is about a case study of remote indigenous education at Maningrida School in Australia. A predominant indigenous population of about 2600 Gunibidji with ten languages lives at Maningrida. The Maningrida Community Education Centre (CEC) is responsible for both the primary and secondary education in the township and the outskirt communities. CEC is a bilingual school with primary school students learning in their own language (Ndjebbana and Burarra) before they can learn English. In 2006 CEC introduced successful secondary programmes. The Report deliberates on two of the programmes, which are Contemporary Issues and Science (CIS) and Junior Rangers (JR).

CIS is based on science, culture and caring for country. Local teachers taught students outdoors when it began in 2005. They planned to identify spiders and other insects in the bush environment. Since its inception, students have managed to identify 45 insect types. CIS is regarded as a good example of a programme that provides an intersection between indigenous and non-indigenous knowledge systems and cultures. Learners use their knowledge of fauna and flora to support their technical learning. The programme also demonstrates innovative teaching methodologies and resources as it engages students hands- on rather than imprison them within the confines of the classroom. Since the introduction of these programmes, students' attendance and performance have improved. This is because the curriculum has been transformed to have cultural relevance by drawing from local resources and knowledge, e.g. linking life in the region to the broader scientific community.

In a Zimbwean case of teacher development, Gwekwerere (2016) relates the benefits of a Dutch sponsored Science Education In-service Teacher Training (SEITT) project. The project aimed at contextualising secondary school Science teaching and learning in which teachers related science to the local realities and developed relevant teaching materials. A Science teacher, Casper participated in a sponsored training programme in Cuba. He learned about teaching Science by relating it to the students' contexts. Gwekwerere reports how Casper came back to contextualise his teaching the concepts of waves and chemical reaction. He (Casper) used a fishing activity that the students are familiar with. He made students imagine one of them throwing a stone into the water on a day they went fishing with their father, not far from the floater on the fish line. Casper then asked students to explain their observation and used it to introduce the wave concept, i.e. water movement and the floater up and down meant that a wave was created; the movement was not caused by the molecules but by the transmitted energy. This meant that the absence of the ripple tank at school does not have to limit the teacher from teaching the concept. The floater and river (ripple tank) provide the required resources.

Santoro, Reid, Crawford, and Simpson (2011) conducted a qualitative case study to examine the professional experiences and career pathways of 50 current and former indigenous teachers in Australia through semi-structured interviews. The study was premised on the idea that non-indigenous teachers do not really know indigenous knowledge, cultures, and identities; hence, they do not know how to teach indigenous students. They, therefore, need to be mentored by indigenous teachers. By involving discourse and thematic analysis, data were analysed through the processes of intercultural dialogue between indigenous and non-indigenous research team. These themes were developed: indigenous ways of knowing, indigenous students' lives beyond the classroom, and building relationships with indigenous students and communities. Under the first theme, for instance, the study revealed indigenous ways of knowing adopted by indigenous teachers, which include hands-on learning, tactile ways, touching and feeling, demonstration and observation, direct experience in the natural world, oral and aural, understanding the particulars from the whole, and learning especially from the elderly community. This shows that teaching from an indigenous perspective thrives on authentic contexts – since Technology teaching is more practical, it can benefit students greatly if approaches like those listed here can be considered.

Nyembe (2015) studied the collard greens (phophoroka in Northern Sotho) and mustard greens (mokhwarepa in Northern Sotho) under the traditional leafy vegetables in Limpopo Province, which are acclaimed as good sources of nutrients and available in rural communities. Seasonal availability and high perishability limit these vegetables' consistent supply and utilisation. Focus groups were conducted with 28 rural and 34 urban indigenous dwellers to gather their views on the consumption patterns of these vegetables and the effect of preservation technology on their nutritional value. Furthermore, the author experimented the blanching and sun drying technology as the dominant indigenous technology to preserve these vegetables to make them available for consumption even in off season. The technology was compared with conventional technology of blanching and oven drying and changes in the colour, texture, nutritional composition, and microbiological content were observed. A five-point pictorial hedonic scale was used. The green colour of these vegetables is important for nutrition. The findings revealed that indigenous technology did not temper with maintaining the colour, whereas the conventional technology was good in preserving the texture. In the end, the author recommended the interfacing of the two preservation technologies. This suggests an effective case study that can boost students' understanding greatly, wherein their learning activities can expose the pros and cons of this food preservation technology. A good thing about this case study is that it combines indigenous and conventional approaches to food preservation.

6 Content for Teaching Technology in "Poorly Resourced Schools"

Relevant teaching resources and physical facilities are needed for quality education of children (Lingam & Lingam, 2013, p. 2161). This section discusses the Technology Education content, taking Food Technology as an example, to contextualise culturally relevant teaching as discussed in the preceding section. I submit that in certain contexts Technology Education curriculum may not include Food Technology as a theme or topic. The point here, however, is to get an idea about contextualising Technology teaching in the stated context – the logic of this example could be transferred to other themes or topics.

In addition to indigenous food systems being important for human sustenance, they also form a treasure trove of knowledge, which plays a role in the well-being and health, environmental sustainability, and cosmic balance of the ecosystem (Kuhnlein in Demi, 2016). From an indigenous perspective, crops are never separated from weeds; there is thus no need to destroy plants as weeds because nearly all plants are either food or medicine (Demi, 2016, p. 3). For example, Fig. 1 shows an indigenous vegetable, which I purchased from women who harvest it from farms in Limpopo Province. These women wait by the roadside to sell it to motorists. When they asked the farmers of European descent to harvest the vegetable, the farmers

Fig. 1 An indigenous vegetable ("morogo wa lerotho" in Setswana)

welcomed the idea that these women would help to unweed their farms. The vegetable was prepared by washing, cooking, packaging and putting it in the fridge. Out of the fridge it is recooked a little bit and mixed with onion and tomato served.

Canada is one of the most advanced countries in the world, ranked number eight on the global scale, yet indigenous people in that country still practice their traditional hunting, trapping, fishing and agriculture, though they face the challenges to access and use their traditional territories for food procurement activities (Demi, 2016, p. 10). This shows the tenacity of indigenous people when it comes to defending their knowledge and technological practices as they have a strong belief in and perceive them as important sources to sustain their livelihoods. Demi (2016) relates four role dimensions of food in indigenous contexts: spirituality, food diversity, harvesting restrictions or regulations and local material.

Table 1 organises content from indigenous and conventional worldviews in a blended or comparable way. This can help students to appreciate the limitations and benefits of foods, their properties and processing and packaging technologies. Most importantly for this chapter, teachers and students will appreciate the value of local resources (food types) and not only look to the more conventional food types. In another study (Gumbo, 2003), I included a case in Zimbabwe where the Consumer Studies students reacted to the external food menus that made their learning difficult. In resolving this matter, these students were asked to list their own traditional food menus. They listed more than 80 types. This caused the transformation of the Consumer Studies curriculum and teaching, which resulted in heightening their interest in learning and improved their performance subsequently. Another quoted case (Gumbo, 2003) is from Namibia. An Examination Monitoring Committee has been established to monitor the balance of local and external content. If an examination question paper does not show a 50/50 balance between these two issues, it is returned to the examiner rework the balance.

Table 1 Food preservation techniques

	Sun	Solar	Oven	Bottling	Freezing/ canning
Method	Expose food directly to the sun or in a shade to remove moisture	Expose food to the sun through covered solar panels to remove moisture	Expose food to high temperature that is constant to remove moisture	Package and freeze fresh food items in an airtight container	Apply heat to food items that are sealed in an airtight container
Input	None	Solar panels	Oven electricity	Refrigerator, electricity, pre-freezing treatment	Heat resistant container electricity/fire
Effect	Causes the highest loss of b-carotene, vitamin A and vitamin C content	Causes the highest loss of b-carotene, vitamin A and vitamin C content	Causes the highest loss of b-carotene, vitamin A and vitamin C content	Retains sensory and nutrient quality, losses are in the pre-treatment phase	Causes the highest loss of b-carotene, vitamin A and vitamin C content
Time	3***	3***	2**	1*	Unknown
Shelf-life	Up to 1 year	Up to 1 year	Unknown	Less than 6 months	Unknown

Adapted from Nyembe (2015, p. 14)
Key: Preservation 1* is faster than 2**, which is faster than 3***

7 An Example of a Lesson About Teaching Technology in "Poorly Resourced" Contexts

This section provides an example about teaching Technology in "poorly resourced" schools with respect to Food Processing and Preservation Techniques. The emphasis is on alternative approaches to allow students to utilise the wisdom of other generations and cultures in order to contemplate contemporary technological developments (Lee, 2011, p. 42).

To give an idea, a teacher may plan the lesson as follows:

Objectives

- To discuss indigenous vegetables.
- To explain in detail a specific indigenous vegetable.
- To logically outline the processing and preservation techniques of this specific vegetable.

Planning the Lesson

The teacher:

1. Identifies an elderly woman who knows about indigenous vegetables and processing and preservation techniques.
2. Asks her if she would mind coming to help teach his/her students about the processing and preservation techniques of these vegetables.

3. Agrees with her on the specific vegetable that will be the focus of the lesson.
4. Establishes the resources that she can use for the processing and preservation, e.g. pots, fire, mortar and pestle, basket, calabash.
5. Also, establishes as to how long it will take her to take the students through the processing and preservation of the vegetable.
6. Finalises the arrangements with her regarding the dates, time, venue, transport, teaching methods (e.g. dialogue, oral presentation, observation), learning activities, etc.

Content

The content will be about the processing and preservation techniques of indigenous vegetables especially a specific type. Its source for this specific task is the invited expert. It also includes some beliefs about the vegetables, regulations related to their harvesting, processing and preservation, processing materials and sources of energy needed to preserve them, processing and preservation knowledge and skills.

Students' Activities

Students can engage in dialogue with the presenter (with the teacher to ensure adherence to the ubuntu principles), listen, or write observation notes in instances where demonstration is used. As a matter of importance, they should note the main points about the processing and preservation knowledge and skills—harvesting, processing, and preservation knowledge and cleaning, boiling, packaging skills, etc.

Assessment

Assessment can take any form, such as oral group presentation, assignments on discussing the processing and preservation techniques of the learned vegetable, homework to interview elders about other vegetables, etc. This lesson may lead to a design project in which students may be given a scenario on the processing and preservation techniques of indigenous vegetables, e.g.

> You have noticed that morogo wa lerotho, in most cases it is not thoroughly washed from dust as part of processing. Furthermore, it is laid on an iron sheet to sundry it in an open so it can be preserved, which makes it susceptible to stealing or vandalised by animals.

The students would have to design alternative processing and preservation techniques to ensure the vegetable's dust-freeness and security.

8 Conclusion

Technology is entirely teachable in the so-called "poorly resourced" environments. The argument in this chapter is that teachers who teach in such schools need to make their teaching relevant by recognising the wealth of resources in such environments which are popularly known as indigenous or rural. Indigenous teachers have a responsibility to unclaw themselves from perceiving technology in western terms only. They should first think local in which case they will be able to identify available technologies which are relevant to local communities and can thus make learning meaningful to their students. By virtue of being part of indigenous communities, these teachers have an advantage to make Technology Education interesting and enjoyable to the students. They do not have an excuse not to decolonise or indigenise Technology Education. They need to expand their scope in teaching Technology, to tapping on local communities' wealth of technological knowledge. That way creative teachers can demystify "poorly resourced" by creating an awareness about local contexts being rich in resources – energy forms, modes of transport, cooking technics, etc. Such teachers also value the elders and local technological specialists who can impart important knowledge that is needed in Technology Education lessons and design projects. As shown in the example of a lesson above, they can include the services of the local experts in different technology fields to co-teach with them.

This chapter also creates awareness in non-indigenous teachers, who may think that there are no other forms of technology besides the conventional forms. It creates a paradigm in these teachers so that they can learn along with their indigenous students who know more than they do about their local technologies. They could allow themselves to be guided by indigenous teachers about how to teach from a culturally responsive manner. Technology teachers and students from the so-called "affluent" contexts can learn a lot from those in "poorly resourced" contexts. In line with indigenous food that is discussed in this chapter, designing an indigenous dish provides an example for designing relevant solutions for local problems/needs/wants. Another example is considered: flooding schools with unilinear technological artefacts which portray western cultural images disadvantages students as they are denied the opportunity to learn about local technologies. Lego data, for example, which are used to inspire ideas about structural design in students, may not be relevant for designing for the popular rondavel housing designs common in most indigenous contexts.

These two types of teachers should, therefore, value a team approach to teaching in which they can be resourceful to each other. Indigenous teachers are not very conversant in conventional (western) forms of technology. Equally, non-indigenous teachers lack the understanding of indigenous forms of technology. A decolonial paradigm could help bridge this gap. Prioritising local resources for the teaching of Technology does not, however, mean shifting the responsibility of resource provisioning away from ministries of education. Such provisioning should, however, not override the value that should be attached to local resources.

The lesson in this chapter is that being "resource poor" can also mean being "technology rich." Acknowledging this wealth of resources can go a long way in as far as transforming the teaching of Technology Education – transforming the entire curriculum in terms of its outcomes, content, assessment; addressing issues of hidden curriculum; respecting indigenous people; and so on.

"Poorly resourced" contexts are actually endowed with locally available resources that can enrich the teaching of Technology.

References

Alloway, N., Gilbert, P., Gilbert, R., & Muspratt, S. (2004). *Factors impacting on student aspirations and expectations in regional Australia*. Canberra, Australia: Department of Education, Science and Training.

Australian Human Rights Commission. (2008). The Social Justice Report: Sydney: Australian Human Rights Commission.

Chambers, R. (2006). What is poverty? Who asks? Who answers? *Poverty in Focus*, 3–4.

Demi, S. (2016). *Indigenous food cultures: Pedagogical implication for environmental education*. Paper Presented at an International Colloquium on global governance/politics, climate justice & agrarian/social justice: linkages and challenges. Kortenaerkade, Netherlands.

Department of Basic Education. (2011). *Curriculum and assessment policy statement grades 7–9: Technology*. Pretoria: Government Printers.

Eady, M., Harrington, A., & Jones, C. (2010). Literacy practitioners' perspectives on adult learning needs and technology approaches in indigenous communities. *Australian Journal of Adult Learning, 50*(2), 260–286.

Fiji Ministry of Education. (2011). *Annual report*. Suva, Fiji: Government Printers.

Gardener, M. (2008). *Education in rural areas: Issues in education policy number 4*. Braamfontein, Johannesburg: Centre for Education Policy Development.

Gumbo, M. T. (2003) *Indigenous technologies: Implications for a technology education curriculum*. Unpublished doctoral thesis, Vista University, Pretoria, South Africa.

Gumbo, M. T. (2016). Pedagogical principles in technology education: An indigenous perspective. In G. Emeagwali & E. Shizha (Eds.), *African indigenous knowledge and the sciences: Journeys into the past and present* (pp. 13–32). Rotterdam, Boston: Sense.

Gwekwerere, Y. (2016). Schooling and the African child: Bridging African epistemology and Eurocentric physical sciences. In G. Emeagwali & E. Shizha (Eds.), *African indigenous knowledge and the sciences: Journeys into the past and present*. Sense: Rotterdam, The Netherlands.

Harris, D. (1992). *Around the developing world*. London: Hodder & Stoughton.

Khoza, R. J. (2005). Let Africa lead. Sunninghill: Vezubuntu.

Kokko, S. & Dillon, P. (2010). Crafts and crafts education as expressions of cultural heritage: Individual experiences and collective values among an international group of women university students. *International Journal of Technology and Design Education, 20*(4), 487–503.

Ladson-Billings, G. (1995). Toward a theory of culturally relevant pedagogy. *American Educational Research Journal, 32*(3), 465–491.

Ladson-Billings, G. (2000). Fighting for our lives: Preparing teachers to teach African American students. *Journal of Teacher Education, 51*(3), 206–214.

Lee, K. M. (2011). Looking back, to look forward: Using traditional cultural examples to explain contemporary ideas in Technology Education. *Journal of Technology Education, 22*(2), 42–52.

Lingam, I. G., & Lingam, N. (2013). Making learning and teaching a richer experience: A challenge for rural Fijian primary schools. *Educational Research and Reviews, 8*(21), 2160–2168.

Ludi, E. & Bird, K. (2007). Brief No. 1: Understanding poverty. Available at: www.odi.org/sites/odi.org.uk/files/odi-assets/publications-opinion-files/5678.pdf. Accessed 21 Mar 2018.

Marchant, H. C. (2009). *Indigenous ways of learning, being and teaching: Implications for new teachers to first nations schools*. Master of Education dissertation. Victoria, Canada: University of Victoria.

Mawson, B. (1999). *In search of the missing strand: Technology and society*. Paper presented at the TENZ conference. Auckland, New Zealand.

Nyembe, S. N. (2015). *Preserving traditional leafy vegetables using indigenous knowledge-based drying technologies to improve household food security in Limpopo Province, South Africa*. Unpublished MEd dissertation, University of Kwazulu-Natal, Durban, South Africa.

Ockenden, L. (2014). *Positive learning environments for indigenous children and young people*. Available at: https://www.aihw.gov.au/getmedia/107b2cd9-88f7-4e8e-8ab3-33718c966ff9/ctgc-rs33.pdf.aspx?inline=true. Retrieved 30 Jan 2018.

Sade, D., Moreland, J. & Jones, A. (2007). *Technology education teacher development in Solomon Islands: Enhancing teachers' perceptions and classroom practices.* Paper presented at Technology Education New Zealand. Auckland, New Zealand.

Santoro, N., Reid, J., Crawford, L., & Simpson, L. (2011). Teaching indigenous children: Listening to and learning from indigenous teachers. *Australian Journal of Teacher Education, 36*(10), 65–76.

Seemann, K. W. (2000). Technacy education: Towards holistic pedagogy and epistemology in general and indigenous/crosscultural technology education. *International Technology Education Research Conference*, Gold Coast, Australia.

Shade, P. (2001). Habits of hope: A pragmatic theory. Nashville: Vanderbilt University Press.

Walker, R. (2008). Twelve characteristics of an effective teacher: A longitudinal, qualitative, quasi-research study of in-service and pre-service teachers' opinions. *Educational Horizons, 87*(1), 61–68.

Pedagogy Involving Social and Cognitive Interaction Between Teachers and Pupils

Niall Seery

Abstract To better support pedagogical decision-making in design and technology, this chapter considers the relationship between cognition and social interactions. This chapter begins with a personal reflection of early practice as a pragmatic way of looking back to move forward. This frames the relationships that exist between constructs of D&T capability, the nature of knowledge and core interactions.

The chapter explores the nature of pedagogical practice from a teacher to pupil and small group perspective. The aim of this chapter is to give a rationale for practice so as to help understand the objective of the pedagogical strategy, at the same time empower teachers in developing new practices. It is hoped that effective practice can be understood when it appears as organised chaos knowing it will become more chaotic as the pupils get older!

Keywords Design and technology · Pedagogy · Cognitive and social interaction

1 Initial Perspectives

The art of teaching is a complex activity that requires a myriad of insights, skills and knowledge to support and advance the learning of others. Describing it as 'art' is a fitting qualification as it is the focus of this chapter to unpack a small yet important aspect of the effective teaching gestalt. At the core of effective teaching is the interaction and exchange between the teacher and pupil and by extension, pupil-to-pupil transactions. Understanding the confounding dimensions to these transactions is fundamental to advancing practice. The complexity of learning is further complicated when you consider technology education, and more specifically design and technology (D&T). Design education as a discipline brings with it its own complexities, most of which are difficult to articulate. Norman and Baynes (2017) highlight this by editing contributions in an attempt to articulate design epistemology.

N. Seery (✉)
Athlone Institute of Technology, Athlone, Ireland
e-mail: nseery@ait.ie

P. J. Williams, D. Barlex (eds.), *Pedagogy for Technology Education in Secondary Schools*, Contemporary Issues in Technology Education,
https://doi.org/10.1007/978-3-030-41548-8_16

While studying to be a teacher, many related anecdotes and experiences resonated with me. These initial perspectives from practice form the contextual framework for the interrelated thematic agendas of this chapter and provide a narrative for considering how to support learning. Having just begun my teacher education programme of study, I recognised the need to understand the complexity of the design and technology agenda and from there I try and conceive a way of translating it to practice. The following section tries to capture some of the elements that formed the basis of my thinking.

I recall being asked on my first teaching practice what would I have done differently if there were no pupils in the classroom? As a student teacher I was challenged with this question, and although obvious, it pointed to the teacher-centred approach that inhibits much of what we value in D&T education. It became a challenge for me to understand the epistemological underpinning of technology education and as such its unique relationship with knowledge. How to develop a pedagogy that could balance the speculative with an efficient development of the individual is difficult when bounded by an emphasis on knowledge. Acknowledging modelling as the critical factor in D&T helps frame and articulate the synthesis between the critical and the speculative, creating a discipline founded on a disposition of enquiry. The dimension of design (as a verb) highlights the kernel of the pedagogical challenge, as the lack of a common language compounds the distance between teachers' expectations and pupils' assumptions about what is evidence of effective learning.

I recall being given some advice from an elderly retired primary school teacher that my job as a teacher was to ensure that at least one pupil in my class understood what I was trying to explain and then I needed to create an environment to allow that understanding to be shared. Although questionable, this advice was thought provoking. The simplicity of the idea was striking and yet the complexity of the execution was apparent. As a student teacher, it made me think about my pedagogical deficiencies – why could I not explain to all and how would the class function? What was the difference in educational transactions between me and the pupils that understood and the remainder of the class? And why should peer to peer differ so much in terms of effectiveness? As a young ideological teacher, I generationally 'connected' with my pupils, so what were the limiting factors that inhibited the teaching and learning? This example speaks to the same underlying pedagogical frailty, that is, communication.

Years later, I was told a story about an in-house (low stakes) assessment where the supervising teacher would become (intentionally) distracted and have to leave the room for a few minutes at the start of the examination. The theory used to justify the practice was that if the pupils could teach and have learned from each other in the length of time the teacher was not invigilating, they would have achieved something that the teacher was not able to do in the entire semester. This example, albeit inherently flawed, interested me and raised a number of questions. What was the qualification of learning? What is the relationship between being able to access information and learning information? What was the environmental change that effected collaboration? Where the outcomes really representative of learning at an individual level? Evidence of interaction has utility but should not be assumed to be useful. In principle, the capacity to create the same engaged conditions within normal practice, mediated by a clear educational objective and where the basis of the learning contract is collaborative, is the pedagogical goal. So how do these social and cognitive interactions manifest in design and technology education?

Imagine an (overly general) design and technology activity for 14-year-olds, *to design a device to enhance the life of an elderly person.* The teacher has previously focused on mechanical movement, basic electronics and material properties and the students are now ready to move beyond the critical foundational knowledge to its application, where the learning outcomes are enveloped by three critical themes:

- *The utility of empathy and experience:* Design activity affords the opportunity to harness the individual's perspective, experience and values that can be mediated through their design agenda. The individualised agenda requires a frame of reference to help the student understand their world. Being able to share in the perspectives and experiences of others is a critical dimension in their design decision-making. As such the embedding of social interactions as a foundational priority for the success of classroom activity is essential to guide a pedagogy that can link design, data and critique. Useful peer-to-peer and teacher-to-student interactions would focus on questions such as: What is the design motivation? Who are we designing for? What is the challenge/problem that the design solution addresses? Has the student experienced this challenge? These interactions will expose the students to other dimensions of the design task that are beyond the direct knowledge and experience of the individual student.
- *The bounded nature of knowledge and skills*: Solely considering curricular knowledge can be a significant limitation for students in developing design solutions (or more accurately design resolutions). A broad knowledge base supports the capacity to explore and develop more conceptual approaches but must be considered in the context of 'relevant' knowledge and how we develop the capacity and environment to facilitate bespoke learning. Furthermore, striking the balance between the student's capacity to imagine (in context) and the probability of successful execution must be negotiated. We can all share examples of when the ambition of the student in developing their idea or project is not aligned with the limiting factors of time and/or resources.
- *The importance of communication:* Ideas can be very 'precious' to the creator, often as they come for very personal experiences or perspectives. This is a confounding variable for design activity as the difficulty in communicating the idea can be clouded by the need for the idea to be valued. The difficulty in communicating an idea or agenda can manifest in a *its stuck in my head* type exchange, where the limitations of natural language or lack of knowledge to communicate inhibit the identification of shortcoming or an ability to guide the progress of thinking and development. Like knowledge, communication (irrespective of medium) may need to expand beyond the immediate frame of discourse and employ analogy, metaphor, simile, etc. to evolve the conceptual thinking. The reference for effective broadening of the communication strategy often begins with peer to peer and 'brings' the teacher's contribution into a slightly more

evolved conception of the design solution, again explicating the social and cognitive interactions required to support meaningful design activity.

The opportunities for students to enact their thinking and practically explore options, alternatives, modifications, etc. sharpen the pedagogical focus on utilising the declarative and procedural knowledge within a conceptual domain. When understanding the practical challenges and the characteristics of student activity in D&T (e.g. the visual, the make, the enquiry and the designerly), it is important to view the core educational transactions from the perspective of knowledge, transactions and collaborations, as we protect the idea that students can be successful in dramatically different ways (Kimbell, 2011). Therefore, the pedagogical challenge is to examine the relations between cognition and social interactions focusing on:

- The nature of practice – e.g. the visual, the make, the enquiry and the designerly
- The relationship between knowledge and authentic D&T activity
- Unpacking the complexity of language in articulating the disposition (relationship between critical and speculative) of enquiry
- The educational transaction and the interactions between student, material and teacher as core to cognitive development

The following sections explore these challenges with respect to qualities and constructs of D&T, and how they relate to knowledge acquisition and collaboration, framed within the definition of educational transaction.

2 Qualities and Constructs of D&T

Highlighting the role of design in technology education, Kimbell (2011) differentiates technological knowledge from scientific knowledge. Scientific knowledge is concerned with facts, whereas technological knowledge is better described as 'provisional' knowledge. It is threshold knowledge that you have at the beginning of a problem/task which you can use to guide your search for additional relevant knowledge. Common knowledge does exist across cultures but is not definitive, and largely this knowledge is more appropriately described as expertise. Therefore, it is perhaps more appropriate to view the benefits of design education from the perspective of human capacities as this amplifies the importance of interactions as a fundamental pedagogical strategy.

Baynes (2013) describes design as a distinct way of thinking, acting and learning. The requirement of imaginative modelling is a unique reasoning capacity that relies on spatial qualities and is fundamental to design education. This view is well supported by Nelson and Stolterman (2002) who view design as being a natural part of human behaviour and is seen as an integral part of everyday life. They emphasise this by stating that humans did not discover fire, they designed it and that the wheel was not accidently discovered, but was also designed (Nelson & Stolterman, 2002). The enactment of designerly activity is the driving force behind human evolution.

Fundamentally it is the unique imaginative capacity that supports us 'seeing the world as it could be'. Comprehensively understanding this capacity so as to affect it is at the kernel of the challenge facing teachers. As Stables (2008) wrote:

> At the heart of early discussions was a belief that amongst the myriad of abilities human beings possess are three that make a particular contribution to the 'designerly: our ability to 'image' in our minds things we have experienced and also that we haven't; our ability to manipulate those images, both in our minds and through externalised actions such as talk or drawing; and our ability – and determination – to utilise imaging and modelling of ideas to create new future realities. (p.8)

There is a contemporary challenge to define the epistemology of design (Norman & Baynes, 2017). In an attempt to articulate a starting position, Baynes (2013) posed the question not 'what' is design but rather 'when' is design? The emphasis on 'when' frames an absolute intent that design education seeks to resolve issues as distinct from solving problems, perhaps it frames the intent of design education as seeking to provide students with the opportunity to develop the skillset to resolve unforeseeable issues. The 'when' protects for a time where we currently have no capacity to predict what society will look like, but also acknowledges that for the pupils' learning frames new insights (which may be only new to them).

The nature of D&T education requires engagement with a dialectic process, where conversations with the self, the medium and with others identify the need for discourse, as a means of pushing thinking and performance forward. The generative process at the centre of the educational transaction in the classroom becomes even more critical as we consider the complexity of design education. Design education supports interactions that see social construction as the cornerstone of negotiating new meaning and as such positions designing as exploratory and conditional. However, building an iterative paradigm of speculation and critique requires clarity of understanding as to what is success (Seery, 2017).

Focusing on the educational transaction, there is an argument that from a design epistemology perspective you cannot 'see' learning; it is outcome focused, so guiding at critical points of the learning process is difficult. This school of thought is framed by the idea that design capability is often impossible to articulate and the innate abilities just 'happen' and cannot be explained by natural language. Alternatively, the view of design education from the perspective of D&T is that articulation of the journey and the critical decisions throughout are representative of design capability. There is little doubt that the relationship between designing and making influences the process. Kelly, Kimbell, Patterson, Saxton, and Stables (1987) propose a model framing the iterative dialectic that highlights the relationship between thinking and external modelling which demonstrates an observational pedagogical model. This model describes the 'private voice' of the learner as they explore and develop their ideas through externalised activity. The deepening of understanding and refining of their conception is aided by sketches, models, prototypes, etc., enabling further insight. However, the variance between the ability to articulate design is accounted for by the inclusion of knowledge. Notwithstanding the importance of dialogue and the central theme of negotiating meaning, this is a particularly interesting concept when considering design education provision.

Pink's (2005) description of a 'conceptual' age creates the need to align carefully our definition of core aptitudes with a contemporary provision.

Given the fluidity of international D&T interpretations and the qualities and constructs that determine success, there are significant implications for practice. In my previous work, I explore the ubiquitous nature of modelling, specifically its relationship with critique. This work referenced the implications of practice when considering the influencers and inhibitors of effective modelling as:

- The learner's reaction to the learning task that employs modelling above other reasoning and logic strategies
- The cognitive faculties that are then activated in response to the modelling act
- The actions and behaviours that are enacted as a result of conceiving and refining a cognitive model and the iterative nature of modelling building towards equilibrium
- The capacity to effectively utilise the modelling process to highlight anomalies, intuitive theories and reason relevance and appropriateness (Seery, 2017 p.266)

Learning is a complex system of interactions, and although the focus of any pedagogical action is moving the individual forward, it is difficult to think about the student in isolation. This chapter is concerned with the face-to-face and small group interactions that are critical to the effective teaching of design and technology. The following section will position knowledge, so as to later consider collaboration and focus on the educational transaction.

3 Foundations of Knowledge Exchange and Threshold Concepts

The current state of thinking in D&T is that there is a fundamental expectation that the traditional views of knowledge boundaries are challenged in design and technology education. Describing the role of knowledge in context, Williams (2009, pp.248–249) notes 'the domain of knowledge as a separate entity is irrelevant; the relevance of knowledge is determined by its application to the technological issue at hand. So the skill does not lie in the recall and application of knowledge, but in the decisions about, and sourcing of, what knowledge is relevant'. The ability of a learner to identify, access, and understand relevant new knowledge will depend to some extent on their current knowledge. The broader a learner's knowledge the easier it will be for the learner to access and acquire new knowledge. Roberts, Archer, and Baynes (1992) acknowledge the importance of meaning making as to be truly involved in learning and that this active enquiry has a 'polysemous quality'. The treatment of knowledge and the associated pedagogical implications highlight the challenge of contemporary provision. From the perspective of the pupil, not knowing the 'answer' is fundamentally aligned with the indeterminacy of design challenges where there is no single 'right' answer or solution and no fixed field of knowledge, where Buchanan (1995) argues that when considering deign there are no definitive conditions or limits.

The relationship and balance between knowledge and the designerly can be somewhat managed in younger design classes, where the need for foundational knowledge is apparent and serves the purpose of a starting position (e.g. the material properties of wood as a consideration for designing a handle) and an explanatory position for further exploration (what else are these characteristics useful for? What characteristics have other materials that also may be useful? etc.). Therefore, the emphasis on the delivery of knowledge associated with a design task is directly related to how permeable the task design is. Considering the role of knowledge as a means of speculating and synthesising new conceptions as a creative endeavour is further enforced when knowledge is used to validate and confirm the viability and utility of the proposed. The more permeable, the less likely students are to rely on a specific knowledge base, therefore relying more on a collaborative process, where there is a negotiated meaning and an evolving knowledge base. In addition, permeable tasks that push the boundaries of the collective knowledge result in a generative process to create the necessary insights, which relies heavily on biological primary knowledge and innate human capacity. The objective is governed by the search, appraisal and application of knowledge in a 'lean' or 'just-in-time' model. From the perspective of practice this can appear to be chaotic, as pupils work on different interpretations of the design task and try to seek out knowledge through an experiential and experimental practice, loosely linked by 'hunch' and 'half-knowing' (Kimbell, 2011). Once understood, the organised chaos can be respected for its sophistication and benefit.

Contextualising design within an educational paradigm, there is recognition that to be capable we have to produce (Kimbell & Stables, 2007). The production of a solution or proposed resolution to a problem is supported by the utility of threshold knowledge. This foundational knowledge is the spark for the critical and speculative dialectic. The relationship between the designerly and knowledge is therefore important. However, the treatment of knowledge and in particular knowledge acquisition must be understood and carefully considered, as the paradigm of transmitting optimal procedures and processes is not the exclusive remit of design education. The speculative is critical as actions are predicated on 'but what if?'. This shifts practice to consider an alternative framing of the boundaries of disciplined knowledge. We may need to consider alternative ways of thinking about knowledge in D&T, but it is important to do so within the broader foundational understanding of how people learn.

Successful learning, defined as a change in long-term memory (Kirschner, Sweller, & Clark, 2006), will always require a knowledge base, and the acquisition of knowledge can be considered from five principles (Sweller, Ayres, & Kalyuga, 2011). The *information store* principle, describes the importance of the long-term memory for storing information. What is of interest is access to the store of others and hence the *borrowing and reorganising* principle suggests that learners 'borrow' information that is stored in the long-term memory of others. This can happen through teaching or collaboration.

When access to information stores is limited or insufficient, the *randomness as genesis* principle explains how information is generated. Primary biological knowledge is used to generate information through problem solving. Heuristics like the 'generate and test' heuristic (Gigerenzer, 2001) are used to generate new knowledge relative to the learners' schema. This principle validates the inclusion of design activity in education and is directly relevant to the contribution of Stables (2008). Kirschner, Sweller, Kirschner, and Zambrano (2018) highlight that in order to avoid combinational overload during the generation of information, limits are imposed by the capacity and duration of our working memory. Therefore, the *narrow limits of change* principle is educationally important as it describes the effect the high element interactivity in new knowledge has on learners capacities to process this new information. This is especially important when considering design education. Kirschner et al. (2018) highlight that collaborative learning may address some of the limitations of working memory, as learners can benefit from the 'collective working memory effect'.

By having too much scope and working within novel problems, there can be too much information and result in no learning. This means how we teach will have to be considered. The *environmental organising and linking* principle trigger can bring previously learned information into our working memory. However, as this is a biological primary process, it is difficult to manipulate in a pedagogical sense but understanding it and utilising it is pedagogically important. From a knowledge perspective, it is recognised that some skills associated with knowledge generation, appraisal and confirmation can be developed by learners through mediated collaborative educational transactions, where the benefits of design education begin to emerge.

4 Interactions and Collaboration

Linked directly to the ideas emerging in this book and specifically chapter "Technology Education: The Promise of Cultural-Historical Theory for Advancing the Field", the relationship between cognition and social interactions call significantly on cultural-historical concepts. The components of Vygotskian theory (Vygotsky, 1987, 1997) align directly with the argument to this chapter. The concept of 'private voice' describes the characteristics of dialogic reasoning the initiates with the 'hunch' and 'half-knowing' (Kimbell, 2011) as students explore the relationship between designing and making. Understanding the nature of knowledge and its function in threshold engagement and future knowledge creation is well articulated by consideration of the zone of proximal development. Additionally, the scaffolding of cognitive development aligns with the teacher-pupil and pupil-pupil interactions that support, encourage, and guide the mastery of new skills and concepts.

Ensuring the quality and sustainability of pedagogy that builds towards permeable tasks and fosters speculative enquiry needs a robust capacity to confirm and qualify created insights. Logic forms a control for declarative knowledge as it emerges and formulates new schema; however, the mechanisms to support this needs to be considered. There are two simultaneous perspectives that are worthy of note. The idea of interaction with materials and processes during speculative enquiry is well framed in the iterative dialectic model described by the work of Kelly et al. (1987). The ability to interact with physical resources and the environment is a critical dimension of D&T pedagogy. Although there is little empirical evidence to suggest the significance of this in relation to cognitive development, it aligns with much of the principles outlined by Sweller et al. (2011). The interaction between inside the head and outside the head is a sophisticated process often described in D&T as 'making'. It is this designing and making that requires the support of teachers (to support and guide) and peers (to affirm, unpack and refine) to externalise thinking governed by the cultural-historical concepts of amplification, mediation, contradictions and dialectics and requires as a fundamental collaborative learning.

Collaborative learning is considered with respect to knowledge creation through a shared language. Kirschner et al. (2018) highlights that if we have evolved to collaborate, then the act of collaboration is biological primary knowledge – meaning it is not teachable. He also argues that based on the development of language, we have strong evidence that humans have evolved to work together. Again, starting with the position of defined knowledge, there are some significant advantages to collaboration. Collaborative education differs from individual learning as we can acquire information that may be difficult to obtain from other means (Kirschner et al., 2018). Particularly when considering design education and the disposition of enquiry, the principle of borrowing and reorganising is useful. It also needs to be noted that collaborative learning may ameliorate some of the limitations of working memory (Kirschner et al., 2018). Considering the breadth of knowledge useful in designing, accessing the insights of the teacher through direct critique, and negotiation and being able to access the long-term memory stores of peers, places collaboration as a critical aspect of the 'chaotic' process.

Although, just because collaboration happens does not mean that it is useful or appropriate, nor can one assume the knowledge exchange and development were relevant, correct or efficient. Popov, van Leeuwen, and Buis (2017) highlight that teachers must ensure peer-to-peer exchanges and small group work are focused on trans-active discourse (i.e. critique, challenging of positions and attainment of synthesis via discussion). The idea of trans-active discourse engages cognitive activities that stimulate knowledge construction.

The complexity of knowledge creation and acquisition puts a significant demand on the cognitive resources of the pupils. Using the example of element interactivity (Sweller et al., 2011) to demonstrate the cognitive implication for knowledge acquisition may be helpful. Learning the colour codes for a resistor has relatively low element interactivity, where the colour code is linked to a resistance value; the lower interactivity is described by the colour representing a numeric value. Although the knowledge in isolation is important, in isolation it is not often useful. Element interactivity increases when you consider the combinations of colours as a representation of resistance value and tolerance, this knowledge increases utility. Higher element interactivity occurs when the selection of a resistor is required for the design of a circuit with specific function. The increased element interactivity increases the degree of cognitive load. This example illustrates the need for foundational knowledge supported by direct instruction and also the application of knowledge developed through an iterative dialectic with environment, teacher and peers. Therefore, in design and Technology education, the synthesis of knowledge forms the kernel of the activity that is then further mediated through social, utility, effectiveness and emotive considerations.

The value of peer-to-peer interaction is amplified when the constituent members of small groups are purposefully selected. The gap between conception and communication is often lessened by virtue of succinct analogy and metaphor mediated by relevant context and situational understandings (peer-to-peer). Kalyuga (2013) emphasises that learners develop experience working in a collaborative environment that supports the development of domain-general group knowledge. This will enhance the ability of each learner in acquiring domain-specific knowledge from the combined effort. Furthermore, this will help frame the evolving heuristics used to explore the unknown (see Seery (2017) for example). Kimbell's (2011) description of 'hunch and half knowing' and using provisional knowledge to develop better insights and applications do not support a determinist pedagogy, as often the value is in the diagnosis of misconceptions, and the tangential agenda that emerged from the creation of new meaning. Schön (1983) describes this as 'a conversation with the materials of a situation' (p.78). The design is situated in enquiry and meaning is grounded in the action and creation of insights, where pupils are actively exploring, experiencing, confirming and affirming their world.

So far, the ideas of collaboration were limited to the discussion on knowledge. Collaborating to share and develop knowledge will not in itself eliminate the pedagogical challenge facing D&T teachers. Linking back to the capacity to articulate design capability emerges again from a collaborative perspective. Roberts et al. (1992) highlight that design does not have a definitive design vocabulary; therefore, being able to explain imagined conceptions of reality without ambiguity is problematic. However, he suggests it would be useful to have a design meta-language that facilitated the principles of collaboration. How do you develop a meta-language for an activity that resides in the minds of your pupils and at best can be framed with respect to a journey and at worst cannot be articulated? To acknowledge that design education lacks a defined epistemology is a positive position to take, and therefore the move to the acceptance of a meta-language is an interesting position. The work of O'Connor (2016) makes a significant contribution to design and technology education and highlights the capacity to describe the processes and functions of education in the context of design activity. His work highlights the development of a common discourse developed over time that sees the convergence of understanding between the intended and subsequently enacted manifestations of design.

The debate between process and product is a critical epistemological dyad. The ability to articulate what is in the mind's eye can be difficult with natural language, as the evidence of capability is in the product and often only apparent 'when you see it'. This position is predicated on a process of learning and development that is more sophisticated and complex that what we can describe from a traditional ontological position. However, the work of O'Connor (O'Connor, 2016; O'Connor, Seery, & Canty, 2018) frames the Experiential, Procedural, Individual (EPI) pedagogical model that using technology-mediated interactions helped the development of a meta-language for design. The EPI Model is a conceptual framework for supporting discourse and helps develop the design of the learning environment, the delivery of the educational transaction and the EPI domains of learning. His work to support discourse in a traditional classroom demonstrated the evolving understanding between teachers and pupils when tacking a design challenge. Using hashtags and a pedagogical framework to link the cognitive processes with social interaction, O'Connor created a rich learning experience. This work managed the experiential, procedural, individual and environmental domains centred on the educational transactions to construct, capture, communicate and cogitate. His work demonstrated empirical evidence of the effects of a pedagogy that mediated the complexity of design and technology education.

5 Educational Transaction: Social and Cognitive Interdependency

Having considered knowledge and interactions independently, there is a need to consider their interaction. This section looks at synthesising the expectations, environment, language and practices that are explicitly aligned with design and technology education. Firstly, we have no absolute measure of designerly ability. The context-specific nature of design makes this nearly impossible. What we can do is identify (as a professional teacher and/or researcher) which characteristics and attributes are important in our own context. But it is the teachers that are in a position to identify variances in ability due to their judgement (interactions with students). So considerations for task design (Sweller et al., 2011), formulating groups (Petty, 2009) and supporting dialogic exchange (Stables, 2017) will be explored so as to clarify effective practice in D&T.

The amalgam of social and cognitive factors can be operationalised by effective pedagogical practice as demonstrated by the work of O'Connor (2016), O'Connor et al. (2018). Dewey (1938) proposed a transactional conception of activity-based education, which describes the teaching and learning experience as a set of transactions 'taking place between an individual and what at the time, constitutes [their] environment' (p.43). Moore developed this concept in the context of technology-enhanced learning and explained the idea of transactional distance is pedagogy made of three dimensions: structure, dialogue and learning (Moore, 1991; Moore & Kearsley, 1996). He explained that structure is the design of the learning experience and communication medium. Therefore, the sequential and strategic development of knowledge-related development and utilisation skills is recognised as critical for design education. The communication of expectation, assumptions and outputs cannot be framed with respect to standards or criteria but a disposition and exemplars. Dialogue refers to the internal or external communication between teachers and pupils. In Moore's theory, the most distant transaction has low structure and low dialogue, while the least distant transaction has high structure and high dialogue. Therefore, it is critical to reduce the social and cognitive space that separates teachers from pupils and pupils from pupils. Although the indeterminism of design is acknowledged, the exchange is absolute. Bandura, Ross, and Ross (1961, 1963) provide a basis for us to consider modelling as a form of critique. They suggest that behaviours are learnt through a self-regulated and considered response that is influenced by 'reciprocal determinism'; this is the relationship between the environment, the behaviour and the person (values, beliefs and cognition).

Having considered the knowledge and the nature of the exchange, the environment becomes a critical focus. The separation of social and cognitive interaction is unhelpful when considering the nature of teaching design and technology. The 'leap of faith' required when conceiving a solution that is 'unique' or 'innovative' requires an environment that supports the translation of what is in the 'head' of the designer (student) to furthering the resolution of the design issue. Redmond and Lock (2006) argue that the intersection between social and cognitive is where learners move beyond the exchange of information to a more reflective (cogitative) and in-depth investigation or analysis. Therefore, the environment needs to facilitate pupils comparing, contrasting and connecting ideas generated from peer discussion, trial and error and speculation. This act must be supported by clear assumptions about the learning activity and clearer expectations that there is no correct answer. The learning environment has to expect the unexpected perspective and trust in the underlying logic that will determine the necessary knowledge or experience.

Critically, how we mediate the educational transaction can significantly impact on the actions and associated outcomes in the learning process. It is arguable that because of the nature of technology and design education, this is amplified. Considering the 'didactic contract' as an exemplification of the importance of a shared purpose and agenda, when engaging with design education, demonstrates the importance of a common goal. Unfortunately, as Brookfield (1995) highlights, we as teachers are often living contradictions.

6 Conclusion

The challenge of attempting to describe the relationship between cognitive and social interactions from the perspective of pedagogy is amplified in the context of D&T. The relationship with knowledge is a complicated idea, where the acquisition of new knowledge is only a partial goal. Outcomes that are exemplary can often only be recognised when they are observed. The capacity to describe the designerly through natural language is a fragile construct. Yet, the objective of this chapter is to try and untangle the contested areas of cognitive development and social interactions through an equally complex pedagogical lens.

However, there are a number of pedagogical considerations that are exemplified and I hope useful in the context of practice. For example, it is essential to ensure that pupils have the time to consider, incubate, explore alternatives and create meaning. Furthermore, pupils should aim to explore resolutions to problems and create new conceptions of reality, safe in the knowledge that D&T is about understanding the world as it could be. Learning tasks should be considered in the context of useful knowledge and foster the speculative and critical dispositions. The learning activity should support the idea of material, process and environment as pedagogues, while peers are positioned as critical references for elaboration and sense checking. I sincerely hope D&T teachers will be encouraged to embrace the chaos that is the exploration of new realities and focus on an environment that celebrates the breadth and ingenuity of human capacity.

References

Bandura, A., Ross, D., & Ross, S. (1961). Transmission of aggression through imitation of aggressive models. *Journal of Abnormal and Social Psychology, 63*(3), 575–582.

Bandura, A., Ross, D., & Ross, S. (1963). Imitation of film-mediated aggressive models. *Journal of Abnormal and Social Psychology, 66*(1), 3–11.

Baynes, K. (2013). *Design: Models of change*. Loughborough, England: Loughborough Design Press.

Brookfield, S. (1995). *Becoming a critically reflective teacher*. San-Francisco: Jossey-Bass.

Buchanan, R. (1995). Wicked problems in design thinking. In V. Margolin & R. Buchanan (Eds.), *The idea of design* (pp. 3–20). Cambridge, MA/London, England: MIT Press.

Dewey, J. (1938). *Logic: The theory of inquiry*. New York: Holt, Rinehart and Winston.

Gigerenzer, G. (2001). The adaptive toolbox. In G. Gigerenzer & R. Selten (Eds.), *Bounded rationality: The adaptive toolbox* (pp. 37–50). Cambridge, MA: The MIT Press.

Kalyuga, S. (2013). Enhancing transfer by learning generalized domain knowledge structures. *European Journal of Psychology of Education, 28*(4), 1477–1493.

Kelly, A. V., Kimbell, R., Patterson, V. J., Saxton, J., & Stables, K. (1987). *Design and technology: A framework for assessment*. London: HMSO.

Kimbell, R. (2011). Wrong ... but right enough. *Design and Technology Education: An International Journal, 16*(2), 6–7.

Kimbell, R., & Stables, K. (2007). *Researching design learning: Issues and findings from two decades or research and development*. Dordrecht, Netherlands: Springer.

Kirschner, P., Sweller, J., & Clark, R. (2006). Why minimal guidance during instruction does not work: An analysis of the failure of constructivist, discovery, problem-based, experiential, and inquiry-based teaching. *Educational Psychologist, 41*(2), 75–86.

Kirschner, P., Sweller, J., Kirschner, F., & Zambrano, R. J. (2018). From cognitive load theory to collaborative cognitive load theory. *International Journal of Computer-Supported Collaborative Learning., 13*, 213. https://doi.org/10.1007/s11412-018-9277-y

Moore, M. (1991). Theory of distance education. In *2nd American Symposium on Research in Distance Education*. Pennsylvania: Pennsylvania State University.

Moore, M., & Kearsley, G. (1996). *Distance education: A systems approach*. Boston: Wadsworth.

Nelson, H., & Stolterman, E. (2002). *The design way*. Englewood Cliffs, NJ: Educational Technology Publications.

Norman, E., & Baynes, K. (Eds.). (2017). *Design epistemology and curriculum planning*. Leicestershire, UK: Loughborough Design Press.

O'Connor, A. (2016). *Supporting discourse using technology-mediated communication: A model for enhancing practice in second level education*. PhD Thesis, University of Limerick, Limerick, Ireland.

O'Connor, A., Seery, N., & Canty, D. (2018). The experiential domain: Developing a model for enhancing practice in D&T education. *International Journal of Technology and Design Education, 28*(1), 85–99.

Petty, G. (2009). *Evidence based teaching: A practical approach*. Oxford, UK: Oxford University Press.

Pink, D. (2005). *A whole new mind: Why right-brainers will rule the future*. New York: Penguin Group (USA) Inc.

Popov, V., van Leeuwen, A., & Buis, C. (2017). Are you with me or not? Temporal synchronicity and transactivity during CSCL. *Journal of Computer Assisted Learning, 33*(5), 424–442.

Redmond, P., & Lock, J. (2006). A flexible framework for online collaborative learning. *The Internet and Higher Education, 9*(4), 267–276.

Roberts, P., Archer, B., & Baynes, K. (1992). *Modelling: The language of designing. Design: Occasional paper No. 1*. Loughborough, England: Loughborough University.

Schön, D. (1983). *The reflective practitioner: How practitioners think in action*. London: Temple Smith.

Seery, N. (2017). Modelling as a form of critique. In P. J. Williams & K. Stables (Eds.), *Critique in design and technology education* (pp. 255–273). Singapore, Singapore: Springer.

Stables, K. (2008). Designing matters; designing minds: The importance of nurturing the designerly in young people. *Design and Technology Education: An International Journal, 13*(3), 8–18.

Stables, K. (2017). Talking with avatars: The potential and impact of design dialogue with an on-screen avatar on the development of learner's design and technology project work. In S. Warner & L. Litowitz (Eds.), *PATT2017: Technology & Engineering Education – Fostering the creativity of youth around the globe* (pp. 1–13). Philadelphia: PATT.

Sweller, J., Ayres, P., & Kalyuga, S. (2011). *Cognitive load theory*. New York: Springer.

Vygotsky, L. (1987). In R. Rieber & A. Carton (Eds.), *The collected works of L.S. Vygotsky: Volume 1: Problems of general psychology, including the volume thinking and speech*. New York: Springer.

Vygotsky, L. (1997). In R. Rieber (Ed.), *The collected works of L.S. Vygotsky: Volume 4: The history of the development of higher mental functions*. New York: Springer.

Williams, P. J. (2009). Technological literacy: A multliteracies approach for democracy. *International Journal of Technology and Design Education, 19*(3), 237–254.

Philosophy of Technology for Children and Youth

Stephen Petrina

Abstract This chapter addresses philosophy in design and technology pedagogy. It problematizes philosophy as a guide and resource for pedagogy and instead explores how children and youth philosophize in a process of designing and making. This chapter provides a brief history of the Philosophy for Children (P4C) movement and questions its neglect of design and technology. In response, this chapter explores the philosophy of technology for children and youth (PT4CY). Philosophy may be defined as a "love of wisdom" but in the real world of designing, engineering, and making, philosophy often reduces to a "love of conventional wisdom." Examples of this are provided along with a research vignette of PT4CY. This chapter concludes with the juxtaposition of disruptive technologies, wherein children and youth are configured as experts, and slow pedagogies, wherein parents and teachers may intervene with spaces and time for philosophizing.

Why do children, overdetermined with gifts, fail to develop into adults that have in their interest a world that the next generation actually needs? We've heard for a century that "children are natural artists" and "natural scientists." In the anthropocene, children are found to be "natural conservationists" and "natural environmentalists." It is often asserted that "children are natural designers," "natural engineers," "natural inventors," "natural makers," and "natural technologists." Increasingly since the 1970s, we are told that "children are natural philosophers." The gifts that children bear in this world are abundant. With a twenty-first-century turn on the eternal truism that "the child is the father of the man and mother of the woman," we consistently resolve that "students know more about technology than their teachers"

S. Petrina (✉)
University of British Columbia, Vancouver, BC, Canada
e-mail: stephen.petrina@ubc.ca

P. J. Williams, D. Barlex (eds.), *Pedagogy for Technology Education in Secondary Schools*, Contemporary Issues in Technology Education,
https://doi.org/10.1007/978-3-030-41548-8_17

(Ebner, 2017). Repeatedly, we are reminded that "children are savvy users of technology before they even start preschool" (Kellahan, 2016). Equally infantilizing is a newfound sentiment that "technology is child's play." Similarly, it is no mystery why we are told that childhood holds a key to innovation and "thinking like a kid" unleashes potential for the creativity desired of entrepreneurs and technology mold breakers.

All these inferences give one pause to wonder, is there anything at all that pedagogy and philosophy can offer children that they do not already have or know? What exactly can education offer children if they are natural designers, engineers, inventors, makers, and technologists? Or worse, do pedagogy and philosophy eradicate or waste these natural gifts? "With the years," says Jaspers (1954), "we seem to enter into a prison of conventions and opinions, concealments and unquestioned acceptance, and there we lose the candour of childhood" (p. 10). Indeed, it has become common sense that schools – especially secondary schools – disrupt and stifle children's natural development and quash their innate gifts of creativity and criticism. Why then is it a paradox that in a transformation of youngster and youth to adult is the loss of the gifts and wisdom necessary to obligations toward future generations (Qvortrup, 2009; Weiss, 1990)?

This chapter addresses the entangling alliance of pedagogy and philosophy in design, engineering, and technology education and focuses on the philosophy of technology for children and youth (PT4CY). Philosophy for children (P4C) generated a range of curricula and pedagogical techniques since the 1970s but has yet to attend to design, engineering, and technology education. Although acknowledging for over a century that children are natural makers *and* philosophers of technology, teachers and theorists of design, engineering, and technology education have not formed an alliance with P4C or developed curricula and methods for PT4CY. One gets an uneasy, false sense of security in scenarios wherein PT4CY is otherwise left to the children and youth alliance with commercial enterprise. The first two sections provide brief histories of philosophy in the schools and P4C. The third section gives an overview of PT4CY, focusing on the void of philosophy of technology in P4C over the past 40 years on one hand and the void of P4C in design, engineering, and technology education on the other (Lipman, 2001/2009; Naji & Hashim, 2017). This section builds on the review of research and provides a variety of leads into PT4CY for advanced development of curriculum and pedagogy (C&P) or instruction (C&I). This chapter concludes by considering Barlex's (2017) challenge to account for disruptive technologies in design, engineering, and technology education practices by asking if this necessitates a counterbalance of slow, soothing pedagogies and philosophies. But for all we hear about natural tendencies toward distraction and "twitch speed," one might just as well propose disruptive, spontaneous, turbulent pedagogies, and philosophies. If children are naturally gifted and suited to philosophy in various ways, why are they ultimately unable to preserve wisdom or transfer this to sustainable design, engineering, and technology education as they age?

1 Philosophy for Children and Youth

In one of his early analyses, James (1876) asserted that philosophy for students "means the habit of always seeing an alternative, of not taking the usual for granted, of making conventionalities fluid again, of imagining foreign states of mind" (p. 178). Half a dozen years later, Dewey (1893/1967) defended arguments against teaching philosophy in the schools if this amounted to "conscious moralizing" (p. 222). Qualifying the argument, he reasoned that if ethics was alternatively defined as "human relationships in action," then it is "not only teachable, indeed, but necessary to any well-adjusted curriculum" of the schools (p. 223). Any "college undergraduate course in philosophy at the introductory level," he conceded, "can be successfully taught to bright high school seniors" (p. 247). Dewey explored a range of definitions of philosophy over his career and eventually acknowledges that, yes, "philosophy is *love of wisdom*" if wisdom is understood as "knowledge-plus" (1949, p. 713). In turn, for this chapter, pedagogy is defined as translating or rendering knowledge-plus teachable and learnable.

In one breath, philosophy is indispensable to pedagogy. The consequences of misapprehending this may be dramatic. "Based on a wrong philosophy, educational research can wreck" a country, Newlon (1923, p. 112) exclaimed with a bit of flair. In another breath, philosophy is entirely dispensable. By most counts, pedagogy does not need philosophy, if it ever did. For instance, over the past 200 years, philosophy has only sporadically been offered as a course in the schools and philosophers seldom write about curriculum design. Historically, philosophy served various roles, ranging from "handmaid to theology" to "queen of the sciences," and by the twentieth century its place in schools was basically reduced to service or questioned as inaccessible. "There are those who claim that philosophy itself has ceased to have any unusual or even worth-while function to fulfill in the modern world," an analyst sarcastically reported in the depths of the Great Depression (Schilpp, 1935, p. 231). He continued: "The day of empirical science spelled the doom of philosophy as surely as it spelled the doom of religion and mythology" (p. 231). Still, educators were challenged to accommodate James's insights into its potential for students as well as the tendencies of children to make critical observations or explore deep questions and theological problems.

The problem of philosophy in the schools was persistent across the world. In UNESCO's (1953) survey of *The Teaching of Philosophy*, only a few countries reported on courses in the schools and fewer on technology as a subject for philosophy. The most robust was the French system, wherein "*lycées* and *collèges* (secondary schools), the last year of study is devoted to philosophy (in the philosophy class) or includes compulsory courses in philosophy" (Canguilhem, 1953, p. 53; Goldstein, 2013). Similar conclusions can be drawn from the American Philosophical Association's (APA) (1958) study of "The Teaching of Philosophy in American High Schools." The APA countered excuses "that boys and girls of 15, 16 and 17 are intellectually too immature to understand and profit from the study of philosophy" but also cautioned that "high school teachers are, for the most part, simply incompetent to teach philosophy" (pp. 95, 97). If separate courses were unfeasible, integrated, or "interstitial," philosophy was a solution. Given newfound student interests if not newfound students, the inclusion of philosophy in the schools increased through the 1960s the world over. The resurgence in "humanities" courses helped the cause of philosophy at this time. Surveying departments of education across the United States (US), Glass and Miller (1967) asked "whether philosophy or any [encompassing] course (Humanities, Great Books, etc.) is taught in schools of their state" (p. 228). About 57% responded yes while 37% said no, as they were either unaware or certain these types of courses were not taught. Despite a brief run of 3 years (1975–1977), *The Journal of Pre-College Philosophy* signified the emphases on pedagogy in the 1960s and 1970s. But perhaps the most noteworthy signs were Lipman's (1976) Institute for the Advancement of Philosophy for Children (IAPC) (est. 1974) in New Jersey, *Metaphilosophy*'s special issue on P4C (Bynum, 1976), and Lipman, Sharp, and Oscanyan's (1977) *Philosophy in the Classroom* (Ayim, 1980).

2 Philosophy for Children (P4C)

P4C is based on a figure of the child philosopher (Kohlberg, 1968; Piaget, 1931) and children's inquisitiveness or propensity for wonder and problem posing, often prefaced with "why?" It began with *Harry Stottlemeier's Discovery* (Lipman, 1971/1974), a children's book (grades 5 and 6, 10–12-year-olds) drafted in 1969 and revised for field research in 1970–1971. "Logic," "philosophy," and "syllogism" do not appear in the text but are ever present as Harry Stottlemeier (aka Aristotle) and friends reason through how statements can be twisted into truths or falsehoods. For instance, Harry's friend Tony exclaims that if a machine's parts were all small, "that wouldn't necessarily mean that it was a small machine. The parts could be light, and still it could be a heavy machine. So what's true of the part doesn't have to be true of the whole" (p. 66). *Pixie*, a P4C book published in 1981 for 9- to 10-year-olds, explores ethics and freedom. Home alone with her older sister, Pixie sings "free, free, free! Everything's possible!" But she's reminded that "there are family rules, and they stay the same whether Mom and Dad are here or not" (Lipman, 2001/2009, p. 38). Following reading aloud sessions in class and questions about the book, the children are then challenged to discuss statements such as "family rules remain the same, whether or not adults are present" and "we are free if we think we're free" (p. 39). Sharp (2017) asserts that fundamentally, P4C "does not tell the child what to think: ultimately that is up to the child" (p. 26). Challenging philosophical concepts are addressed, she affirms, "but ultimately they have to make up their own minds whether in this particular circumstance lying or divorcing or stealing was the right or wrong thing to do" (p. 26).

By the mid-1990s, Lipman authored eight P4C books, and a range of children's and youth literature were used as an alternative or complement to the IAPC materials (Murris, 2016). P4C had diffused through 41 countries, from Argentina to

Zimbabwe (Lipman, 1997). Schools gravitated to P4C as *critical thinking* became a major goal for educational systems (Facione, 1990). And it was relatively easy and inexpensive, given, as Lipman (2005/2017) maintained, "the teacher needs only one novel for each child, as well as an instructional manual" (p. 8). Despite the saturation of lives with electronics since the mid-1990s, P4C and spinoff PwC (Philosophy with Children) practitioners have overlooked pedagogy to challenge children's thinking about technology. For instance, a section dedicated to "Specialized Uses of Philosophical Dialogues" in a P4C book does not contain any examples of technology as a case study for children and youth (Naji & Hashim, 2017, pp. 67–89). The more expansive *Philosophy in Schools*, with 25 chapters and 300+ pages, offers little to nothing on technology (Goering, Shudak, & Wartenberg, 2013). Similarly, Gilmore's (2016) *Kids Can Think* offers an adequate backdrop but then omits technology from the 24 "scenarios for the classroom" that follow. Design, engineering, and technology education educators readily isolate Lipman's comment that children need only a text for engaging with philosophy as a sure sign of the problem with the pedagogy. A counter is that design, engineering, and technology education has not taken up P4C despite access to children's literature awash with thematic content of their subject (Axtell, 2017). If the "Emperor's New Clothes" provides a model of the child critic, what is in this story that could help us draw out the technology critic from the savvy child?

3 Philosophy of Technology for Children and Youth (PT4CY)

If pedagogy is rendering knowledge-plus teachable and learnable, then of course it is inseparable from philosophy *and* technology. Dewey (1916, p. 386) at one point defined philosophy as "theory of education in its most general phases" but he also defined it as "generalized theory of criticism" (1929, p. ix). Theory, for Dewey, was an articulation of insight and understanding. Albeit elegant in its simplicity, his definition of technology as "intelligent techniques" is limited given a translation into "smart technologies" (1930/2004, p. 218).

Inasmuch as P4C overlooked technology, with rare exceptions, both design, engineering, and technology education and philosophy of technology have overlooked P4C (Pritchard, 1991). Since the 1960s, science, technology, and society (STS_1) and science and technology studies (STS_2) inspired some effort in the pedagogy of philosophy of technology for schools but a reality check is needed. In British Columbia (BC), the STS_1 course (*Science and Technology 11*) for high schools had little interest and was decommissioned in 2018. As it was, neither "philosophy" nor "philosophical" appear in the combined 150 pages of the original and revised "integrated resource package" (IRP) for teachers (BC Ministry of Education, 1995, 2008; Nashon, Nielson, & Petrina, 2008). In turn, BC's (2016) new "Philosophy 12" elective omits technology. In *Teaching about Technology*, de Vries (2005) offers an introduction to philosophy of technology with hopes that teachers will design C&P for their schools. Similarly, *Philosophical, Logical and Scientific Perspectives in Engineering* provides a scope of activities and analyses that could be readily applied to high school courses (Sen, 2014). In sum, we have yet to meet the challenge of pedagogy for PT4CY.

A promising initiative in PT4CY is the "Philosophy Short Course" developed by Ireland's National Council for Curriculum and Assessment (NCCA) (2016) for junior grades in Irish high schools (Canavan, 2014). Currently in the Philosophy course, content for the "Philosophy of Science and Technology" strand is a bit light and tilted toward science. Guiding questions include "Does technology always advance human wellbeing?" and "Will technology be able to save our fragile earth" (p. 19)? "We will need people who are prepared to ask, and answer, the questions that aren't Googleable," a reporter remarked (Blease, 2017).

While education entails helping or challenging students to think, Kohlberg and Gilligan (1971, p. 1072) and Kitchener (1990) cast doubt on assertions that children 10 years and younger *think philosophically*. Kitchenor stipulates that "to think philosophically one must be engaged in... *critical thinking about a philosophical issue*" (p. 425). Thinking philosophically also involves raising burning and puzzling questions yet "one must also be able to think the puzzle through to the end, to advance tentative answers to it, to subject proposed solutions to criticisms" (Kitchener, 1990, p. 419). Doubts and technicalities aside, Mitcham's (1994, p. 1) primary question can be reframed: what does it mean for children and youth to *think philosophically about technology*? What is a Socratic design, engineering, and technology education classroom, lab, or workshop? Clearly, at this point, we cannot say what characterizes this thinking or Socratic design, engineering, and technology education pedagogy.

The upshot of a void of PT4CY is we can assemble curriculum to balance western philosophy of technology canons. Van Norden's (2017a, 2017b) *Western Philosophy is Racist* and *Taking Back Philosophy* are symbolic of an intensification of critiques of undergraduate and graduate philosophy courses. African philosophers' efforts to decolonize curricula via "conceptual liberation" are enlightening for PT4CY initiatives (Wiredu, 1984, p. 35). These philosophers have been especially attentive to the nuances of conventional wisdom and "spontaneous philosophy" (Jacques, 1995, pp. 232–233). The imperative here is extending the spontaneous philosophy of technology of children and youth the world over beyond common sense and conventional wisdom.

4 Conventional Wisdom of Technology

If we provisionally interpret knowledge-minus as *belief* and knowledge-plus as *wisdom*, how might we render design, engineering, and technology education wisdom teachable and learnable? However much we are challenged to design C&P for "Traditional Ecological Knowledge and Wisdom," we are doubly challenged by Traditional Technological Knowledge and Wisdom (Stables & Keirl, 2015; Turner,

Ignace, & Ignace, 2000). How might we distinguish between the "wisdom of technology" and the "conventional wisdom of technology" (Lower, 1987, p. 1149)?

Upon introducing the concept, Galbraith (1958) defined "conventional wisdom" as "ideas which are esteemed at any time for their acceptability" or as understandings we accept because we are accustomed to them (p. 6). These are sometimes referred to as "old adages," truisms, or what Ellul (1968) calls commonplaces: "living beliefs, formulas that were repeated and used by everybody as criteria for judgment" (pp. 4–5). An example is "The Machine is a Neutral Object and Man [or Woman] is its Master" (pp. 226–235). "It is a fearful thing to attack this commonplace," he warns, "for it is the base, the foundation, the cornerstone of the whole edifice" upon which the average person elevates "technology, its glories, and its achievements" (p. 226). The neutrality of technology, keeping it under human control, raises implications of "technological determinism" as a recurrent theme in philosophy of technology (Dusek, 2006). As conventional wisdom, this is often stated "technology is neither good nor bad… it is how it is used" (Kranzberg, 1986, p. 545; Richardson, 1974, p. 5). A manifestation is "guns don't kill people; people kill people," repeated since the late 1960s. The reality is first of all, says Ellul, "there is not *one* machine but hundreds of machines" (p. 228). Who actually controls technology as a "network of all the machines," he asks? Ellul has students beginning with logic and questions of "which technologies?" and "who are the humans in control" (Lafrance, 2016)?

Equally entrenched conventional wisdom is "technology is a tool" – "just," "merely," or "only" "a tool." Ascended as high advisors or redeemers, philosophers once reveled in this conventional wisdom: "technology is merely a tool; the direction of its use must be determined by social and political philosophy" (Chen, 1950, p. 130). To what extent do millennial computer and network specialists repeat this conventional wisdom of technology? Dean (24 years old) says "technology is neutral" while Ray (29 years old) confirms that "technology's neutral.... It is just a tool. A gun is not evil because it can be used to kill" (quoted in Tapia, 2003, p. 498). When asked by talk show host Donny Deutsch whether new devices and apps were reinforcing crass individuality and antisocial behavior in young people, Gates (2006) spun the question. "Technology is just a tool," he answered, "to let you do what you're interested in." Melinda Gates (2013) in turn repeated this conventional wisdom in a commencement speech. Microsoft's (2014) infomercial during the Super Bowl then raised the stakes on the question "What is Technology?" Today, a student might inquire whether their design, engineering, and technology education course might better be titled hoplonology, organology, or toolology, the study of tools (Canguilhem, 1947/1992; Montagu, 1976, p. 270). A professor might still complain that if we design a course for design, engineering, and technology education, why not "develop a course in "pencil literacy" which would include learning what pencils are made of, how to sharpen them, and perhaps how to sign one's name" (Papert, 1996a, p. R01)? A critical theorist might leap to instrumental rationality: "In a socialist system the worker maintains [her and] his dignity and self-respect, while under capitalism [she or] he is just a tool or instrument to be exploited" (Nettler & Huffman, 1957, p. 53). This conventional wisdom of technology takes for granted that we know what a tool is or does. Logically, if "a doll is a tool" then "technology is just a doll" (Bronstein, 2017, p. 143)? There is a reservoir of examples and implications PT4CY (Petrina, 2017).

Conventional wisdom of technology also includes "necessity is the mother of invention," "technology is advancing," "technology is technology," and "technology is natural," or "technology is a natural part of children's lives" (Petrina, 1992). An urgent challenge for PT4CY is conventional wisdom, not wee wisdom or juvenile wisdom, as Piaget (1931) implied. What additional adages do students and teachers introduce into classrooms?

5 Research Vignette

Our PT4CY research participants (aged 7–13) indicate that their spontaneous philosophy of technology ranges from mundane to extremely sophisticated (MacDowell & Petrina, 2020). Some are quick to characterize technology as devices but their unusual descriptions also suggest they are giving serious thought to what technology means. For example, Jovan sees technology as something new and superior while Dan disagrees:

Dan: [interrupting] it's like saying I invented paper, and it's a technology, but in twenty years from now it's not a technology. We still use paper don't we? It's still something you use.

Jovan: Yeah, but it's not technology anymore. Technology is when you discover a thing for the first time.

Dan: Yes, but I find that technology is the same. Right now, you would say a computer is technology, right?

Jovan: This is a new computer [points to an iMac] and it is now the technology. The old one is not technology anymore.

Dan: I agree, but I think the old things are still technology, cuz you still use them. If this [iPod] is five years old, would you throw it in the garbage cuz it's five years old and it's not technology anymore? Technology is something you use as a form of like [pauses] as a tool. Like, let's say, fire.

Jovan: You know what, you are confusing electronics and technology. Technology is the *new* thing.

Dan: People are still using fire right?

Jovan: Yeah, but it's not technology. You are confusing technology. It's not the thing that you use. Technology is an abstract thing. It's the thing that is first, the best thing.

Dan: Well, you are basically saying that technology is a new invention. I find you are not saying that technology is technology.

Jovan: [talking excitedly] Technology is the new thing, the best thing in every capacity, every time. It's not just a thing – it's an abstract thing.

Dan: This subject is really weird. Like in a good way [smiles].

In another interview, Marie describes the problem of the ontology of technology with an insightful alternative to the black box concept. Technology is like a chicken egg, she explains, "cuz you don't know what's inside growing and it's like, 'how did this chicken come out of an egg?' If you didn't know about that then you'd think someone must have made the chicks."

In *Brain Gain: Technology and the Quest for Digital Wisdom*, Prensky (2012) observes that "technology-based wisdom is something we teach to all our children, starting at a very young age." Yet he seems to mean conventional wisdom of technology. *Three Little Pigs*, he writes, "teaches that those who are wise use better technology…. The wise pig employs the more advanced technology" (p. 47). Drawing the wrong conclusions but on the right topic, *Brain Gain* helps keep open a question of whether technology offers wisdom other than conventional wisdom.

6 Disruptive Technologies and Slow Pedagogies

Nearly each day we hear about the "breakneck speed of technology" and get reminded that kids "operate faster than any generation that has come before" (Prensky, 2010, p. 11). Kids and technology are fast and impulsive while pedagogy and philosophy are slow and contemplative, conventional wisdom holds. Pedagogy and philosophy's slow adoption of kids' and new technologies' spontaneous adaptation to one another is proof positive, we are told (Prensky, 2010, pp. 9–10). Philosophers and teachers grew up pulling wagons around, just like medieval children, while kids now "sitting in their classes grew up on the 'twitch speed' of video games" Prensky, 2001, p. 4). Ancient philosophers and teachers time traveling to our contemporary classrooms "might be puzzled by a few strange objects" but "could quite easily take over the class," it is said (Papert, 1993, pp. 1–2). We often marvel at the achievements of kids and technology in spite of the laborious nature of pedagogy and philosophy. Kids and technology roll with Zuckerberg's (2010) wisdom, "move fast and break things," whereas pedagogy and philosophy are preoccupied tinkering with what cannot be fixed.

Barlex's (2017) C&P of "disruptive technologies" for design, engineering, and technology education and PT4CY is refreshing and unique juxtaposed against volumes offering the C&P of "disruptive students." For example, Barlex challenges students to distinguish between conventional wisdom (a drone or nanobot is just a tool) and deeper insights into disruptive technologies. Design, engineering, and technology education and PT4CY are challenged to complement turbulent, disruptive pedagogies, including racing outside to remotely control drones, with slow pedagogies, such as asking students "what do you think needs disrupting?" and providing scenarios to develop sophisticated critiques (p. 225). Another option is weighing consequences of a potentially disruptive technology. The Nuffield Foundation, for instance, encourages students and teachers to identify how or why "winners and losers" are persuaded to accommodate disruptions. Indeed, the challenge is acknowledging that design, engineering, and technology education is practical *and* philosophical.

7 Conclusion

This chapter confronted a bifold problem: The adultification of children and infantilization of technology (Lafrance, 2016). An untimely convergence means the duty to teach technology is progressively passed to "children." Since the early 1980s, it became increasingly difficult to distinguish whether the sages of technology are cyberpunks of fantasy or children of reality (Leary, 1988). When Papert (1996b) was asked, tongue in cheek, if a 2-year-old was smarter than mom and dad, he answered "we're trying to hurry along children to think like adults, whereas we'd do much better if we got more adults to think like children" (p. 100). As Turkle (1984, pp. 29–63) envisioned, with artificial intelligence (AI) the burden of wisdom is further lifted as machines relish the role of new, youthful philosophers. With emphases on contradicting the love of conventional wisdom, this chapter noted a relative absence of technology within P4C and philosophy in secondary schools. Is it not time for children and youth to study *and* do philosophy of technology?

References

American Philosophical Association. (1958). The teaching of philosophy in American high schools. *Proceedings and Addresses of the American Philosophical Association, 32*, 91–137.

Axtell, C. (2017). Critiquing literature: Children's literature as a learning tool for critical awareness. In P. J. Williams & K. Stables (Eds.), *Critique in design and technology education* (pp. 237–254). Dordrecht, Netherlands: Springer.

Ayim, M. (1980). Pre-college philosophy. *Canadian Journal of Education, 5*(2), 15–22.

Barlex, D. (2017). Disruptive technologies. In P. J. Williams & K. Stables (Eds.), *Critique in design and technology education* (pp. 215–235). Dordrecht, Netherlands: Springer.

BC Ministry of Education. (1995). *Science and technology 11*. Victoria, Canada: Author.

BC Ministry of Education. (2008). *Science and technology 11*. Victoria, Canada: Author.

BC Ministry of Education. (2016). *Philosophy 12 (Draft)*. Victoria, Canada: Author.

Blease, C. (2017, January 9). Philosophy can teach children what Google can't. *The Guardian*. https://www.theguardian.com/commentisfree/2017/jan/09/philosophy-teach-children-schools-ireland

Bronstein, M. (2017). *The Lonely Doll* series. In T. Jones (Ed.), *Toy stories: The toy as hero in literature, comics and film* (pp. 143–156). Jefferson, NC: McFarland.

Bynum, T. W. (1976). What is philosophy for children? — An introduction. *Metaphilosophy, 7*(1), 1–6.

Canavan, F. (2014). Philosophy in Irish schools. *Studies: An Irish Quarterly Review, 103*, 331–343.

Canguilhem, G. (1947/1992). Machine and organism. In J. Crary & S. Kwinter (Eds.), *Incorporations, Zone 6* (pp. 44–69). New York: Zone.

Canguilhem, G. (1953). The teaching of philosophy in France. In UNESCO (Ed.), *The teaching of philosophy* (pp. 53–72). Paris: UNESCO.

Chen, T. H. (1950). The guidance of foreign students. *Journal of Higher Education, 21*(3), 126–131. 166.

de Vries, M. J. (2005). *Teaching about technology: An introduction to the philosophy of technology for non-philosophers*. Dordrecht, Netherlands: Kluwer.

Dewey, J. (1893/1967). Teaching ethics in the high school. *Educational Theory, 17*(2), 222–226, 247.

Dewey, J. (1916). *Democracy and education*. New York: Macmillan.
Dewey, J. (1929). *Experience and nature*. Chicago: Open Court.
Dewey, J. (1930/2004). What I believe. In D. Capps & J. M. Capps (Eds.), *James and Dewey on belief and experience* (pp. 215–225). Urbana, IL: University of Illinois Press.
Dewey, J. (1949). Experience and existence: A comment. *Philosophy and Phenomenological Research, 9*(4), 709–713.
Dusek, V. (2006). *Philosophy of technology: An introduction*. New York: Blackwell.
Ebner, T. (2017, March 21). When students know more about technology than their teachers. *Education Week Teacher*. Retrieved https://www.edweek.org/tm/articles/2017/03/21/when-students-know-more-about-technology-than.html
Ellul, J. (1968). *A critique of the new commonplaces* (H. Weaver, Trans.). New York: Knopf.
Facione, P. A. (1990). *Critical thinking*. Newark, DE: American Philosophical Association.
Galbraith, J. K. (1958). *The affluent society*. London, UK: Hamish Hamilton.
Gates, M. (2013). *Duke commencement*. Retrieved from https://www.gatesfoundation.org/Media-Center/Speeches/2013/05/Melinda-Gates-Duke-Commencement-2013
Gates, W. (2006, May 9). Interview. *The big idea* [Television broadcast] (D. Deutsch, Host).
Gilmore, R. (2016). *Kids can think: Philosophical challenges for the classroom*. Lanham, MD: Rowman & Littlefield.
Glass, C. A., & Miller, R. I. (1967). Humanities courses in secondary schools. *Educational Theory, 17*(3), 227–235.
Goering, S., Shudak, N. J., & Wartenberg, T. E. (Eds.). (2013). *Philosophy in schools: An introduction for philosophers and teachers*. New York: Routledge.
Goldstein, J. (2013). Neutralizing Freud: The lycée philosophy class and the problem of the reception of psychoanalysis in France. *Critical Inquiry, 40*(1), 40–82.
Jacques, T. C. (1995). Is there an African philosophy? *Centennial Review, 39*(2), 231–264.
James, W. (1876, September 21). The teaching of philosophy in our colleges. *The Nation*, 178–179.
Jaspers, K. (1954). *Way to wisdom*. New Haven, CT: Yale University Press.
Kellahan, K. (2016, October 29). It's never too early to get appy. *Sydney Morning Herald*. Retrieved http://www.smh.com.au/business/workplace-relations/its-never-too-early-to-get-appy-20161026-gsbmv9.html?deviceType=text
Kitchener, R. F. (1990). Do children think philosophically? *Metaphilosophy, 21*(4), 416–431.
Kohlberg, L. (1968). The child as a moral philosopher. *Psychology Today, 2*(4), 24–30.
Kohlberg, L., & Gilligan, C. (1971). The adolescent as a philosopher: The discovery of the self in a postconventional world. *Daedalus, 100*(4), 1051–1086.
Kranzberg, M. (1986). Technology and history: "Kranzberg's laws.". *Technology and Culture, 27*(3), 544–560.
Lafrance, A. (2016, March 22). What is a robot? *The Atlantic*.
Leary, T. (1988). The cyber-punk. *Mississippi Review, 16*(2/3), 252–265.
Lipman, M. (1971/1974). *Harry Stottlemeier's discovery*. Upper Montclair, NJ: IAPC.
Lipman, M. (1976). Philosophy for children. *Metaphilosophy, 7*(1), 17–39.
Lipman, M. (1997). Special report to the APA committee on pre-college instruction in philosophy. *Proceedings and Addresses of the American Philosophical Association, 70*(5), 105–109.
Lipman, M. (2001/2009). Philosophy for children: Some assumptions and implications. In E. Marsal, T. Dobashi, & B. Weber (Eds.), *Children philosophize worldwide: Theoretical and practical concepts* (pp. 23–46). Frankfurt, DE: Peter Lang.
Lipman, M. (2005/2017). The IAPC program. In S. Naji & R. Hashim (Eds.), *History, theory and practice of philosophy for children* (pp. 3–11). New York: Routledge.
Lipman, M., Sharp, A. M., & Oscanyan, F. S. (1977). *Philosophy in the classroom*. West Caldwell, NJ: Universal Diversified Services.
Lower, M. D. (1987). The concept of technology within the institutionalist perspective. *Journal of Economic Issues, 21*(3), 1147–1176.
MacDowell, P. & Petrina, S. (2020). Philosophy of technology for children and youth II. In *STEM in Education 2020 proceedings* (pp. 1–6). Vancouver, BC: EDCP

Microsoft. (2014). *What is technology?* [Television commercial]. Retrieved from https://www.windowscentral.com/microsoft-super-bowl-commercial-2014

Mitcham, C. (1994). *Thinking through technology. The path between engineering and philosophy*. Chicago: University of Chicago Press.

Montagu, A. (1976). Toolmaking, hunting, and the origin of language. *Annals of the New York Academy of Sciences, 280*(1), 266–274.

Murris, K. (2016). The philosophy for children curriculum. *Studies in Philosophy of Education, 35*, 63–78.

Naji, S., & Hashim, R. (Eds.). (2017). *History, theory and practice of philosophy for children: International perspectives*. New York: Routledge.

Nashon, S., Nielson, W., & Petrina, S. (2008). Whatever happened to STS? *Science & Education, 17*(4), 387–401.

National Council for Curriculum and Assessment (NCCA). (2016). *Philosophy short course*. Dublin, Ireland: Government of Ireland.

Nettler, G. N., & Huffman, J. R. (1957). Political opinion and personal security. *Sociometry, 20*(1), 51–66.

Newlon, J. H. (1923). What research can do for the superintendent. *Journal of Educational Research, 8*(2), 106–112.

Papert, S. (1993). *The children's machine*. New York: BasicBooks.

Papert, S. (1996a, October 27). Computers in the classroom. *Washington Post*, R01.

Papert, S. (1996b). Testimony of Seymour Papert. In *Educational technology in the 21st century*: Joint Hearing, House of Representatives, 104th Congress (October 12, 1995) (pp. 86–120). Washington, DC: GPO.

Petrina, S. (1992). Questioning the language that we use. *Journal of Technology Education, 4*(1), 54–61.

Petrina, S. (2017). Critique of technology. In P. J. Williams & K. Stables (Eds.), *Critique in design and technology education* (pp. 31–50). Dordrecht, Netherlands: Springer.

Piaget, J. (1931). Children's philosophies. In C. Murchison (Ed.), *A handbook of child psychology* (pp. 377–391). Worcester, MA: Clark University Press.

Prensky, M. (2001). Digital natives, digital immigrants, Part 1. *On the Horizon, 9*(5), 1–6.

Prensky, M. (2010). *Teaching digital natives*. Thousand Oaks, CA: Corwin.

Prensky, M. (2012). *Brain gain: Technology and the quest for digital wisdom*. New York: St. Martin's.

Pritchard, M. S. (1991). STS, critical thinking, and philosophy for children. In P. T. Durbin (Ed.), *Europe, America, and technology: Philosophical perspectives* (pp. 217–246). Dordrecht, Netherland: Springer.

Qvortrup, J. (2009). Are children human beings or human becomings? *Rivista Internazionale di Scienze Sociali, 117*(3/4), 631–653.

Richardson, R. A. (1974). Some social consequences of technological success. *Hastings Center Report, 4*(5), 4–6.

Schilpp, P. A. (1935). An "apology" for philosophy. *Journal of Higher Education, 6*(5), 231–239.

Sen, Z. (2014). *Philosophical, logical and scientific perspectives*. Dordrecht, Netherland: Springer.

Sharp, A. M. (2017). Philosophical novel. In S. Naji & R. Hashim (Eds.), *History, theory and practice of philosophy for children: International perspectives* (pp. 18–29). New York, NY: Routledge.

Stables, K., & Keirl, S. (Eds.). (2015). *Environment, ethics and cultures: Design and technology education's contribution to sustainable global futures*. Rotterdam, Netherland: Sense.

Tapia, A. H. (2003). Technomillennialism. *Science, Technology, & Human Values, 28*(4), 483–512.

Turkle, S. (1984). *The second self*. New York: Simon and Schuster.

Turner, N. J., Ignace, M. B., & Ignace, R. (2000). Traditional ecological knowledge and wisdom of Aboriginal peoples in British Columbia. *Ecological Applications, 10*(5), 1275–1287.

UNESCO. (Ed.). (1953). *The teaching of philosophy*. Paris: Author.

Van Norden, B. W. (2017a). *Taking back philosophy*. New York: Columbia University Press.

Van Norden, B. W. (2017b, October 31). Western philosophy is racist. *Aeon*. https://aeon.co/essays/why-the-western-philosophical-canon-is-xenophobic-and-racist
Weiss, E. B. (1990). Our rights and obligations to future generations for the environment. *American Journal of International Law, 84*(1), 198–207.
Wiredu, K. (1984). Philosophical research and teaching in Africa. In UNESCO (Ed.), *Teaching and research in philosophy: Africa* (pp. 31–54). Paris: UNESCO.
Zuckerberg, M. (2010, October 14). Mark Zuckerberg, moving fast and breaking things [interview with Henry Blodget]. *Business Insider*. Retrieved http://www.businessinsider.com/mark-zuckerberg-2010-10

Synoptic Review

David Barlex

Abstract This chapter considers chapters 2 to 17 and comments on each chapter separately. Each commentary identifies some of the key elements in the chapter, considers the pedagogy related to these elements and discusses other pedagogies that are relevant and might be supportive of associated teaching and learning, where appropriate links have been made between chapters.

1 Introduction

The strategy in writing this synoptic review has been to treat each chapter separately, identify some key elements within each chapter, consider the pedagogy related to these elements and discuss other pedagogies that are relevant and might be supportive of associated teaching and learning, where appropriate links have been made between chapters.

2 Considering Chapter "Technology Education: The Promise of Cultural-Historical Theory for Advancing the Field"

Having outlined a range of learning theories and described their relevance to technology teaching, Marilyn Fleer presented a vignette describing a sequence of lessons in which students considered the dilemmas posed by the introduction of

D. Barlex (✉)
University of Exeter, Exeter, UK
e-mail: david.barlex@btinternet.com

P. J. Williams, D. Barlex (eds.), *Pedagogy for Technology Education in Secondary Schools*, Contemporary Issues in Technology Education,
https://doi.org/10.1007/978-3-030-41548-8_18

autonomous vehicles. This vignette illustrated a sociocultural approach to learning, in this case about autonomous vehicles and their possible impacts on society (see page 32 for more detail). There is little doubt that there is the possibility of considerable learning *through* tackling the tasks in the way described. These are not trivial activities, and the teacher has clear and demanding learning intentions for the students; they will not only learn about developments in a new and emerging technology, but they will also learn how to consider the implications of deploying this technology and identify possible consequences. Exploring such issues will always be speculative to some extent because the students are being asked to consider both 'what might be' and also 'what we want to be'. Shannon Vallor (2016) argues that our growing technosocial blindness, which she refers to as *acute technosocial opacity*, makes it increasingly difficult to make the ethical decisions leading to lives worth choosing and lives lived well. Hence, engaging young people with such speculation is an important aspect of general as well as technology education.

It is important to note that the learning intentions will be predicated on previous learning; what we might term learning *for* the task. It is generally agreed that small group and class discussion are powerful means of achieving learning (Hattie, 2012) but such learning does not happen without clear instruction with regard to the ground rules for such discussion. Mercer, Wegerif, and Dawes (1999) identified the following set of ground rules for such small group discussion which they name exploratory talk.

- All relevant information is shared
- The group seeks to reach agreement
- The group takes responsibility for decisions
- Reasons are expected.
- Challenges are accepted.
- Alternatives are discussed before a decision is taken.
- All in the group are encouraged to speak by other group members.

Barlex and Welch (2009) have considered the use of exploratory talk in technology education and asked three questions:

- To what extent would students' ability to engage in exploratory talk enhance their ability to make design decisions?
- In what ways would students' engagement in exploratory talk change their design decisions?
- How can students in design and technology classrooms be taught to use the ground rules of exploratory talk?

Prior learning *for* the task will be required for some of the elements embedded in the tasks set by the teacher. For example, 'design a road system that would support autonomous vehicles' a task embedded in 'design the road rules and environmental features for self-driving cars in your community'. How might this design be developed and presented? What might students need to be taught about simple maps, grids, scales and keys and acquire the drawing skills necessary to present their design ideas in an attractive, easy to interpret format that reveals their appreciation of the impact that autonomous vehicles might have in their community. Should this teaching be done via direct instruction calling on a constructivist approach before tackling the design task? Or should the teacher allow time for students to work in groups, within the design task, to research the required knowledge, understanding and skills on a ‘just-in-time’ basis calling on a social constructivist approach. Is there a middle way in which the teacher organises a few pertinent short instruction sessions within the task that enables the students to dig deeper on an as needed basis? Deciding on the balance between learning *through* a task and learning *for* that task is an important part of pedagogy to inform curriculum planning. Insufficient learning *for* the task and the students are unable to learn *through* the task. Too much learning *for* the task and there is the possibility of learning *through* the task diminishing as students become disenchanted because they do not see its relevance to the upcoming larger task.

3 Considering Chapter “The Case for Technology Habits of Mind”

Janet Hanson and Bill Lucas develop in some detail the components of technology habits of mind (THOM) taking into consideration the notion of habits of mind itself and drawing upon the habits of mind already established for mathematics, science, engineering, visual arts and creativity. They present a compelling case for its consideration with regard to enhancing the position of technology as a contributing discipline to STEM learning where its position is often marginalised for a variety of reasons. John Holman (Royal Society, 2007), when he was the National Director of the Programme, summed up this disparity in status of the STEM subjects well with the following graphic device.

In school:

STE**M**

In the world outside school:

S**TE**M

John argued that there should be a better balance across the STEM subjects in school. In addition, Janet and Bill argued that the reduction of technology education to simply ‘making things’ which excluded the process of designing in response to needs and wants to develop prototypes of worth from different and wide-ranging stakeholder perspectives can be challenged through using THOM. They suggest that by using THOM it is possible to develop pedagogy for teaching technology that embraces its breadth far beyond, but still including, the teaching of making.

They present a vignette concerned with biomimicry and describe the teacher’s approach to engaging her students with THOM (see page 57). Biomimicry is a strategy that uses ideas from nature as inspiration for creating and developing design ideas. It was mentioned specifically by Dick Olver, Chairman of BAE Systems, one of the UK's biggest companies, when he was arguing for a modern approach to design and technology (Olver, 2013) and as a result included as an example in the non-statutory guidance in Key Stage 3 English National Curriculum (Department for Education, 2014). Although it is relatively straightforward to explain biomimicry to learners and provide examples of its use by professional designers, it is less straightforward to enable learners to use it as a strategy in their own designing. Hence, the teacher in the vignette is attempting something not easily achieved. She requires the learners to observe the way different seed types increase their time in the air by spinning and to identify possibilities for producing such spinners using paper and card. There is certainly plenty of opportunity for refining and improving initial design prototypes so that they spend longer in the air. The vignette does not include any details of how the resultant spinners might be incorporated into a useful design but it is not difficult to imagine how the suggestions might be included in 'flying' toys or perhaps as a means to distribute seeds across fields when farmers are trying to develop meadows as part of their land use. The explicit use of the THOM to scaffold a learner so that his diffidence is overcome and he learns to become confident in tackling new things provides a specific example of how THOM might be used to develop confidence across a wide range of activities. Hence, a small amount of time in considering what the paper spinners might be used for would call upon the THOM imagining.

4 Considering Chapter "Making the Invisible Visible: Pedagogies Related to Teaching and Learning About Technological Systems"

Jonas Hallström and Claes Klasander discuss the research into teaching systems as part of technology courses and from this derive four pedagogies for enabling students to make sense of systems. These are as follows:

- Interface pedagogy – starting with the human–system interface and work outwards to identify and understand the elements of the system and how they work together
- Holistic pedagogy – starting with an overview of a given system and exploring it to reveal the subsystems, their interaction and importance
- Historical pedagogy – starting with a particular system and exploring its development by considering its origins
- Design pedagogy – in contrast to the above which concern analysing existing systems students are challenged with designing a system

See page 73 for more detail.

In developing how these might be used, they identify a useful view of progression using the concept of systems horizon shown in Fig. 1.

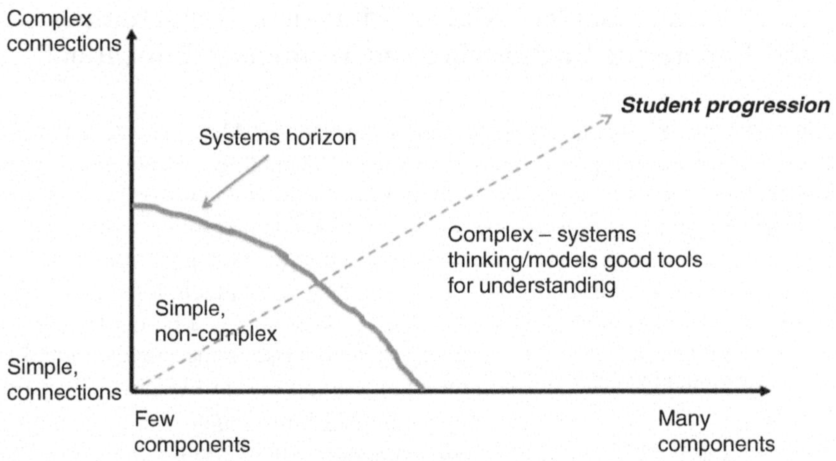

Fig. 1 Student progression into systems thinking

This enables the teacher to move students from considering simple artefacts with few components and simple connections which does not require systems thinking to considering more complex technological outcomes which in order to be understood require systems thinking and an appreciation of system concepts.

I wonder about the extent to which some of the system concepts might be taught by direct instruction and this learning used in the three analytical pedagogies they identify. These concepts include subsystems within a system, system boundary, flow of information, materials and energy within systems, feedback, lag, etc. To overload students with detailed information about such concepts would be counter-productive, but introducing them as vocabulary to be used in the suggested analysis activities might enable learners to become fluent at using systems thinking. The provision of a glossary of terms with simple definitions to which students could refer or be referred to by their teacher would aid this process.

Claes and Jonas consider that the design of sociotechnical systems will be too complex in educational settings. However, I wonder if there are opportunities for this in the realm of digital design. Developing an interactive website is now a relatively straightforward task and any such site is part of a wider sociotechnical communication system. Identifying a group that need information about a particular issue, providing such information and enabling communication between members of the group through the design of a website would be an interesting if somewhat simple example of designing a sociotechnical system. That was how Facebook started. Teaching young people to view the world in which they live through a systems lens is an important part of technology education. It helps them make sense of the very complicated and interconnected world in which they live where many of the systems on which they depend are invisible. The pedagogy suggested by Jonas and Claes will help teachers reduce this invisibility.

5 Considering Chapter "Maker Education: Opportunities and Threats for Engineering and Technology Education"

Gerald van Dijk, Elwin Savelsbergh and Arjan van der Meij provide a balanced view of the threats and opportunities afforded for technology and engineering education (ETE) curricula by engaging with maker education. On the one hand there is little doubt that young people involved with their local maker movement experience a rich creative environment and learn to use particular technologies, but on the other hand the informal setting and level of choice they have precludes the deliberate teaching of previously identified knowledge or skill. This is a particular issue for those educational systems in which there has been a rise in the interest in and significance of knowledge rich curriculum and the importance of teachers identifying and teaching specified substantive and disciplinary knowledge.

The serendipity of learning within maker spaces would seem to have little place in this context.

Yet the attractiveness of the maker space environment is seductive. In the USA, DARPA, for example, supports Maker Activity with young people through its Manufacturing Experimentation and Outreach (MENTOR) programme. This focuses on engaging high school-age students in a series of collaborative design and distributed manufacturing experiments. The goal is to encourage students across clusters of schools to collaborate via social networking media to jointly design and build systems of moderate complexity, such as mobile robots, go-carts, etc., in response to prize challenges.

The authors use maker education to identify principles for strengthening Engineering and Technology Education curricula (see page 96 for details); what they term a hybridisation approach. Within this, it is the role of the teacher in the classroom interaction with the students that make use of the approaches developed in maker spacers that is significant. The research agenda offered by the authors is intriguing in that one of its proposals is to observe maker education practice in its own right, as opposed to searching for hybridisation, with a view to identifying those features of practice that might be relevant to current ETE curricular. This has resonance with Niall Seery's chapter, *Pedagogy Involving Social and Cognitive Interaction Between Teachers and Pupils*, in which he explores the social and cognitive interdependency between teachers and learners and argues for its deliberate development.

6 Considering Chapter "Signature Pedagogies for Designing: A Speculative Framework for Supporting Learning and Teaching in Design and Technology Education"

Kay Stables describes the complexities and demands of designing very clearly. To my mind, expert designers have considerable substantive and disciplinary knowledge. They know lots about materials, their properties and applications, manufacturing methods, ways to achieve functionality of many sorts and they know how to deploy this knowledge through designing. In addition, their disciplinary knowledge equips them to explore the contexts in which their proposed designs will be used such that their suggestions meet the requirements of stakeholders. They often suspend judgement and avoid becoming definitive too early in the process and intuitively use case-based reasoning from their considerable experience to develop unexpected and provocative solutions.

Yet we expect young people at school to design! This is why Kay's suggestion of pedagogies of designing is so important in that they provide the teacher with the means to support iterative designing within a design task on an ongoing basis. The vignettes are instructive (see pages 112 and 113 for detail). They indicate how the teacher may intervene with particular pedagogies at key moments throughout the designing in an as needed just-in-time basis. Of course, there will be learners other than Rebecca or Abdul in the class who will need appropriate pedagogic advice. Hence the teacher will need to be highly aware of each pupil's progress and intervene accordingly which makes for very demanding teaching.

I wonder if there is a curriculum development solution to this problem in that the various pedagogic tools available might be used as the basis for planning a sequence of design tasks. The tasks would be devised such that particular tools would be expected to be used within each task and that over the sequence of tasks the learners would become au fait with the entire suite. The tools themselves are described in the glossary to Kay's chapter on page 115. Such an approach would avoid making the teaching overly demanding but it would be important to ensure that the devised tasks were genuine design tasks with an appropriate degree of authenticity although deliberately contrived to require the use of particular pedagogical tools. A class of learners moving through the design task sequence would be adding to their disciplinary knowledge and some of them might decide in a particular task to use tools that they had learned about earlier. It would be at the teacher's discretion as to whether to allow this or insist on the use of tools assigned to be used during the task in hand. There are arguments for both positions. Insisting on using the pedagogical tools assigned gives the teacher insight into whether the learner has understood the new tool and is able to use it. Allowing the learner to use a previously learned tool gives insight into the effectiveness of that learning and supports the learner's sense of self-efficacy. A discussion with the pupil as to these pros and cons is a way forward.

7 Considering Chapter "Pedagogies for Enabling the Use of Digital Technology"

Debi Wynne makes a strong case for teaching digital technologies, partly economic utility – lots of occupations will require the use of such technologies and partly cultural – such technologies are rapidly becoming ubiquitous and it is difficult to make sense of the world if you do not understand something about their nature, limitations and how to use them. As an experienced design and technology teacher,

Debi knows well the pros and cons of various approaches and is sensitive to the students' attitudes to learning new knowledge and skills and their expectations with regard to likely success or failure. She is also clear about the features of digital technologies that are particularly relevant and useful for developing students' confidence and competence: declarative knowledge and strategic knowledge are of particular importance; procedural knowledge much less so.

With all this in mind she develops a set of important factors to consider when developing a pedagogical approach to digital technologies which if used over time are likely to engage interest, increase confidence and resilience, and develop the ability to choose and use appropriate software. In many design and technology curricula, the teaching of CAD figures large and following Debi's advice will enable teachers to help students to make progress, but it is important to ask to what ends will the CAD be used in pursuing designerly intention. CAD linked to CAM has considerable potential and it is possible for students to design and make simple shapes and forms that they would find difficult if not impossible to make using conventional hand and machine tools. It is worth considering the worth of some of these items. Many items produced by school laser cutters or 3D printers are tchotchkes, decorative items of little worth kept small, to reduce cost and simple to allow the production of CAD files by novices. Simple fixings and components to be integrated into a more complex design might be of more worth although perhaps not as immediately attractive.

It is worth noting that CAD is not a substitute for designerly imagination and intuition. There is the problem that the nature of the software and what it can do easily may overly influence the nature of any resulting design and limit the creativity of the designer. Experienced designers have reported that they leave the use of CAD until as late as possible in their designing to avoid this problem (Carr, 2015). CAD alone can produce images of designerly intention without the possibility of realisation. In the case of designing without making tasks this may not be a bad thing as it enables interesting ideas to be explored in a variety of forms. Such designs can then be imported into presentations in which the worth of the design intention is justified. With regard to designing and making the design and technology will need to use the sound pedagogic advice provided by Debi in the light of the values that underpin that activity.

8 Considering "Developing a Pedagogy of Critiquing as a Key Dimension of Design and Technology Education"

Steve Keirl makes a powerful argument for the act of critiquing to permeate the design and technology (D&T) curriculum and provides an interesting set of examples of critiquing activities from the South Australian D&T curriculum along with possible responses. But he issues a stern warning that simply following such examples, by rote as it were, is a betrayal of the essence of critiquing. Such an approach to critiquing would lead to it becoming humdrum, what is expected and bland. This must be avoided at all costs. But within his robust and assertive view of critiquing lies a particular problem.

It is clear that the 'successful' interaction of science, technology and capitalism, whilst providing considerable wealth for the few and an increase in living standards for the many, comes at a great cost in terms of the nature of lives currently being lived, the behaviour of the planet in response to the impact of technological activity, and the future existence of human and other life on the planet. Exposing these consequences through critiquing can easily lead to young people seeing themselves as powerless against forces over which they have no control and feelings of despair.

The pedagogical challenge for critiquing is not only to raise the issues but also to actively consider what solutions should be developed and how these might be deployed. Such pedagogies that engage with critiquing can be a means to develop a 'future worth wanting'. As Steve points out such considerations will involve ethics and it is encouraging to read the work of Shannon Vallor (2016) in this light. Taking the thinking of three ancient scholars, Aristotle, Confucius and Buddha as her starting point, Shannon makes suggestions for technomoral wisdom to enable us to move towards a future worth wanting. In particular, she focuses her thinking on social media, surveillance and robotics as examples of new and emerging technologies that are already disrupting our lives and identifies ethical positions that enable critiquing to become a change agent. As I write there is a wonderful example of critiquing as a change agent being provided by Greta Thunberg (2019), unheard of a few years ago but now a key figure in energising critiquing leading to action across the planet.

In his concluding cautions, Steve warns of the pernicious constraining curriculum constructs such as STEM, training for jobs and computing. I understand the need to resist a utilitarian view of education and appreciate that support for STEM in particular is often justified in these terms. But it does seem obvious that the funds of knowledge in mathematics and science have both revealed the current climate crisis and will be essential in developing responses to mitigate its effects. Approach with caution by all means but conversations with colleagues across the STEM subjects with particular regard for their role in critiquing will, I think, become increasingly important.

9 Concerning Chapter "Question Think Learn: A Pedagogy for Understanding the Material World"

This chapter by Belinda von Mengersen and Terry Wilkinson is nothing less than a clarion call for D&T teachers to help their students meet the ecological request of our time. They describe a well thought through and research informed approach to engaging young people with the highly problematic issues surrounding the way that materials are chosen and used and the consequences, often hidden, of this. They question the 'technical issue only' paradigm of teaching about materials: what matters is their fitness for purpose and deliberately widen the scope of reasons for choosing, or not, a particular material for a particular purpose. This has resonance with the view of Barlex and Steeg (2017) who argue that learning about materials should involve not only their properties (intrinsic and working) but also their source, footprint (during extraction, use and disposal) and their longevity if they are not renewable. Belinda and Terry widen this consideration to include the human as well as environmental consequences of material choice.

There are clear links here with Steve Keirl's championing of critiquing as an essential element of D&T teaching and learning (see chapter "Developing a Pedagogy of Critiquing as a Key Dimension of Design and Technology Education") but Belinda and Terry focus intently on the issue of material choice and the consequences of such choices. They acknowledge that there are no easy answers and that simple positions such as 'paper is good, plastic is bad' are both flawed and unhelpful. Of course they are arguing for teachers to challenge the prevailing paradigm of most young peoples' lives in the developed world: consumption is good, your identity is intimately linked to what you purchase, it is important to get a job that pays well and enables you to acquire goods and services that show you are a success. Although there is a growing awareness that 'business as usual' is no longer an option with regard to the way we live our lives overturning the consumerist ethic is a tall order. The pedagogy suggested by Belinda and Terry accepts this challenge and acknowledges the need to develop in learners a willingness to find out more, listen to and consider the views of others, be constructively critical of the status quo in the search for a future worth wanting (Vallor, 2016). This entails spending significant time in activities that are not geared to the sometimes hectic pursuit of practical capability through designing and making. For some pupils, this 'doing with a concrete outcome' is a welcome relief from other subjects in the school curriculum and they might well react adversely to lessons that are less obviously active and more reflective. Hence, we should not underestimate the pedagogical challenge embedded in the approach that Belinda and Terry are advocating. The challenge for those developing curricula that adopt their suggestions will be one of balance in enabling the design and making activities pursued by young people to engage with a deeper understanding of the material world. Their approach requires equal weight being given to developing both technological capability and technological perspective avoiding their separation such that both contribute to a holistic interpretation of the design and technology curriculum.

10 Considering Chapter "Pedagogy for Technical Understanding"

Torben Steeg and David Hills Taylor unpack and clarify the uses of technical understanding that are significant for learners in design and technology:

- To support and inform learner's design decisions
- To develop novel (for the learner) and interesting ways for things to work
- To inform making
- To help learners understand what might be going wrong when something isn't working

- To help learners repair something that isn't working
- To enable learners to improve the design of something so that it works better
- To develop technological perspective by helping them understand the technologies they engage with in every-day life

They declare their support for constructivism and constructionism as important and research justified theories of learning that are of particular relevance to design and technology noting that there are many different ways to take these theories of learning into account when devising pedagogy. They illustrate this point by identifying three approaches, each illustrated by a case study:

Reconceptualising scientific knowledge
Product analysis
Systems thinking

A particularly strong feature of the tools developed by Torben and David is that they take into account what cognitive psychology and neuroscience tell us about how learning happens. These complement the pedagogic tools identified by Kay Stables and presented in the Appendix to her chapter (page 115).

Some criticise current D&T as a school practice as being concerned mainly with 'product styling' (Lewis, 2019) as opposed to genuine product design with technically functioning internal workings. The chooser charts that David and Torben regard as enabling the reconceptualisation of scientific knowledge can also be used to present knowledge from a range of sources in ways that support design decisions and these would make a welcome addition to the tool kit.

Chapter "Making the Invisible Visible: Pedagogies Related to Teaching and Learning About Technological Systems" by Jonas Hallström and Claes Klasander explores this thinking in terms of making the invisible visible and as a way of conceptualising and simplifying the workings of hidden technologies in which we are embedded. Some of these technologies operate at the microlevel through programmable interface controllers linked to sensors and actuators of various sorts and, as Torben and David show it is perfectly feasible for learners to design and make items that include embedded intelligence through a systems thinking approach. An interesting exercise for aspiring curriculum developers would surely be to use the many suggestions in this book to develop pedagogic tools which can provide an a la carte menu of pedagogy from which teachers may choose to meet the needs of their learners.

11 Concerning Chapter "Capability, Quality and Judgement: Learners' Experiences of Assessment"

Richard Kimbell begins his chapter making a distinction between 'understanding that might properly be the domain of science, and *capability*, that is the province of design & technology' clarifying his view of capability as follows:

> By capability, we mean that combination of ability and motivation that transcends understanding and enables creative development. It provides the bridge between what is and what might be. (Kimbell, Stables, & Green, 1996 p 25)

Richard's point on pedagogy as duality is well made identifying the two roles of the teacher as both supporter and critic; a feature of enabling creativity that was revealed in a study to identify conditions to support creative outcomes in both art and design and technology in the secondary school (Barlex, 2007). His emphasis on the need for the learner to be a genuine reflective practitioner is also significant. Teachers using the 'at any time in the moment assessment' of current design ideas through a thumbs up/thumbs down approach is a sound strategy which can inform and feed into more formal timed reviews of progress.

The discussion of ideas of the nature of what constitutes quality between teachers and learners is an essential plank in Richard's argument if there is to be a development of a shared understanding. Making judgements about learner's achievements through comparative pairs, judgement has revealed that teachers and learners do in fact agree on what constitutes quality, and this does not require the use of atomised criterion statements. In fact the use of such statements is counterproductive. All this supports Richard's position that achieving capability in design and technology involves the learner in a journey in which the teacher plays a variety of roles in which there has to be mutual trust.

In supporting learners in their journey towards ever more competent capability, the teacher has to wrestle with the just in time or just in case dilemma. In the case of developing a steerable skateboard, the learner was exploring the pros and cons of a rack and pinion device. Did the teachers teach about mechanical systems prior to the learner beginning their skateboard project (just in case learning) or did the teacher rely on the learner being able to learn this for themselves on an as needed basis (just in time learning)? Most learners will be keen to improve their capability and they will almost certainly ask what they need to do differently next time. Here the teacher will need work with the learner to unpack the elements of quality that provide their shared understanding of holistic quality in order to provide specific guidance to which the learner may respond.

12 Considering Chapter "Technology Education Pedagogy: Enhancing STEM Learning"

John Wells and Didier Van de Velde provide interesting descriptions of how integrative approaches to STEM education are being developed in their respective jurisdictions. John teases out the differences between interdisciplinary and transdisciplinary approaches within technology education in the USA, whilst Didier points to integrative STEM as being an elective subject for young people in Belgium. Interestingly, the examples they describe are from very different fields.

John describes the challenge of designing a photo-bioreactor undertaken by graduate teachers in training to illustrate the integrative STEM education approach (see page 230). This is a demanding challenge, probably not typical of school design challenges and requires the use of equipment not found in most school technology departments. This however does not detract from the powerful learning that has to take place for the challenge to be tackled successfully. Some of this learning might take place before the task is tackled, learning for the task, and be supported by guided instruction. Some will have to take place during the task in response to the needs of the task. This provides a very powerful model of the sorts of learning that need to take place in the high school classroom.

Didier describes an activity much more typical of school design challenges but none the less significantly demanding for learners and teachers. It required development of a prototype electric vehicle that had to behave in a certain way plus a consideration of the possible impacts of autonomous vehicles on society (see page 226). Interestingly, the work revealed tensions between the requirements of developing understanding and the production of a high-quality prototype.

Both studies reveal the importance of teachers in any attempt to develop integrative approaches to STEM. In John's case, the knowledge base for the photobioreactor design task is probably outside that of most technology teachers so teachers wanting to become involved in this sort of work will need significant subject knowledge enhancement professional development. In Didier's case, the involvement of teachers from physics and technology revealed different priorities – both of value. So again there is the need for professional development in this case to enable teachers to appreciate the legitimate yet different priorities of colleagues and work together to find ways to resolve any differences these might create in the classroom.

The implications for STEM education in which the subjects deliberately try to overcome a single silo approach to teaching through integrative approaches are clear. Unless teachers have the necessary professional development time to reflect on the potential costs and benefits of such approaches and become creative collaborators in both devising and teaching appropriate schemes of work the likelihood of integrative STEM being successful is small. Is it worth the effort in putting resource into this professional development? There is a strong sign from Didier's study that it is. Mathematics teachers were initially sceptical as to the benefits of the interdisciplinary approach but found that as a result of engaging with the prototype electric vehicle project their students were better able to recognise the importance of different representations (table, graph, formula) and the possibilities for modelling and predicting system behaviours.

13 Considering Chapter "Teaching Problem-Solving in the Digital Era"

In his chapter, Moshe Barak challenges several of the prevailing norms of design and technology education practice. First, he situates what young people might design and make firmly in their technological world and the many changes that are likely to happen in that world within a short space of time. Second, he suggests that what they design and make should move beyond the small, often ingenious, electro-mechanical devices still common in many schools and embrace digital functionality. Third, he argues that the primary driver for such activity should not be meeting needs but the invention of innovative products that meet as yet unidentified needs and open up new markets. To some this radical approach may seem to be giving free reign to the worst aspects of neoliberal capitalism and should be resisted but others may see it as an opportunity to democratise innovation and give young people the chance to become reflective innovators in developing a world in which they want to live.

Moshe supports a curriculum based on his challenge to orthodoxy by suggesting a range of teaching methods that support inventive thinking and problem solving and argues that it will also be important to develop computational thinking. He provides some interesting examples. He raises some very important points with regard to the devising of student assignments in his brave new world of product/service innovation citing a task taxonomy that distinguishes between three levels of student assignment: practice, problem solving and projects (see page 261 for details). He suggests that this approach is necessary to counter the argument put forward that constructivist-oriented instructional methods such as discovery, problem-based and inquiry-based learning fail because they involve only minimal guidance during the process and as a result do not achieve learning. He acknowledges the severe limitations of minimal guidance and uses the idea of progression through his task taxonomy as a means of overcoming this.

Moshe's approach has some resonance with the Nuffield Design and Technology approach of guided instruction through preliminary, short resource tasks providing the necessary learning to be successful in capability tasks (Longman, 1995). There is also resonance with the findings of research carried out in England that creative outcomes are produced by a combination of two sets of features. The first set involves teaching relevant knowledge, situating the task in a relevant context, providing appropriate stimulus and encouraging students to be reflective, whilst the second set requires that the teacher steps back and allows the students to take the intellectual risks necessary to be creative whilst at the same time managing that risk taking so that it is neither too small, leading to little of no creativity, or too large leading to failure (Barlex, 2007). It is worth noting too that such approaches to problem solving in the design and technology curriculum may be seen as composed of two sorts of learning: learning *for* the task in which guided and/or direct instruction play a key part and learning *through* the task in which the learning *for* the task is deployed. In the deployment of such learning, it is likely that the overall learning becomes refined and is further embedded in long-term memory in forms related to its active use.

14 Considering Chapter "Pedagogical Approaches to Vocational Education"

John Williams provides a description of vocational education and how its low status is dependent particularly on the way success in vocational courses is framed in terms of achieving particular individual occupation-specific competences. The administration to record success is easy and straightforward and appeals to government who can point to 'training success to meet economic requirements'. Overall, this gives the system considerable inertia in resisting any change and results in a limited instructional pedagogy which those responsible for providing vocational education have adopted more or less without question. The paradox is that this pedagogy does not develop the so-called 'softer' skills that many in industry say are essential for the modern and future workplace. Given the almost certain influence of greater automation on the workplace with many routine technical operations being carried out by intelligent machines or algorithms (Frey & Osborne, 2013), the need for such skills will become ever more paramount.

A recent twitter conversation sums this up rather well. Brian P Hogan tweeted: *I've hired a few people in my time. I've never been disappointed by hiring someone who is less technical but is a great human who cares for others. I can teach tech pretty easily. But boy have I seen people who are strong technical people with toxic attitudes destroy a team.*

Oliver Caviglioli replied: *Have you heard of that saying in business: we hire people for their technical skills and fire people for their lack of communication skills?*

John presents a framework for reconceptualising vocational pedagogy involving social practice, discourse community, signature pedagogies, and practice architectures (see page 274) and in discussing these highlights the increasing importance of developing a pedagogy that transcends traditional boundaries, meets complex workplace requirements and is not based on pre-specified statements of measurable, competence and standardised content. He notes that changing the situation will not be easy but makes the key point that the revised aims of vocational education will embrace many of the aims of general education. This in no way decries the need for vocational education to enable technical competence but does indicate that this, while necessary, is not sufficient.

I think this has particular relevance to the status of design and technology in the secondary school from two perspectives. First, in that it is important to see design and technology as a key element in the general education of all students to 16+ years and labelling it as a vocational subject detracts from this possibility and leads to it having a low status. In England, this often results in those learners who are academically successful being advised not to study the subject after the age of 14 years. Seeing vocational education as having significant general education components would decrease the pejorative effect of labelling design and technology as a vocational subject. Secondly, the range of soft skills that are taught within design and technology are significant and from this point of view the ways these are taught might inform the pedagogy used in vocational education.

15 Considering Chapter "Teaching Technology in "Poorly Resourced" Contexts"

Mishack Gumbo presents a compelling case for the possibility of teaching technology in the so-called 'poorly resourced' situations. Arthur (2009) has argued that technology may be considered as the exploitation of phenomenon described, explored and explained by science. If one considers the rich vein of knowledge acquired and exploited by indigenous peoples as science, then it makes perfect sense to see this as the basis for a technology curriculum. One might argue that such knowledge lacks a theoretical base, but this is more than mitigated by the deep understanding of natural phenomenon acquired over many generations living successfully in a particular environment which is sometimes hostile to human habitation. This knowledge and understanding is embedded in the day-to-day activities of indigenous people and underpinned by their holistic approach to how best to live with one another in their particular situations.

Hence Mishack's argument that there is an abundance of technology teaching resources at hand in rural environments if teachers would only open their eyes to see them is very strong. Using the practices of an indigenous community in growing, preserving, storing and cooking food is particularly appealing as this is often underpinned by a requirement to live in harmony with nature. It is noteworthy that many young people in the so-called developed countries have little knowledge or understanding about the origins of their food and the communities in which they live lack any expertise in food production. Barlex (in press) in exploring how food technology might be taught in secondary schools in England has suggested that a component should be growing and harvesting vegetables.

Paradoxically there is some resonance between Mishack's approach and that advocated by Moshe Barak (see Chap. 13). Moshe argues that the digital and information revolutions of the last half-century have impacted on almost every aspect of our lives so much so that youngsters today think, shop, spend leisure time and learn differently from how they did 20 or 30 years ago. As a result, technology educators need to make technology studies as relevant as possible to students' daily lives. This seems to me to be parallel to Mishack's argument. The only difference is the technological milieu in which the students are immersed. This parallel should go some way to convincing non-indigenous teachers who think that there are no other forms of technology besides the conventional forms and the worth of adopting Mishack's suggestions. There are of course the problem-facing young people from rural settings who move to urban settings where the technological environment in which they find themselves will be significantly different from that they have experienced before. However, I think it is possible to argue that they will be better equipped to adapt to their new situation if they have had a technology education of the sort suggested by Mishack than if they had had none justified on the grounds of a lack of resources.

16 Considering Chapter "Pedagogy Involving Social and Cognitive Interaction Between Teachers and Pupils"

Niall's contention that to the uninitiated a technology workshop in which young people are tackling a range of designing and making projects appears chaotic with doing taking precedence over learning will come as no surprise to those of us who have taught in such situations. But we know from our experience that this is not the case and that underlying this apparent chaos is a network of social interactions which 'grows' the knowledge available to the students enabling them to make difficult decisions about the details of their emerging and as yet unresolved design proposals. This knowledge is not evenly distributed amongst the students; serendipity plays it part in who knows what but the skilful teacher orchestrates the social interaction to ensure that the workshop is a place in which communication between students is the norm, invariably on task and beneficial. This requires the teacher to develop trust between herself and the students and between the students. Establishing the technology workshop as such a collaborative creative community of practice takes time and the teacher will need to nurture this over a significant period of time. Within this, there is a place for specific teaching of matters relevant to making sound design decisions.

I would argue that this is essential if the students are to be able to find out more on an as-needed basis. Such teaching forms a springboard for independent activity and if ignored puts the students in a position where they have to learn everything relevant to a designing and making task from scratch which is difficult and inefficient. As students become adept at finding out more for themselves such specific teaching becomes less imperative but it is probably worth revisiting important ideas at regular intervals in the light of the learning that has taken place across the class as a result of their designing and making. Imagine asking students, 'Okay, now what more do we know and understand about X or Y given our experience and conversations in the designing and making project we've just tackled?' This isn't an easy task, but it would make explicit learning that is taking place within the class and provide the opportunity to make this more widely available. It could also be used to convince those who have a limited view of the learning that takes place in technology education lessons, often seeing it as limited to making with little if any cognitive gain.

In England, it is ironic that the current assessment arrangements set out to deny this powerful learning through social interaction. The contextual challenge set by awarding organisations requires students to explore a context, identify a need or want to address and then design, make and evaluate a prototype solution. But as this is for individual assessment purposes, the student is required to operate in the company of, but in isolation from, his/her peers. One can only imagine the frustration of students whose teacher has developed a creative community of practice in her classroom when faced with this situation.

17 Considering Chapter "Philosophy of Technology for Children and Youth"

Stephen summarises the developments of philosophy for children and young people in schools noting the lack of engagement with the philosophy of technology. This lack of engagement is, he asserts, a two-way street in that technology education has rarely engaged in philosophy, and he questions the conventional wisdom that technology is just a tool and whether it is used for good or evil will depend on the intentions of the depending on the user. In doing this, he argues for young people to engage with the nature of technology.

This is a difficult territory as the intrinsic nature of technology is a contested territory. Some philosophers of technology are pessimistic with regard to its nature and the impact it will have on humans. Jacque Ellul (1964) would fall into this category. Others such as Kevin Kelly (2010) are much more optimistic and believe that it will ultimately be beneficial to humans although both camps see that technology and those technologies which make it up as having trajectories with significant momentum making technology seem autonomous (Hughes, 1994).

The research vignette Stephen presents shows two students actively struggling with the nature of technology, grappling with the idea that technology is always what is new, echoing to some extent Douglas Adams (2002) in the set of rules he devised to describe our reaction to technologies:

1. Anything that is in the world when you're born is normal and ordinary and is just a natural part of the way the world works.
2. Anything that's invented between when you're fifteen and thirty-five is new and exciting and revolutionary and you can probably get a career in it.
3. Anything invented after you're thirty-five is against the natural order of things.

(p.95)

If teachers are to help young people engage with the nature of technology it is important that they themselves have significant knowledge of the philosophy of technology. It is unlikely that many of them will have encountered much in the way of this in their initial teacher education or subsequent professional development. Help is at hand in the shape of two recent publications. The first is Reflections on Technology for Educational Practitioners, edited by John Dakers, Jonas Hallstrom, and Marc de Vries (2019) in which the work of 14 different philosophers of technology is explored and their implications for the school technology curriculum are discussed. The second is Technology and the Virtues by Shannon Vallor (2016). This is an exploration of the ethics needed for us to develop 'technomoral wisdom' grounded in the philosophies of Aristotle, Confucius and Buddha. Shannon acknowledges the acute technosocial opacity with which we are faced. We simply do not know and cannot with any accuracy predict how the future will unfold with regard to technological developments. But, she argues we need a robust ethical framework to decide what to do with and about technology if we are to move towards a future worth wanting.

This is clearly heady stuff, and at first sight, the busy technology teacher might well give a shoulder shrug and decide to give all this a miss, saying, 'I've got more than enough to do as it is.' This would I think be a mistake as the sort of learning about technology through a philosophical approach that Stephen is advocating is essential if humans are to be in control of technology as opposed to being controlled by it.

18 In Conclusion

What are we to make of this wide range of pedagogy suggested in this book? The diagram in Fig. 2 illustrates some of the features that inform student learning in technology education. The features are arranged in a tetrahedron indicating that each has an interaction with the other three features and that if any one feature is removed then the overall structure will become unstable and collapse leading to a lack of learning.

The nature of curriculum will clearly be highly influential. The students will bring to their lessons a range of pre-dispositions, knowledge, understanding skills and values, and these will influence the ways in which they respond to the curriculum and what it is that they learn. The pedagogy that might be employed in teaching the curriculum will clearly play a significant role. And the teacher will have played a role in devising and interpreting the curriculum, deciding just how best to engage the students being taught and choosing which of the many and varied pedagogy to use in the light of the curriculum and the students being taught. Which particular pedagogy to choose and how to deploy it sits at the heart of teacher professionalism? There is no single right answer to such a complex conundrum, and what works well in some situations might not work at all in others. This book has been written in the hope that it will provide teachers with a sufficiently broad menu of conceptually and reasoned-based options that they will in their different jurisdictions with differing technology education requirements be able to choose and justify pedagogy that is appropriate for and successful in their particular and unique situation.

Fig. 2 Features that might inform student learning

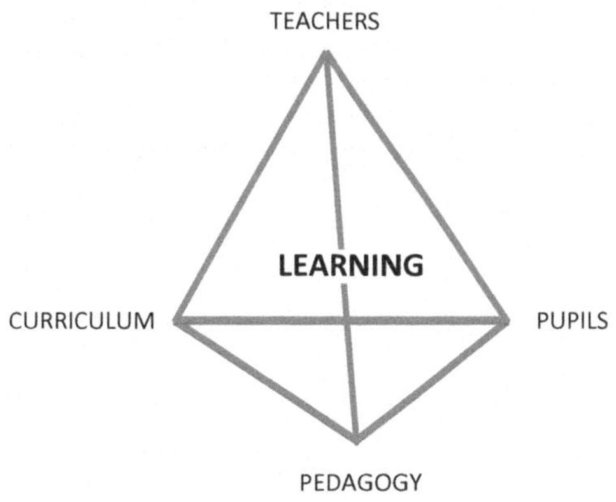

References

Adams, D. (2002). *The Salmon of doubt*. New York: Ballantine.

Arthur, W. B. (2009). *The nature of technology*. London, England: Allen Lane.

Barlex, D. (2007). Creativity in school design & technology in England: A discussion of influences. *International Journal of Technology and Design Education, 17*, 149–162. Klewer, The Netherlands.

Barlex, D. (in press). A curriculum developer's perspective on the place of food in the secondary school. In M. Rutland & A. Turner (Eds.), *Food education and food technology ... an international perspective*. Singapore: Springer.

Barlex, D. & Steeg, T. (2017). Re-building Design & Technology in the secondary school curriculum. Available at this URL: https://dandtfordandt.files.wordpress.com/2017/05/re-building-dt-v21.pdf. Accessed 10 March 2020.

Barlex, D., & Welch, M. (2009) *So we're going to have this huge spike here? Pupils' exploratory talk while designing and making*. Presented at The Design and Technology Association Education and International Research Conference 2009 A Platform for Success.

Carr, N. (2015). *The glass cage*. U.K.: Penguin.

Dakers, J., Hallstrom, J., & de Vries, M. (Eds.). (2019). *Reflections on technology for educational practitioners*. The Netherlands: Brill Sense.

Department for Education. (2014). *The English National Curriculum*. Available at this URL: https://assets.publishing.service.gov.uk/government/uploads/system/uploads/attachment_data/file/381344/Master_final_national_curriculum_28_Nov.pdf. Accessed 5 October 2019.

Ellul, J. (1964). *The technological society*. New York: Vintage.

Frey, C.B. & Osborne, M. A. (2013) *The future of employment: How susceptible are jobs to computerization?* A working paper Available at this URL: https://www.oxfordmartin.ox.ac.uk/downloads/academic/future-of-employment.pdf. Accessed 5 October 2019.

Hattie, J. (2012). *Visible learning for teachers*. Oxon, UK: Routledge.

Hughes, T. (1994). *Technological momentum*. In M. R. Smith & L. Marx (Eds.), *Does technology drive history?* MA: MIT, Massachusetts, US.

Kelly, K. (2010). *What technology wants*. New York: Penguin.

Kimbell, R., Stables, K., & Green, R. (1996). *Understanding practice in design and technology*. Buckingham – Philadelphia: Open University Press.

Lewis, T. (2019). Emeritus Professor of Technology Education Sheffield Hallam University. private communication.

Longman. (1995). *Nuffield design and technology teacher's guide*. Harlow, England: Longman.

Mercer, N., Wegerif, R., & Dawes, L. (1999). Children's talk and the development of reasoning in the classroom. *British Educational Research Journal, 25*(1), 95–111.

Olver, R. (2013, 1 April) Quoted text from a conference in March 2013, reported in *the Guardian*. Available at this URL: http://www.theguardian.com/education/2013/mar/31/school-curriculum-cookery-horticulture-before-technology. Accessed 5 October 2019.

Royal Society. (2007). *S-T-E-M working together for schools and colleges*. Unpublished.

Thunberg, G. (2019). *No one is too small to make a difference*. UK: Penguin.

Vallor, S. (2016). *Technology and the virtues*. New York: Oxford University Press.